Hattie Lovisa Borden Weld

Historical and Genealogical Record of the Descendants

Hattie Lovisa Borden Weld

Historical and Genealogical Record of the Descendants

ISBN/EAN: 9783337379018

Printed in Europe, USA, Canada, Australia, Japan

Cover: Foto ©berggeist007 / pixelio.de

More available books at **www.hansebooks.com**

HISTORICAL AND GENEALOGICAL RECORD

Of the Descendants
as far as Known

OF

RICHARD AND JOAN BORDEN

Who Settled in Portsmouth, Rhode Island, May, 1638

WITH

HISTORICAL AND BIOGRAPHICAL
SKETCHES OF

Some of their Descendants.

COMPILED BY

HATTIE BORDEN WELD.

HISTORICAL AND GENEALOGICAL
RECORD
OF THE
BORDEN FAMILY.

INDEX OF PORTRAITS.

		PAGE.
1.	REV. PARDON G. SEABURY	33
2.	GAIL BORDEN	143
3.	DR. JOSEPH BORDEN	145
4.	THOMAS RICHARDSON BORDEN	147
5.	JOHN BORDEN	149
6.	JUDGE JAMES W. BORDEN	151
7.	DAVID WALLACE BORDEN	153
8.	BENJAMIN BORDEN	155
9.	ISAAC PENNINGTON BORDEN	157
10.	JOHN ALLEN BORDEN	161
11.	SAMUEL W. BORDEN	163
12.	EDWARD PARKER BORDEN	165
13.	RICHARD BORDEN	167
14.	JEFFERSON BORDEN	169
15.	SIMEON BORDEN	175
16.	NATHANIEL BRIGGS BORDEN	177
17.	GAIL BORDEN	187
18.	THOMAS H. BORDEN	189
19.	PASCHAL P. BORDEN	191
20.	JOHN P. BORDEN	193
21.	JUDGE RHODES BORDEN	203
22.	WILLIAM J. BORDEN	207
23.	THOMAS JAMES BORDEN	225
24.	SIMEON BORDEN	233
25.	NATHANIEL B. BORDEN	237
26.	HON. FREDERICK W. BORDEN	239
27.	HENRY LEE BORDEN	247
28.	JOHN GAIL BORDEN	249
29.	PRES. W. A. OBENCHAIN	267
30.	DR. MILBANK JOHNSON	289

KEY TO TRACE THE GENEALOGY.

The number and name of the father or mother of each family in direct lineage from No. 1, Richard Borden, is placed at the head of the list of their children. Take the number of this name, trace it back to the next generation, until it is found in the list of children under another parent, and so on through all the generations. For instance, take any name, say the last one in the book, No. 2965, Mary Louise Stover. The name and number of her mother is No. 2892. Georgie Hulse McLeod. Tracing this number back to the next generation we find it among the children of No. 2560, Maria Louise Borden. No. 2560 is found under No. 1827, James Cole Borden, which in turn is found under No. 967, Arnold Borden, then No. 967 is found under No. 490, Seth Borden, and so on through the eleven generations back to No. 1, Richard Borden. Where no surname is given it is always Borden. It is omitted to avoid repetition.

"No fad of the present day can be more effective in bringing the American people into close sympathy than genealogical study pursued in a liberal spirit."

PREFACE.

The compiler of historical genealogy depends on four sources for his information: First, the accounts which history has preserved concerning the deeds of his family; second, the early records of land transactions, marriages and deaths; third, the accounts which various branches have kept of themselves, and, fourth, that vague and unreliable field of tradition.

The sources of history from which information has been gleaned concerning the Borden family are many, notably:

Austin's Genealogical Dict. of Rhode Island.
Early Annals County Kent.
Hosiers History of Kent.
Ireland's History of Kent.
Dr. Jameson's Dictionary.
Wedgwood's English Etymology.
Chaucer, Vol. III, page 406. The word "Bourdon."
Lower's History of English surnames, origin of "Burden."
Lewis' Topographical Dictionary, Eng.
Archives of France—Bourdon Coat of Arms.
Peterson's History of Rhode Island.
Fowler's Historical Sketches.
Savage's Genealogical Dictionary.
Fowler's Genealogical Dictionary of Rhode Island.
Records of Colonial Assembly of Rhode Island, June 24, 1684.
Massachusetts Historical Collection.
Arnold's History of Rhode Island; also mention of bounty given by the General Assembly of Rhode Island to William Borden for the manufacture of sail cloth.

Withers' History of the Settlement of West Virginia. Granting of a large tract of land by Gov. Gooch to Benjamin Borden.

Archives State of North Carolina. Mention of William Borden's services during the Revolution.

Fowler's History of Rhode Island. Capture by British of Richard Borden (3rd generation).

Ellet's Women of American Revolution. Heroism of the wife of Joseph Borden (3rd generation) at Bordentown, New Jersey.

The first Bordens in America were members of the Society of Friends. The Records of the Friends were faithfully kept, concerning the marriages, births and deaths of its members, and access can be had to them today. Glad should every Borden be that his ancestors were Quakers.

Tradition has furnished its share in weaving family romance, and

PREFACE.

while it cannot be drawn upon without an interrogation point, yet it has its place in family history.

An accurate dependence upon facts, has been the aim of the editor of this Historical and Genealogical Record of the Borden Family, and much information, purely traditional, has been discarded. Facts could only be obtained from living members of the family, aside from those furnished by History and Record, and no effort has been spared to gather them. Doubtless much information has escaped even vigilant effort. Some Bordens to whom we addressed our communications, did not care enough about family history and genealogy to reply. Others knew nothing beyond the names of their fathers or grandfathers. Search for genealogical data under such circumstances is not encouraging.

But the grand majority of Bordens have been eager and willing helpers in this work, and deserve our hearty thanks, for without their assistance this Record would have been impossible. Notable among these good helpers were Judge Rhodes Borden of San Francisco, Miss Sarah Borden of Philadelphia, Mr. James E. Borden of Eatontown, New Jersey; Mrs. Robert M. Parker of Chicago, Mr. Charles E. Borden of Wilmington, North Carolina, and Mrs. M. E. L. Minnich of Philadelphia. We also owe much to a manuscript written by Judge James W. Borden of Ft. Wayne, Indiana, and loaned by his son, Mr. William Borden of New York, and to Mr. J. O. Austin of Providence, R. I.

For material aid, advice and encouragement the editor wishes especially to acknowledge her indebtedness to Dr. Milbank Johnson, Mr. H. L. Borden, Mrs. M. J. Munsill and Mr. A. B. Church.

Notwithstanding all this assistance this work would have been far from complete without the aid of the manuscript of Reverend Pardon Gray Seabury. Upon this work he spent many years of his life and died leaving it unfinished and unpublished. From the Seabury manuscript we have taken most of the biographies of the early Bordens and the history of the family before coming to America. We give the author credit by signing quotations from his manuscript with an S.

This valuable manuscript was in the possession of Miss Charlotte A. Seabury of New Bedford, Mass., daughter of the author and a direct descendant of Richard Borden. Rev. Pardon G. Seabury brought this genealogy of the family down to about 1865, especially working on the Bordens of Fall River, Mass. Our arrangements of the genealogical tables and generations differ somewhat from his, yet all the information of his large manuscript has been embodied in this work, together with the result of years of labor on the part of the editor.

HATTIE BORDEN WELD.

1040 W. 36th St., Los Angeles, Cal.
July 1st, 1899.

THE
BORDEN
COAT OF ARMS.

BORDEN COAT OF ARMS.

The Borden Coat of Arms is described as follows: "The field azure, a chevron engrailed, ermine, two bourdens or pilgrim's staves proper in Chief; and a cross-crosslet in base, Or (gold).

"Crest: A lion rampant above scroll (Argent) on his sinister foot, holding a battle axe proper.—Motto: 'Palma Virtuti;' and above the Crest 'Excelsior.'

This description is taken from the London and Middlesex Heraldic Offices, 44 High Hilborn, W. C., London, certified by Thomas Morine. (Certificate No. 25,472) and sealed with the seal of the Heraldic Office.

This officer states as follows: "It appears from the works of Berry, Gwillim, Burke and others, on Heraldry, and the books of family Crests, that there are twelve Coats of Arms under the name of Burdon, Burden and Borden; and all bear pilgrims' staves, variously emblazoned, and the Crests are uniformly a lion rampant.

"The battle axe and hautboy are the same as the palmer's staff, or take the place of it often, in heraldry—thus the Coat of Arms of one branch of this family, is three battle axes, and of another three hautboys, and as many crosses-crosslet; we think, however, that there can be no doubt, but the Bourdon or Pilgrim's staff, is the proper device for the shield, and that a lion rampant is the proper Crest belonging to the Coat of Arms of the family.

Signed

[SEAL.]   "THOMAS MORINE."

IN HERALDRY.

"Field" is the full surface of the shield before any of the emblazonments are put on—in this case blue.

"Chevron" is the rafter-like figure which occupies the lower part of the shield. White in this case.

"Engrailed" means scalloped around the edges of the emblazonment. In this case around the edge of the chevron.

"Ermine" is one of the four furs used in markings in heraldry. In this case it is the white with black tips through it, which covers the chevron.

"Proper" means in its natural color and appearance; and when applied to the bearings on the shield, means that the bearing is shown in its natural color and appearance.

"In Chief" means in one or more of the "Chief Points" on the shield.

There are nine points on a shield, and the three nearest the upper edge are called the "Chief Points."

The point in the middle, and near the upper edge of the shield, is called "Center Chief."

The point on the right of the shield, and near the upper edge of the shield, is called the "Dexter Chief."

The point on the left of the shield, and near the upper edge of the shield, is called the "Sinister Chief."

The pilgrim staves on the Borden shield are in the Dexter Chief and Sinister Chief Points. In some of the descriptions of the Coats of Arms of this family there is the word "Gules" after the words "Pilgrim's Staves proper," which would mean that the staves should be shown in Red.

"Gules" means Red.

"Cross-Crosslet" means a cross with small crosses on each end of the larger Cross.

"In Base" means in the Base Point of the shield (the lower part of the shield).

"Or" means gold.

"Argent" means silver.

"Rampant" means standing upright.

"Passant" means in the act of walking.

"Couchant" means the act of springing.

These are used in heraldry to describe the positions of animals used as emblazonments upon shields.

"Sinister" means left. Sinister foot means left foot.

"Dexter" means right.

THE NAME BORDEN.

"A learned writer, who lived during portions of the 16th and 17th centuries, remarks that "About the year of our Lord one thousand, surnames began to be taken in France," and that in England, "they came into use about the time of the Conquest, or a little later, under Edward the Confessor, who was all Frenchified." And again he remarks: "Whatever may be advanced in favor of an earlier adoption of surnames or family distinctions in particular cases, it is certain that the practice of making the second name of an individual stationary and transmitting it to his descendants came gradually into use in England during the eleventh and three following centuries. By the middle of the twelfth century the people began to feel that persons of rank should bear some designation in addition to their baptismal name. An instance occurred about this time which shows the state of feeling which then prevailed in England. The wealthy heiress of Baron Fitz-Hamon, being addressed by Robert, Duke of Rochester, on the subject of matrimony, a natural son of Henry I, objected to the marriage, declaring,

"It were to me great shame
To have a Lord withouten his twa name."

The King, on hearing of the lady's objection to his son, immediately removed it satisfactorily to the lady and her friends by adding as a surname Fitz-Roy, so that his name became Robert Fitz Roy (the son of the King). His name, Camden remarks, has since been borne by many of the illegitimate offspring of British kings.

The practice of adopting hereditary surnames from manors, towns and other localities originated in Normandy some time previous to the Norman conquest, as is evident from the celebrated Battel Abby roll, which gives the surnames only of the companions and commanders of Duke William.

But we are not to suppose that the adoption of the preposition de, or de-la prefixed to any name in England proves that the person bearing it is of Norman origin; as the Saxons who preceded the Normans in England used the same designations, following the example of their conquerors. In some cases the Normans chose to retain the surnames derived from their ancient patrimonies in Normandy, and in others they substituted those taken from the estates given them by the Conqueror of his successors in England. In a few instances the de and de-la was retained, but generally it was dropped from surnames about the time of Henry VI, when Esquire was given to the heads of families and Mr., or gentleman, was substituted among the younger sons. Brevity, probably was the cause for the omission of the particles usually prefixed to ancient surnames. It is generally supposed that the practice of borrowing surnames from patrimonial states became common about the close of the tenth century, particularly in Normandy, and the contiguous portions of France. Chiefly of this class are the names in that far-famed, though apochryphal document called "The great roll of Battel Abby," which contains a list of the principal commanders and companions in arms of William, the Duke of Normandy. Under the feudal system the great barons assumed as surnames the proper names of their seignories; the knights, holding under them, did the same; and those holding under them followed their example, which was imitated by everyone possessing a landed estate, however small. And so extensively were the different civil divisions of Normandy represented in the army of the Conqueror that Camden asserts with the utmost confidence that "There is not a simple village in Normandy that has not surnamed some family in England." And as William remodeled the government of England, turning out the old aristocracy and replacing them by his own barons and others of his officers, the succeeding aristocracy were of Norman origin with but few, if indeed, any exceptions. Hence the Roll which contains the names of the commanders and companions in arms of the Norman Conqueror, has become an object of deep interest to multitudes of people in England and America.

The history of the origin of this roll, as given by Guilliam Tayleur (William Taylor), a Norman historian, is as follows:

"On the morning after the Battel (of Hastings), very early, Odo, Bishop of Baieux, sung a mass for those that were departed; after that the Duke, desirous to know the estate of his battel and what people he had lost therein and were slain, he caused to come unto him a clerke that had written their names when they were embarked at St. Valeries and commanded him to call them all by their names; who called them that had been at the battel and passed the seas with Duke William." This account was taken from Foxe's Acts and monuments by M. A. Lower and published by him in his work on the origin of surnames, Vol. 2, p. 168.

In commemoration of this great victory the Conqueror caused an Abby to be erected on the field of battle at Hastings, and the roll which contained the names of his surviving commanders and companions in arms was deposited there in the custody of the monks. Here it remained with other valuable historical memorials until the time of Henry VIII, when the abbeys and monasteries throughout the kingdom were forced to surrender their charters to the government and they were deposed by the King. "Three montns after the surrender of the Charter of Battel Abby to Henry VIII, the site on which it stood and the lands around it were given, by him, to Sir Anthony Browne, Baronet, ancestor of the Vis-Counts Montague. This family sold the mansion and its appurtenances to Sir Thomas Webster, Bart., whose descendants still possess it, but resided afterwards at their other seat, Cowdray House, near Midherst. Thither this famous document was carried. Cowdray House was destroyed by fire 1793, when the roll was presumed to have perished with everything else of value which that lordly edifice contained." M. A. Lower's essay on family nomenclature. Of this interesting document (Vol. 2), several copies have been made—two only have I seen—one was copied from the original by John Leland, the topographer, with the approbation of Henry VIII, while the Abby was still in his possession. From the known character of Leland, the excellent opportunity afforded him for obtaining an accurate copy, and the care and scrupulosity manifested by him to secure an exact transcript of the original Lower does not hesitate to give his copy the preference over all others. Leland's copy contains 494 names, about 100 less than that of Holinshed. The reason for this difference between the two copies seems to be that Leland has thoroughly sifted this document and rejected as spurious some names which had been added to it by the monks who had it in charge, and had made these entries to gratify the ambition of some aspiring families. They were publicly charged with this breach of trust, and none was better qualified to detect them than John Leland.

The names in the roll given by Leland are arranged in couplets in such a manner that the last syllable in the second line should correspond in sound with the last syllable of the first line, thus forming a species of rythm, probably to assist the memory of the clerk who called the names, and to attract the attention of those who were expected to answer the call. The number of syllables in each line were intended to be equal as far as practicable. It should be remembered that the names are Norman French, and should be pronounced according to the rules of that language. That this was the original arrangement of the Conquerer's muster roll, there seemed to be no good reason to doubt. The appearance of the copy, with its dots and marks, which are without meaning to us, are faithfully copied by Leland with the names, showing that he meant to make an exact transcript of the original, and of course, we cannot suspect him of altering the arrangement of the names, but he has given them to us in the same order in which he found them on the original roll. I will here transcribe a few of them taken promiscuously from the roll, which will give the reader a correct idea of the peculiarity of this arrangement:

Geneville et Giffard,	Graundin et Gerdoun,
Someray et Howarde,	Blundel et Burdoun,
Baudin et Bray,	Glauncourt et Chaumont,
Sauluan et Say,	Baudwin et Deaumont,
Maoun et Mainard,	Gallofer et Gubioun,
Banestre et Bekard,	Burdet et Baroun,
Bealum et Beauchamp,	Jardyn et Jay,
Loverak et Longechamp,	Fourges et Tay,

Of these thirty-two names more than one-half are still extant in this country, with slight modifications in the mode of spelling them, which will occur to the minds of most persons who read them. Thus, Giffard becomes Gifford, by changing the a for an o. Howarde has dropped the e; Baudin is now Boden, this name occurs now in Salem and New Bedford; Maoun is Moon; Mainard is Maynard; Banestre is Banister; Bealum is Beale; Beauchamp is the same; Loverall is Lovel; Gerdoun is Gorden; Blundel is Blondel and Blendin; Burdoun is Burden and Borden; Baudwin is Baldwin; Beaumont is the same. Gallofer is Gallop, a very common name in Connecticut and New York; Gubion is Gibeon; Burdet is Burdett still; Baroun is Barron and Le Barron; Jardyn is Jardin and Jorden; Jay and Tay the same.

Our ancestors came to this country bringing with them many of the surnames of these bold adventurers. It would be strange indeed if their descendants should look with indifference upon the record which proves their family identity, as far as a similarity in surnames can do. The question is often asked "in what part of Europe did your family name originate?" Those who find it in the roll of Battle Abbey may rest assured that their ancestors came from Normandy, and by looking over an old map of that country, may, perhaps, designate the district or village from which they came, and to which they are indebted for the surname which they bear. This must be an interesting matter to many people in this country, as much so to them as it will be to the Bordens, for whose information I have prosecuted this investigation.

BURDOUN.

Among the extracts already given from "the great roll" I have selected the above name as the undoubted type of Borden. It occurs there as Blundel et Burdoun. The first is now spelled Blondell, and was the surname of the Prussian minister resident at Washington 1865. The other I consider as one of the forms which the name Borden has taken in this instance through the ignorance of the age in which it was written. For I find among different writers at various periods a marked difference in spelling names. Holinshed spells this name Burdon in his copy of this roll. Leland does the same, but the u in the second syllable was added by the clerk for the sake of euphony Gerdoun and Burdoun giving similar sounds which would not be done by Gerdoun and Bourdon.

Camden remarks that "too much stress should not be laid upon uniformity in spelling ancient surnames, as ignorance of letters and their exact sounds prevailed almost universally among the people of all nations during the dark and middle ages, when there could not have been any common standard of orthography to which those who wrote could appeal, as authority. Hence every scribe wrote according to his own views." And it often happens in perusing old deeds and other ancient writings that the same names are often written differently in the same document. This is particularly true in England, where one of their writers says that he had seen an old deed where the purchaser's name was spelled, by the same person, three different ways. People of the olden time seem to have aimed at strength and not brevity in their mode of spelling, and have inserted the greatest number of vowels possible between their consonants. And the attention of modern Philologists has been severely taxed to remove the supernumeraries and thus to simplify the language.

I feel quite sure that the Bordens and Burdens, originated in Normandy—that their original surname was Bourdon and that they are a branch of the numerous families of that name still existing in France. I have found their name variously spelled as Bourdon, Burdon, Burden, Borden, Bourden, Berden, Birdin and Barden. In each case the consonants which are the root of the name are the same and they occur in the same order in each name, which shows a common origin to them all. The diversity in the three last names occurred at an early day in England, as each have a corresponding coat of arms, differing entirely from those that precede them.

If we now return to the assertion of Camden (and we can have no better authority, nor safer guide) "that there is not a village in Normandy which has not surnamed some families in England," we may, by following his advice, be enabled to trace the Bourdons or Bordens to the identical village from which they have derived their surname. And in doing this I have not been disappointed. I find by several literary works that there is, at the present time, in the northern part of France, the ancient Normandy, an ancient village which still bears the name of Bourdonnay. "It is in the Department of Muerthe, and the chief place of a Canton in the district of Chateau-Salins, four leagues northeast of Luneville." This description of this locality is copied from Rees' Cyclopedia, in which may be found a notice of several of the Bourdons who have obtained in France some eminence as artists. Of these one bore the name of Bordone, a pupil of Titian. He executed many memorable works as monuments of his extraordinary abilities both in Italy and France. He was born in 1513 and died 1588.

Another was Amomet Bourdon, a physician of Paris, who published works on anatomy 1678. And a third was Sebastian Bourdon, a painter of great note, born at Montpelier 1616. His picture of the Crucifixion of St. Peter in the church of Notre Dame in Paris is said to be a fine work of art. But, being a Calvinist and exposed to persecution on that account from the Catholics, he went to Sweden and painted for Queen Christina. Bourdon was one of the twelve persons, who in 1648, commenced the establishment of the Royal Aacademy of Paris of which he became a Director. He died 1671. See these names in Rees' Cyclopedia.

These names of distinguished men show the truth of Camden's remarks respecting the surnames of Normandy as being derived from Norman villages. He also shows the existence of Bourdons in France, and that there is even a difference in spelling this name. Bordone is Italian.

I much regret that our sources of information are so limited on this subject. I would have been glad to present here something relative to the early history of Bourdonnay, but I have not any authority to consult. But there is still one fact more within my reach, which is of great importance in indentifying the Borden name under all its various forms, except those already noticed both in England and America, with the Bourdons of Bourdonnay, as the common centre from which they have all emigrated, and this is the coat-of-arms of that village. In the Encyclopedia of Diderot and D'Alembert, Vol. 5., -.388 (Published 1781), under the article "La Bourdonnaye" I find the following description of the coat-of-arms of that village. "De Gueales, a trois bourdons de Pelerine, d'argent, 2 & 1." That is in English, Gules, 3 pilgrim staffs, two and one, argent. This is one of the simplest forms of a coat-of-arms, and its simplicity proves its antiquity. And what has surprised me is that it has been retained by most of the Burdons, Burdens and Bordens of England. It is true that some have added different crests and other devices as embellishments and equivalents, but nearly all that I have found in works of heraldry agree in the main with this of Bourdonnay. I have examined about twenty coats-of-arms in different works and find that eight are identical with that of Bourdonnay, 3 pilgrims staves; 3 have substitutes, 3 hautboys, one 3 halbards and several others battle axes, all of which are equivalent to the bourdon or pilgrim staff in heraldry. And these coats-of-arms belong to families now residing in various parts of England, and Scotland, whose names, as we have seen, are variously spelled, but in a manner which shows very clearly that they have all been derived from the same word, Bourdon, a pilgrim's staff. History and heraldry therefore unite in showing one name and one peculiar symbol, from which these various surnames and coats of arms have derived their origin in common; and it is equally certain that such an amount of mutual resemblance can point to no other conclusion than this, "that the Bordens originated at Bourdonnay in Normandy, that they went into England with William the Conqueror, and, after the victory of Hastings, October 14, 1066, settled down upon the lands given them by the new King of England." I think this conclusion is so firmly established as to exclude all reasonable doubt on the subject.

Heretofore, the British heraldic writers have uniformly asserted that the Borden surname was of Anglo-Saxon origin. And their explanation is thus expressed: "The word, Bour is Saxon and means a cottage; den is a frequent termination in the ancient British or Celtic of the names of localities and surnames, but it originally implied a situation in a woody valley. And a recent writer in the Edinburgh Review, April, 1855, p. 371, remarks that the word, den, is found in no other Germanic language than the Anglo-Saxon and was adopted into that language from the old British or Celtic. These two words united form the name of Bourdon." Well, what if they do? The question at issue is not how the name can be formed, but how and where did it originate? I will take the French word Borde, a cottage, and by simply adding the consonant n for the terminal letter and we have the name Borden perfectly formed without resorting to the old Celtic, Teutonic or Saxon languages or the boars of the wildwood. I see no good reason why this is not as correct a method of deriving this name as that of the English writers on the subject. It has this to give it the preference, the name being French points out the country from which the Bordens emigrated. The statement of English writers on the etymology of surnames often appear to me puerile and sometimes ridiculous, and in no case are they more so than in their attempts to prove the Bordens to be of Anglo-Saxon origin. They were of the Conquerors of England and not of the Conquered in England. And I was pleased to find one English writer who has, at this late day, admitted the fact. Lower in the third edition of his essay on English surnames says, the singular name of Burden must be a corruption of Bourdon, a palmer's or pilgrim's staff—a very appropriate sign for a wayside inn; Lower, Vol. 1,-.205. By this admission Lower abandons the ground on which most English writers have founded their fanciful theories concerning the Borden name, and agrees with us in ascribing to them a Norman French origin. And as history and analogy teach us the same thing concerning them I shall rest the case here so far as their nationailty is concerned.

The Bourdons rallied around the standard of their chief, and under his orders rushed to the battlefield of Hastings and did what they could to win a crown for their sovereign, William, Duke of Normandy. The Normans were successful, and in a short time were comfortably settled in the country which their courage had won. History gives no intimation that any of them returned to Normany after the Conquest, but it is natural to suppose that many of the Normans in France joined their friends in England, and history confirms this fact.

The great Roll contains the name of one Burdoun commander, who survived the battle, but the slain are not enumerated, nor can I find any intimation as to the number of followers which this one commander led to the field. The presumption is that they consisted of several hundred persons, selected from his own vassals and tenantry, each bearing the surname of their Seignior or Lord, according to the custom of their country. This view gives us necessarily the impression that there were many persons in the army of the Conqueror bearing the name of Burdon, who shared in the common danger and received their proportion of the spoils. This idea is confirmed by the fact that the name of Burdon, in all its various forms, was widely disseminated throughout England and Scotland at an early day. Among the coats of arms I notice Bourdons of Scotland, Burdon of Castle Eden, Durham County, a descendant of Thomas Burdon of

Stockton-Upon-Tees in the time of Edward IV; Burdons of Nottinghamshire and Cumberland; Burdens of Lincoln County; Bordens of Kent County, and various other counties. Lewis, in his topographical dictionary, gives two small towns by the name of Burdon. The first is in the parish of Bishopwearsmouth, northern division of Easington ward, County Palatine of Durham, 3¾ miles S. S. west from Sunderland; the second is Great Burdon, and is in that part of the parish of Haughton le Skene, which is in the southeastern division of Darlington ward, County Palatine of Durham, 2½ miles N. E. by east from Darlington; these towns were small and contain but a small number of inhabitants.

The same writer under the head of Borden remarks: "Borden, a parish in the Hundred of Milton, Lathe of Scray, County of Kent, 2¾ miles west by south from Sittingbourne, containing 650 inhabitants. The living is a vicarage in the archdeanery of and diocese of Canterbury, rated in the King's book £8, 10, and is in the patronage of J. Musgrave, Esq. The church dedicated to St. Peter and St. Paul is an ancient edifice, comprising three aisles, and three chancels, with a square tower at the western end. There are some Roman bricks mixed with the flint stones in the building, and cemented with mortar in the composition of which pulverized cockle shells have been used. The chief entrance is under a Saxon or Norman arch, and there are peculiar specimens of architecture in other parts of the edifice. A British, and several relics of Roman antiquity, together with a great quantity of round stones, like cannon balls, have been found in the neighborhood.

A more particular description of Borden has been furnished me by Judge James W. Borden. He remarks: "The County of Kent has long been considered the garden of England, and in no portion of it is the scenery more diversified or more beautiful than in the parish of Borden and its vicinity. When the Normans conquered England the Saxons of Kent were permitted to retain many of their ancient customs, and among them was the law that divided the land equally among all the sons. Consequently the farms in this parish are small comprising arable, meadow, pasture lands with orchards and hop grounds, handsomely interspersed with woodands, consisting mostly of oak, elm, hazel, birch, ash and chestnut, of which latter there are a great number in this vicinity, and among them are found the sweet chestnut, supposed by historians to have been introduced into England from the south of Europe by the Romans. The soil is generally a rich loam and the surface undulating. On ascending some of the hills in the **south part of** the village the verdant pastures and coppice woods present a most enchanting view."

"When Julius Caesar invaded England he cut a road through the woodlands of Kent from the place where he landed on the English Channel to a camp which he established at or near the place where London now stands. This road passed through the parish of Borden and the village of Borden was built beside it. The great Roman highway, afterwards built by them and since called Watling Street by the English, leading from Canterbury to London through the cities of Chatham and Rochester, was located on the north side of the parish of Borden, some distance from the road made by Caesar. The remains of a Roman camp and entrenchments were visible in that vicinity as late as the time of the civil war, called "the war of the Roses." Roman coins, arms, pottery, tiles, bricks and other vestiges of that nation, have been ploughed or dug out of the ground in this parish. And as late as 1846 some Roman coins and medals were found at Sutton Barron, about half a mile from the parish church." Hasted's notices of the Churches of Kent. -.35.

Tradition says that the ancient Druids had a place for heathen worship in an oak and chestnut grove on the spot where the parish church of Borden now stands; and that soon after the Saxons drove out the Britons from Kent, the Romish priests who went to England with Austin, the Monk, caused a temporary edifice to be erected about A. D. 636, which remained until the present parish church was built on the same site, A. D. 1134. Irelands History of Kent, Vol. 4, p. 36. "This church is a curious old building and bears characteristic marks of a very remote antiquity; and what is very remarkable, there is in the tower or steeple a chimney which appears to be coeval with the foundation of the structure."—Ireland's History of Kent. But whatever may have been the origin of this church it should be borne in mind that prior to the Norman conquest there was not a parochial church in England. The parish system was introduced there by William the Conqueror in the early part of his reign, and the parishes were organized on the same plan with those of his native country. So that the emigration from Normandy preceded the organization of the present system of parishes and the division of lands. And we should naturally expect these Norman institutions to bear the names of the conquering party, and not those of the vanquished, whenever indivadual names were appropriated by them. Some of the Bordens settled in Kent and are reported to have held large estates there soon after the conquest; and there seems to be no good reason to believe that the parish of Borden did not derive its name from them. Any other suposition is a mere conjecture suggested by an inkling of some British writers on surnames to go behind the record to explain the origin of names from some old Celtic and Saxon roots or, mere fanciful resemblances in words or things.

Camden in his Brittania Magna, Article Benenden, in the lathe f Scray (near Borden) says "Godric, a Saxon, was the original proprietor of this place. His descendants took the name of Bennenden, and the parish was called so from them." This parish I think joins that of Borden on the east; they are both in the lathe of Scray and the archdeanery of Canterbury, and have both derived their names in a similar way. Camden further says: "Some of the followers of the Conqueror adopted the names of the estates they received from him, while others chose to retain those they had received in their own country." The Bordens appear to have been of the latter class, and by this course have preserved their family connection with their Norman ancestors, and have continued this line of descent to the present time.

The earliest notices I have seen of the Bordens residing in Kent County are contained in some extracts made from Hasted's History of Kent. They were made by a competent person in England at the request of James W. Borden, Esq., of Fort Wayne, Ind., and sent to me for insertion in this work, for which he has my sincere thanks.

These extracts relate to four different persons of the name of Borden, who lived at different periods, in the parish of Borden in Kent. The first relates to Simon. "It appears from ancient records that Simon de Borden of Borden Manor, sometimes called Borden Hall or Court, resided here in the reign of King John, and was among the beneficiaries of the parish church." This John, surnamed Lackland, was the brother and successor of Richard Coeur de Lion, who fought so bravely against the infidels in Palestine. John was crowned King A. D. 1199, and died A. D. 1216 in the 17th year of his reign. Hasted's History of Kent, folio 4th, Vol. 4, p. 33 and 34.

Again, "The family of Bordens were possessed of good estates in this part of Kent, and were distinguished persons among the landed gentry of the county." Philip de Borden is mentioned in the Chartulary of the Abby of St. Radegund as having contributed yearly from his manor of Borden for the support of that Abby, H., Vol. 6 p. 74, Octave Edition.

Osbert de Borden is recorded in a charter of the times of Henry III as having pastured sixty sheep for the monastery of St. Sexlurgh on the Island of Sheppy." Henry III ascended the throne of England A. D. 1216, and died A. D. 1272. Another of this family is mentioned as having pastured a large flock of sheep for the same monastery in the reign of Henry IV, who was crowned King A. D. 1399, and died A. D. 1412. Hasted's History of Kent, Vol. 2, p. 567, folio edition.

These historical notices of the successive owners of the Borden manor in Kent extend from the year 1199 to 1412 embracing a period of two hundred and thirteen years commnecing one hundred and thirty-three years after the conquest and terminating only ninety-seven years before the accession of Henry VIII to the throne, A. D. 1509, who abolished all the monastic establishments in England. It would have been more interesting to us if these extracts had embraced the history of the Borden Manor from its first occupation by the Bordens down to the time of Henry VIII, or even to 1635. But enough historic testimony has been introduced to show that there has been a continued occupancy of this manor by the Bordens for 213 years, and that they have during that time contributed to various religious establishments located in that neighborhood. This is certainly showing a clean record of them, and one which will not cause a blush of shame upon the countenance of any of their descendants, who still live and bear their name. Their history is not written in the blood of their fellow men; their course is not traced by long marches through devastated farms, fortified camps and bloody battle-fields, but in the congenial pursuits of agriculture and the unostentatious works of Christian charity. This exhibition of their kind and generous feelings toward others, and their liberality in aiding to sustain the institutions of religion for so long a period shows them to have been a family of no ordinary stamp and fully justifies the encomium bestowed upon them by Hasted, already quoted, that they, the Bordens, "possessed good estates in this part of Kent and were distinguished persons among the landed gentry of this county." This eulogium of Hasted of the Bordens in Kent reminds me of another of Camden in his "Brittania Magna," more extensively applied to all the inhabitants of Kent. He remarks: "The country people and town residents of Kent retain the spirit of the old English above other counties, bearing good minds one to another, being ready to afford a respect and kind entertainment to strangers, and less inclined to resent injuries." From what I have already stated concerning the owners of the Borden manor, I think those who may peruse this narrative will agree with me in applying to them all the excellent traits of character described in this extract from Camden's work.

Camden and Hasted, whose works I have frequently quoted, are two of the most reliable authorities in investigations of this kind to be found in England. They do not deal in surmises and conjectures, or fanciful theories, but state such facts as they have good authority for. Neither of them displays any disposition to maintain any new theory, nor any desire to embellish their works by an attempt at eloquence. They simply follow the leadings of truth, and in all their allusions to the Bordens of Kent their testimony can be safely relied upon by us all. No better authority can be found in England.

The village of Borden is in the parish of Borden and contains the old parish church already described. It contains about one hundred houses and stands on a high ridge of land which affords a fine view of the surrounding country. The railroad from London to Dover passes through Chatham, Rochester and Borden before reaching Canterbury. The Borden station is 39 miles from London, 15 miles from Canterbury, and one mile south of Borden village. This village may be found on a map of Kent County in Camden's Brittania Magna, published 1720, also on maps of Kent County and railway maps through Kent of a recent date.

BORDEN EMIGRANTS TO AMERICA

These have not been so numerous as we might suppose from their wide dissemination throughout most of the States of the Union, nor were they as numerous as some modern writers have asserted. In the Colonial Records of Rhode Island Thomas Borden is said to have received a bushel of wheat from the town of Portsmouth April 26th, 1640, but the name on the old record is Thomas Gorton, who, the same year was appointed to tend the Pocasset, since called Howland's Ferry. It is recorded in the same work that Samuel was admitted a freeman of Warwick 1655, but there was no man of this name in the state at that time. There was a Samuel Gorton in Warwick who may have had a son, Samuel, Jr., old enough to be a freeman. Richard Borden of Portsmouth had a son Thomas, who, in 1640 was about eight years old, and a son Samuel, who, in 1655 was but ten years of age. Neither of these could be the persons named. It is also said in the same book that Robert Borden was chosen assistant when it should have been Richard, who did perform that service.

George Burden, a shoemaker, who came over with Gov. Winthrop in 1630, joined the church in Boston and had lands for eight heads laid out for him at Mount Walliston. But becoming involved in the controversy respecting Ann Hutchinson he was disarmed. He was not banished, however, with the leaders of her party; but returned to England with his family about 1640.

I have also found the name of John Bordinghe among the first settlers of Pennsylvania; he probably traveled northeasterly from Bourdonnay to Holland to reach America, as his name bears evident tokens of having had a residence in that region. Had he tarried a short time in England on his way hither, his name would have undergone a revision and he would have arrived here as Mr. Burden, Burdon or some other synonym. But as I can learn nothing of his family history, I will take leave of him here.

I have also met with the name of Sarah Borden in Drake's Annals of Boston. She was one of the unfortunate persons who were persecuted for her religious opinions, being of the Society of Friends. She was arrested and imprisoned in Boston with others of both sexes for a long time, and was at last released by the governor, not willingly, but by an order from King Charles II, who had heard of this outrage upon humanity, and determined to correct it. The historian remarks: " Some of these persons had been confined in the Boston jail for two years." The general jail delivery took place in 1661. Who this Sarah Borden was is not stated. Mary Malins, released at the same time, was of Newport, R. I., and Richard Borden's daughter Sarah, born May, 1644, was the only Sarah Borden we know of in this country at that early day.

Bryant Borden is said to have married Elizabeth Lewis, daughter of John of Saugus. But as nothing more is known of him or his family there is no use in pursuing him any farther. These are all the Borden names I have been able to find among the earliest records of emigrants to this country excepting Richard and John Borden." S.

RICHARD AND JOHN BORDEN.

"From a careful and very extended examination of the early record of the Borden family these two persons appear to have been the pioneers in the work of emigration to this country, if not the only persons of that name who came over early and were the ancestors of the numerous Borden families that are scattered throughout every part of the Union. I have examined lists containing a vast number of persons, emigrants and settlers, in various secitons of the country from 1620 to 1800 and I have not met with any new family claiming their descent from any other source than Rhode Island, where Richard lived and died. The children of Richard commenced the movement westward before the death of their father in 1671, and every succeeding generation has done the same, and the tide of emigration still sets in the same direction. And though many of them can no longer trace their genealogies to their progenitor, Richard, they can all remember the place from whence their forefathers started, and their sufferings by the way.

But it is now time to notice some inquiries respecting Richard and John, which have been often presented to me by various persons. The first question: Were Richard and John brothers? To this 1 reply that during all my investigations concerning their history I have not found any positive proof that they were brothers; still, I have met with nothing in relation to either which has led me to think that they were not brothers. And my impression is that they were so. This opinion is founded on the tradition concerning them still prevailing at Borden among the descendants of their relatives, that Richard and John were brothers and the sons of John Borden. That they went to Wales and married wives there (each of these wives bore the Welsh name of Joan), and after some years returned to the neighborhood of Borden, with the intention of emigrating to America. This tradition is confirmed by another among the descendants of Richard in Rhode Island, which states that our Borden ancestor came from Wales where they had married their wives. This report was current among the old people of this vicinity many years ago, and has never been contradicted. They had the best opportunity to gather information respecting their ancestors, and lived and died near the very spot where they did; it seems to me that great confidence is due to their statement on this subject.

The association of Richard and John first in the neighborhood of Boston and afterwards in Portsmouth, R. I., serves to strengthen our convictions of the truth of this relationship between them, as stated in the tradition on both sides of the Atlantic Ocean.

THE BORDEN FAMILY.

Again it is asked What could have induced them to remove from Kent County, the garden of England, and settle among the mountainous crags of Wales and the dense forests of America I answer, it is not an easy matter to go back several centuries and assign the reason for human conduct under trying circumstances, where there are no landmarks left by them to guide us in our researches after truth. But whoever will make himself familiar with the history of their times would never think of proposing this question. Let us look at the situation of living in Borden as late as the sixteenth and the early part of the seventeenth centuries; we shall then see that their position was exceedingly annoying and perplexing. At that time the Archbishop of Canterbury ruled with despotic sway, both in the church and in the state, holding the power of life and death over every individual in every community. Borden was only fifteen miles from Canterbury, and belonged to this Deanery, so that the Bordens lived in the immediate vicinity of this great persecuting power. And there appeared, neither on the earth nor in the heavens, any sign of coming good, no token of future prosperity to cheer the spirits and stimulate a man to brave the dangers of his present position or lead him to hope that this dark cloud which hung over his destiny would ever be dissipated. This was a trying period for the inhabitants of Kent particularly, where safety could alone be purchased by a blind submission to the arbitrary dictates of a church Hierarchy, from whose fatal grasp no situation in life was so elevated and none so humble or low as to secure the possessor from prison or the flames. The minions and tools of the persecutors were thickly scattered through every grade of society whose sole business it was to prefer complaints against individuals who were obnoxious to their employers, and they were tried and condemned on the testimony of these miscreants, and burned at the stake with the approbation of the Arshbishop of Canterbury and by his order.

Some persons may think that this is an overdrawn statement. I will here insert a short list of these executions in Kent County alone for the years 1555-56-57-58, premising that they are taken from Camden's Brittania, published 1720. His character already noticed is a sufficient guarantee of their truth.

List of persons publicly burnt at Canterbury and its vicinity, all in Kent County, and about fifteen miles from Borden.

1555—July 12, five persons; August 30, six persons; September 6, five persons; October 1, one person at Canterbury. Three persons at Dartmouth August 30, three persons at Uxbridge.

1556—Five persons at Canterbury; June, 6 persons at Canterbury; April 1, two persons at Rochester, seven miles from Borden.

1857—Five persons at Canterbury; seven persons at Maidestone, June 18, 1557 near Borden.

1558—November 20, six persons at Canterbury.

THE BORDEN FAMILY. 29

These are by no means the only human sacrifices offered upon the altars of Moloch. But these towns and cities are all in the vicinity of Borden and lie on each side of it and form but a small portion of Kent; and these executions must have produced a profound sensation among its inhabitants. A deep feeling of insecurity must have been the result of this display of religious bigotry and Satanic power, and no doubt many a young man just entering upon the theatre of life secretly determined to seek a new position, where he might avoid this destructive agency and settle down in a quiet and peaceful community. Wales was the only spot on the Island of Great Britain or in all western Europe, where such a place could be found. And is it strange that the parents of Richard and John, who were born probably about A. D. 1570, the most exciting period of the sixteenth century, should have retired to Wales, where their family would be protected by the laws as well as by the crags and ravines of this mountainous country? It was protection from the dangers of persecution and not a rich, prolific soil that they sought; it was liberty of conscience, and not affluence, which they longed for and which led them to take this direction. The Welsh, though a part of the English nation had never allowed the standard of the Pope of Rome to be planted in their territory, and were ever ready to receive and protect all who fled to them to avoid this persecuting power. But when in 1630, Governor Winthrop, with two thousand emigrants sailed to America with the approbation and assistance of the government and under their protection, the news spread throughout England with lightning speed, and gave a new direction to the thoughts, feelings and designs of the terror-stricken people of that country. The idea of placing a great gulf between them and the great persecuting power captivated their hearts, and they immediately prepared to follow their countrymen to the wild lands of America, for these could be tamed.

There is one other point in relation to Richard and John Borden to which my attention has been called. Some persons have asserted as an evidence of the intolerance of the government of Massachusetts, that these two men were forced to leave that colony in consequence of their connection with the followers of Ann Hutchinson. To this I have to reply that I have investigated this matter thoroughly and have been unable to find anything which implicates them in any of the transactions of that stormy period; neither do I believe that there is a particle of truth in the statement. They came from Wales, where they had been associated with a people, who, being free from religious oppression, could readily estimate the advantages of such a position, and therefore must have deeply sympathized with those who were harassed and banished for their religious opinions. As men they must have looked upon this controversy with feelings of sorrow and regret. It was natural that they should do so, and probably they were not the only persons in Boston or New England that indulged such feelings.

My reasons for believing them innocent of this charge are these:
1. They had but recently arrived in the country and neither of them had been admitted to the rights of citizenship, nor even allowed by the old town of Boston to be an inhabitant.
2. Neither of them appear to have been at any time members of any church either in England or America at that early day.
3. They had no connection with the church in Boston where the difficulty with Ann Hutchinson originated and to which it was confined.
4. They were not of the number of suspected persons who were arrested and disarmed and some of them banished.
5. There is no historian of these transactions who accuses them of any complicity in these religious, or to speak more correctly, these anti-religious demonstrations.

There was a George Borden, a Hutchinsonian, who was arrested and disarmed—no doubt as guilty as his associates—but he was neither banished nor punished in any other way, but of his own choice returned to England with his family about 1640, and left Richard and John Borden to bear the reproaches which should have fallen upon him, if, indeed, any were due. And certainly there could have been no reason for the government to use any harshness toward Richard and John. I am inclined to believe that this story, which has come to the ears of some of our Borden friends, had its origin as above stated, and that Richard and John went to Rhode Island as other settlers did from motives of policy. They all came to this country pilgrims and strangers, seeking a country and a people where they might quietly settle down with their wives and little ones and enjoy those blessings which they hoped to find in the society of friends and neighbors bound together by their mutual trials in the old country at their mutual dependence on each other's sympathies in the new. Their short residence in Boston had disappointed their fond anticipations, and they had now to make a new choice, and in doing this, they chose to cast their lot with the persecuted party and thus, by their action, testified to the world their opinion of this whole controversy. They saw nothing in the exiled which disqualified them for citizenship—nothing which affected their moral character, or could lead them to fear any evil to themselves or their families from their mutual association in the same community. Had the Bordens been Baptists they would most likely have joined Roger Williams at Providence; had they been of the reigning order, they could have safely remained in Boston. But being neither, they embraced this opportunity to unite with those who were suffering for conscience sake, hoping thereby to establish one colony in this country which would protect it citizens in the enjoyments of their civil and religious rights. The leaders of the Hutchison party nineteen in all, purchased the Island of Aquidneck, now Rhode Island, from the Indians, by the advice and assistance of Roger Williams. They signed an agreement which was to serve as the basis of their future government March 7, 1638, but neither Richard nor John Borden were of this number. This compact was probably made and signed in Boston. Richard signed the civil compact made by the settlers October 1, 1638, but John's name does not appear on it. All of which shows that they were settlers with this party only, and not otherwise attached to it. They went to Rhode Island to secure a comfortable home in an excellent situation, where they were surrounded by water, accessible at every point by water communication and abounding, at that time, in game, particularly in deer, and the choicest varieties of fish.

Having attended to these various insinuations respecting the two Borden emigrants and found them without any foundation in truth, I will now proceed to the consideration of the history of each separately. I will commence with John first, as he disappears at a very early day and little is known of his descendants, if he has any:

JOHN BORDEN.

"The first authentic information we have of this person is when he applied to the commissioners of emigration for liberty to go to America. This course was made necessary by the arbitrary conduct of Laud, Archbishop of Canterbury, who had procured a law to be made in 1633, by which no person was allowed to leave England without the permission of the government. The object of this law was to prevent dissenting clergymen and persons suspected of disloyal sentiments, from leaving England at all. To obtain a permit, it was necessary for the applicant to present to the Commissioners of Emigration a certificate from the minister of his parish, certifying his conformity to the rules, regulations and doctrines of the church, and also another, from a justice of the peace in the same locality, of his loyalty to the government, and no fifth monarchy man. And in order to insure the success of this scheme, Archbishop Laud, the real instigator of this project, was placed at the head of this commission. This act alone caused great anxiety in the minds of those who were contemplating a removal to the new world. And every expedient which could be thought of was put into requisition to circumvent the tyrant.

John Borden succeeded, through the influence of friends and relatives, probably, in obtaining the requisite documents, and was granted a permit to emigrate to America. Copy of this permit: "May 12, 1635. In the Elizabeth and Ann, Roger Cooper, Master; The underwritten names are to be transported, per certificate from the Minister of Benenden, Kent, of their conformitie to the orders and discipline of the Church of England,'" and it was customary to add to this: "and have taken the oath of allegiance," and also, "he is no subsidy man, nor a fifth monarchy man." But in most cases these were omitted, I suppose because a conformity to the orders and discipline of the church was considered a true test of loyalty except in the case of susjected persons, when they were all required. The family of John Borden named in this permit were thus entered on the list of passengers: John Borden, aged 28 years; Joan Borden, aged 23 years; Matthew Borden aged 5 years and Elizabeth Borden, aged 3 years. The ship sailed about the 20th of June, 1635, and arrived in Boston in the fall of the same year. John probably remained in the vicinity of Boston until he went to Rhode Island, where he next appears. But his stay here was short and the notice of him on the old records of Portsmouth barely show that he came upon the island. That record is this: "John Borden and Daniel Willcox were chosen on the grand inquest (jury) March 15, 1643." As there appears to have been no other John Borden known in this country at that time, except John the son of Richard of Portsmouth, who was then not three years of age, this must have been John, the companion of Richard. His employment on Rhode Island, the time of his sojourn there, and his departure are equally unknown. But in 1651 his name appears again in a list of persons who worked on the mill-dam at New London July 31st. Miss Calkins, in her history of that place says: "John Borden remained in this vicinity two or three years and then disappeared." She further states that John Borden, supposed to be a son of the preceding, was admitted to be an inhabitant of New London, January, 1662, and the same year he married Hannah Hough, the daughter of Deacon William Hough of that city. The children by this marriage were: John, Samuel, Hannah, William, Sarah and Joanna. These children were all baptized at New London, Joanna on the 11th of January, 1680. John Borden lived at Lynn, Conn., but later in life he removed to New Haven, where he died ,1684. And this short notice is all the information I have obtained of these two persons. The children of the second John, probably had descendants, but as I know not the course they took, I can trace them no further. Three sons starting out at that early period, might, under ordinary circumstances, have presented a large number of descendants at this time. But we have found no Borden in America who traces his family back to John." S.

REV. PARDON GRAY SEABURY.

FIRST GENERATION.

BORDEN.

FIRST GENERATION.

1. JOHN.

2. JOHN.
3. RICHARD, born in 1601, died May 25, 1671.
"It has been said that Richard did not come over to New England with John, but waited until he received a letter from him, and came the year following, 1636. If it was so, this circumstance serves, in some good degree, to explain why his name is not found on the list of passengers in 1635. The commissioners found all their labours to prevent the emigration of obnoxious persons had been a complete failure, and, becoming satisfied by the experience of 1635, that they could not attain their object, they gradually relaxed in their efforts, and at last ceased to enforce the law, and suffered emigration to flow to New England unnoticed. On this account we have no list of passengers, or rather a very meagre one, to guide us in the case of Richard in fixing the year of his arrival in this country. He may, however, have come in the same ship with John by an arrangement with the captain, as was done in multitudes of other cases. In 1635 the ship Abigail, of London, is known to have landed in this country ninety persons more than were entered upon her passenger list in London, and a large portion of them were dissenting ministers and their families, and if it was so easy for the most obnoxious persons in England to avoid the surveillance of the government, I see no reason why Richard Borden might not have done the same, and thus have arrived in Boston in the Elizabeth and Ann with John and his family.

Of Richard's early history no more need be said. When the proposition of forming a settlement on Rhode Island was made to him, he entered into it with all his heart, and to it he devoted all his energies. All the necessary arrangements having been completed, the pioneers moved forward like the advance guard of an army to select the route and prepare the way for those who were to follow them, by removing obstructions, building temporary bridges across the rivulets that impeded their way or provide rafts on which to cross the larger streams, and also, charged with the duty of erecting suitable cabins for the reception of their wives and children upon the island. All this required stout hands and willing hearts, and, in this case, as in most others of a similar character, the labour fell upon a different class of men than the leaders of the Hutchinson party, or of its rank and file, most of whom never came to Rhode Island at all.

The place first selected for the settlement was about half a mile southeast from Bristol Ferry, at the south end of a pond that opened into Mt. Hope Bay, which the settlers dignified by the name of Portsmouth Harbor. The pond still retains the name of the town pond, and ebbs and flows as it did then. The town spring has not ceased to send forth its crystal stream, as in days of yore, to gladden the hearts of men, notwithstanding the crowd of settlers have turned their backs upon it, and left it alone in its glory. To the northeast of the spring a neck of land extends about two miles, which was nearly separated by creeks, marshes and the town pond from the rest of the island. This strip of land, called by the natives Pocasset Neck, was set off by the settlers as a common by running a fence from the south end of the pond to a cove on the east side of the island. This common was called the fenced common, to distinguish it from the lands outside to the south and west of it, which were all common; and the north point then received the name of common fence point, which it still bears, though the reason for its name ceased soon after it was given, and it is now a matter of wonder with many how this name could have originated. These different objects enumerated point out the location of the first settlement upon Rhode Island and the birthplace of Matthew, the third son of Richard Borden, who was born May, 1638, and shows very nearly the time when the first families arrived there. His birth, and those of Richard's other children, born on the island, have been handed down to our times by the records of the Friends' Monthly Meeting at Newport, which further tell us that Matthew Borden, son of Richard, was the first child born of English parents upon Rhode Island. It will here be noted that the birth of Matthew occurred so early in 1638 that it must have been at the place of their first settlemnt and not at Newtown, nor yet on the homestead of Richard, since known as the MacCorrie Farm.

In 1639 the settlers concluded to change their location for another about one and one-half miles farther south, on the east side of the island which they called Newtown. There they laid out house lots for a numerous settlement, but the speedy division of the island into farms soon absorbed all the population then in Portsmouth, and the settlement at Newport this year attracted a large portion of the emigrants to that locality. So that Newtown has remained, as the lawyers sometimes say: "In statu quo," until recently it is beginning to put on the appearance of a neat, quiet, prosperous little country village. It has a Methodist and an Episcopal Church, a post office, and bids fair to become all that its original founders anticipated— only they were about two and a half centuries ahead of the times in their anticipations. This town was not laid out on the narrow, contracted, miserly plan of modern speculators. To every citizen was meted out a lot of five acres on which to place his cottage, cabbage and turnip yard, etc. I see by the record that this was the size of the lot granted to Richard Borden June 10, 1638, at the first station, and I think his lot at Newtown was the same. It was afterward built upon by his son, John, and is still held by his descendant, William Borden.

Richard was one of the men who were appointed to survey the town lots ,and subsequently, 2nd day of the 11th month, 1638, he was appointed on a committee to lay out all the farming lands in Portsmouth. He had previously signed the civil compact October 1, 1638.

The Freeman's oath, which he signed at that time being as follows: "I, Richard Borden, being in God's providence an inhabitant within the jurisdiction of this commonwealth, do freely acknowledge myself to be subject to the government thereof. And therefore do here swear by the great and dreadful name of the Everlasting God that I will be true and faithful to the same with my person and estate, as in equity I am bound, and will also truly endeavor to maintain and preserve all the liberties and privileges thereof, submitting myself to the same. And further, that I will not plot or practice any evil against it, or consent to any that shall do so, but will timely discover and reveal the same to lawful authority now here established for the speedy prevention thereof. Moreover I do solemnly bind myself in the sight of God, that when I shall be called to give my voice touching any such matter of this state in which freemen are to deal, I will give my vote and suffrage as I shall judge in mine own conscience may best conduce and tend to the Publike weal of the Body, so help me God in the Lord, Jesus Christ."

During Richard's life the town and state records show him to have been a prominent man among his contemporaries in the town and colony. He was frequently called upon to fill important stations. He was a commissioner for Portsmouth for the years 1654, 1655, 1656, 1657—the last year with William Almy, another emigrant from Benenden. He was chosen assistant, or Senator, 1653 and 1654, and September 12, 1654, he was chosen General Treasurer of the Colony, to fill a vacancy. And if we include in our estimate of him the time and labour spent in surveying the town lands and in the performance of the various other duties assigned him by his townsmen, we must regard him as an active, intelligent business man, who would be honored and respected in any community at that time, or at the present day. Indeed, he seems to have entered heartily into all the plans for the improvement of the town and colony with a just appreciation of the responsibility of those who are legislating for posterity and not for partisans of the day; and by a strictly conscientious discharge of his duty toward all, he secured the entire confidence of his fellow citizens.

It is not known at what time he became connected with the Society of Friends; but as all his children were brought up in this connection it must have been at a very early day. But it is certain that he was one of the founders of that society in Portsmouth, and by his activity and pious zeal did much to extend its influence and promote its prosperity. He was also an advocate for peaceful and gentle intercourse among neighbors, and did all in his power to reconcile the differences between the settlers on Rhode Island and those of the plantations at Providence, wishing to bring both parties under the same general government. At first Portsmouth and Newport acted together; and Providence and Warwick had done the same, each party having a separate government. But finally commissioners were chosen by each party, Richard being appointed for Portsmouth, and in a short time a union was effected under the name of "The Colony of Rhode Island and Providence Plantations," according to the designation contained in a charter obtained by Governor Coddington from Charles I. This old charter has been the basis on which that government has rested from the time this union was formed, until the present constitution was adopted.

But the attention of Richard was not confined to what was passing in the town and colony in which he lived. Great changes were constantly occurring in various sections of the country; new settlements were forming every day, and new grants of large tracts of land designed to form new colonies were made by the King of England, furnishing new openings for settlements or speculation, which kept the people in a feverish state of excitement. Richard was fully informed of these transactions by his son Francis, who had established himself at Shrewsbury in East Jersey, and was induced by him to purchase two shares in a land company in that territory for the purchase of the township of Shrewsbury. The Friends generally throughout New England took a lively interest in this colony on account of the liberal constitution adopted by Sir George Carteret, the owner, and Philip Carteret, the Governor of the new colony, which secured to them and all other persons the free exercise of the rights of conscience, and they hoped to make it a place of refuge to all who were persecuted in the other colonies. This purchase was made near the close of Richard's life, as the grant from the Duke of York to Sir George Carteret bears date June 23, 1664, and Richard died May 23, 1671.

Richard seems to have passed away suddenly at a time when he had not arranged for the disposal of his widely-extended property. When it was announced to him that he had but a short time to live, he requested some of his neighbors to be called in as witnesses, and on their arrival he proceeded to make a "nuncupative will" by declaring what disposition he wished to be made of his property. These declarations were noted down in the presence of the witnesses, and though never revised by the testator, were approved by the Council at Portsmouth July 11, 1671, and established as his lawful will. If the case had not been so urgent, due reflection would have led him to make a more equitable distribution of his property among his children. As it was, the four older sons got nearly the whole and the three younger only forty pounds each, and the three daughters were left almost unprovided for, as also was

FIRST GENERATION.

his widow. But we may charitably hope that these omissions were duly attended to by his son Matthew, the executor, and that the wants of all were abundantly supplied.

The following obituary notice of Richard Borden is copied from the Record of the Friends Monthly Meeting at Newport:

"Richard Borden of Portsmouth, R. I., being one of the first planters of Rhode Island, lived about seventy years and then died at his own house, belonging to Portsmouth. He was buried on the burial ground given by Robert Dennis to the Friends, which is in Portsmouth, and lieth on the left hand of the way that goeth from Portsmouth to Newport, upon the 25th day of the 3rd month, 1671," old style; June 5, 1671, new style. Joan, the widow of Richard, survived him eighteen years and died July 16, 1688 two years after the death of John Alden, who is supposed to have been the last of the Mayflower's company. She lived long enough to see all her children fully confirmed in what she believed to be the truth and in dying she must have had a happy consciousness that they would do honor to their parental training and cordially unite with their friends in all their plans for the support of religious institutions and the promotion of sound morals among the people at large. She died at the age of 84 years, 6 months. Reckoning back from the dates given us by the Friend's record, Richard was born about 1601, and Joan February 15, 1604.

I have endeavored to place Richard prominently before the minds of his descendants as their ancestor, whose wisdom has placed them in a country abounding in all the blessings of God's providence which can make life desirable and where they and their associates can make their influence felt in the government of the nation. If their liberty is ever trampled upon by the feet of tyrants, it will be because they have become unfaithful to the trust committed to their charge and despised the noblest birthright ever committed to mortals." S.

From the records at Portsmouth, Rhode Island, we find that Richard Borden was in the year 1638 admitted an inhabitant of the Island of Aquidneck, having submitted himself to the government that is, or shall be established, 1638, May 20—He was allotted five acres.

1639, January 2—He and three others were appointed to survey all lands near about, and to bring in a map or plot of said lands.

1640—He was appointed with four others to lay out lands in Portsmouth.

1641, March 16—Freeman.

1653, May 18—He and seven others were appointed a committee for ripening matters that concern Long Island, and in the case concerning the Dutch.

1653-54—Assistant Treasurer.

1654-55—General Treasurer.

1654-56-57—Commissioner.

1661, September 6.—He bought of Shadrack of Providence, land in Providence near Newtokonkonut Hill, containing about 60 acres.

1667—He was one of the original purchasers of lands in New Jersey from certain Indians.

1667-70—Deputy.

1671, May 31—Will made by Town Council of Portsmouth on testimony concerning the wishes of deceased. Ex. son Matthew. To widow Joan the old house and fire room, with leanto and buttery adjoining, and the little chamber in new house, and porch chamber joining to it; half the use of great hall, porch room below, cellaring and garret of new house for life. To her also firewood yearly, use of thirty fruit trees in orchard that she may choose, liberty to keep fowls about the house not exceeding forty, and all household goods at her disposal. She was to have thirty ewe sheep kept for her, with their profit and increase; fifty other sheep kept to halves, three cows kept and their profit, and to have paid her yearly a good well fed beef, three well fed swine, ten bushels of wheat, twenty bushels of Indian corn, six bushels of barley malt and four barrels of cider. To son Thomas all estate in Providence, lands, goods and chattels (except horse kind, he paying his mother Joan yearly a barrel of pork and firkin of butter. To son Francis, lands in New Jersey. To son John all land about new dwelling house of said John Borden, etc. To son Joseph, £40, within two years after the death of his mother. To son Samuel £40, half in six months after death of father and half in six months after death of mother. To son Benjamin £40 within four years after death of mother. To daughter Mary Cook, £5. To daughter Sarah Holmes, £40, within six months after death of mother. To daughter Amy Borden, £100 at age of twenty-one. To granddaughter, Amy Cook, £10 at age of eighteen. To son Matthew, whole estate after payment of debts and legacies, and if he die without issue said estate not to remain to any brother older. Inventory, £1572, 8s. 9d., viz: 200 sheep, 100 lambs, 4 oxen, 9 cows, 4 three-years, 5 two-years, 7 yearlings, 5 calves; horseflesh in Providence, £60. Four mares on the island, £20, horse £7, 10s; 6 colts, and other horseflesh at New London, £8. Thirty swine, 11 pigs, negro man and woman, £50; 3 negro children, £25; turkeys, geese, fowls, Indian corn, rye, wheat, oats, barley, pease, 2 cheese presses, 6 guns, pewter, 2 swords, 2 feather beds, 2 flock beds, hat case, silver bowl, £3; cider, £2; money, £11; goods, £16; tables, form, settle, chairs, warming pan, books, £10.

SECOND GENERATION.

SECOND GENERATION.

2. RICHARD.

4. THOMAS, b, 16—, d. Nov. 25, 1676.

"Thomas Borden of Providence was the eldest son of Richard and Joan Borden of Portsmouth, R. I. He was born in England and came to this country with his parents, but the date of his birth is unknown. He was very young when he came to this country, and his boyhood and youth were spent in Portsmouth. In 1655 Richard, Thomas and Francis Borden were admitted freemen by the general court. About this time his father made several purchases of real estate in the western part of Providence upon and around Nucokonecut, now called Nutecognet Hill, about two and a half miles northwest of the town of Providence. The purchases of Richard, Thomas and Mercy Borden in this neighborhood comprised a large tract of land, much of which is still in the possession of the descendants of their daughters, the male line from them having become extinct. Thomas being the oldest was the first to occupy and enjoy this large estate. Soon after taking possession of it he was married to Mary Harris, the daughter of William Harris of Providence, who was one of Roger Williams' party that first settled in Providence, and who is reported to have been one of the four persons who accompanied him in his flight from Salem. This estate is now included within the lines of Jonson. Among the earliest purchases made by Richard Borden was one from Shadrach Manton of sixty acres of land, dated September 6, 1661. This deed was not recorded until December, 1773, having lain neglected one hundred and twelve years. Two other deeds for sixty acres each conveyed to him the farm of David Field dated November 7th and 8th, 1662. These dates give us the date of Thomas Borden's settlement in Providence.

There are but few notices of Thomas Borden in the Providence records. His occupation was that of a farmer, and to this he devoted his whole attention; but when called by his associates to the performance of other duties he yielded to their wishes. In 1666, 1670 and 1672 he appears to have been one of the deputies from Providence, and when the fears of all New England were aroused, and the Indian warwhoop filled all hearts with terror and dismay, Thomas Borden was selected May 5th, 1675, as one of the Assistants to the Governor of Rhode Island. He served in this capacity with so much satisfaction to his constituents that he was chosen again for the year 1676. But he was not permitted to see the final triumph of his friends; he died at his post before the conclusion of this bloody struggle. He had lived in this country about forty years, and was probably not more than six years old when he came here. This would make his age at his death forty-six years. He died November 25th, 1676. His wife died March 22, 1718. In the absence of other evidence of his character and standing we can judge of him only by the estimation in which he was held by his contemporaries. He appears to have enjoyed their confidence in times of peace as a legislator, and in the great struggle for national existence, he was selected by them to advise with and aid the Governor in the performance of the arduous duties imposed upon him by a bloody and exterminating Indian war. We may, therefore, conclude that he was the man for such a time, and a man in the right place, where his country most needed his services.

Thomas probably had but a short notice of his approaching dissolution, and his will was expressed in a few words. He gave one-third of his property to his wife during her life and the other two-thirds were to be divided equally between his three sons, and the widow's third, after her death, was to be divided equally between the three sons.

The registry of the births of Thomas' children has been preserved on the records of the Friends' Monthly Meeting at Newport. It is complete with one exception—that of Richard, the eldest son. But his identity is fully established by two deeds which he made which will be referred to under his own name more fully." S.

1677, April 23—Thomas Borden's will was ratified by Town Council, on the testimony of Walter Newberry and Robert Malins as to the declaration of his mind and will. Exs. father-in-law, William Harris, and brother, John Borden. He divided his property between his wife and his sons. Being asked what he would give his doughters he said his father-in-law, William Harris, had promised to make his daughter's portions as good as he (Thomas) gave his sons. He desired that his "brother Joseph, might have his sons Joseph and Mercy, and that his daughter that was at John Borden's should remain there, until of age" His will was declared at Portsmouth, during a temporary stay there, occasioned by the Indian wars.

1718, April 28—Administration on Widow Mary's estate was granted to Thomas Harris, at the request of Mr. Richard Borden and Lieutenant Mercy Borden, sons of said Mary. Inventory of said estate being £95, 10s., 6d.

5. FRANCIS, born in England, 16—died 1703.

"Francis was born in England, and was the youngest child of his parents when they arrived in America. His childhood and youth were spent in Rhode Island, where he early acquired a knowledge of land surveying, in which operation he acted as an assistant to his father until he was fully qualified to pursue this business as a profession.

SECOND GENERATION.   45

About this time, 1665, great efforts were made by the agents of Sir George Carteret to induce emigration from the New England states to East Jersey, and similar efforts were also made by the agents of Lord Berkley to attract them to West Jersey. These two provinces had been taken from the Dutch and Swedes by Charles II, King of England, and conveyed to his brother, the Duke of York, March 20, 1664, and on the 23rd of June, 1664, the Duke of York conveyed them to the above-named gentlemen. The name to these provinces was taken from the Island of Jersey, lying in the British Channel near the coast of France, which was so bravely defended by Sir George Carteret against the combined attacks of Cromwell's fleet and army. For this noble daring he was rewarded with knighthood and a province which bore the name of the Island he had so gallantly defended. They adopted a very liberal constitution as the basis of the new government, and made the terms of settlement so favorable that many persons from New England accepted the invitation and hastened to remove their families to the new country. Indeed these proprietors, at the very commencement of their undertaking, adopted into their constitution, full liberty of conscience to all and an equality of privileges. These two propositions were posted up in all the principal towns in New England, and in connection with the low price of land drew the attention of many people. The Friends, who had suffered so much from persecutors, took a deep interest in this enterprise, and favored it to the extent of their ability, hoping soon to see their own society firmly established in this new land of freedom.

Among others, Richard Borden, of Portsmouth, purchased two shares in the East Jersey lands, which, at his death, he gave to his son Francis, who had settled at Shrewsbury, E. J., at the beginning of the settlement. He had set up his business here as surveyor and land operator, and probably found employment for all his time for many years.

The notices of Francis Borden are very meagre, but enough appears to show that he was a highly respected member of the Friends' Society, and that his associates were men of the highest rank. After settling in New Jersey he does not appear to have had any business operation in Portsmouth, R. I., as his name never occurs again upon the records of Portsmouth, after 1692, and then only as a resident of New Jersey.

But he soon formed the acquaintance of a very distinguished man which gave a new turn to his employments. The grant of Pennsylvania to William Penn, was made by Charles II, March 4, 1681. In the fall of 1682 William Penn, the Governor and proprietor of this new colony, came to America with a company of his friends and landed at New Castle on the Delaware, October 24. During his residence here he visited New York, New Jersey and Delaware. Previous to his arrival he had become interested in the East and West Jersey lands which had been bought by two Friends of Lord Berkley and the heirs of Sir George Carteret, and resold to two companies in England, consisting at first of twelve persons each, but afterwards increased to twenty-four each. This business caused Governor Penn to spend some time in New Jersey, and as he found Francis Borden fully posted in all things relating to that colony, and landed interests generally, he became very much attached to him and wished to avail himself of his superior knowledge relating to the management and disposition of wild lands. In short, Mr. Borden entered into the service of the Governor of Pennsylvania and continued in it until Penn's second visit to America. During the first visit his residence was at a place about half way between Trenton and Bordentown, on the east bank of the Delaware. Penn returned to England 1684. His rights being restored to him by William and Mary, 1694, he again visited his colony, bringing his family with him in 1699. He then resided in Philadelphia. He, however, soon found his enemies at work against him in England, and as it seemed necessary for him to remain there permanently, he returned in 1701 and arrived there in December.

It has been asserted in print that Francis Borden was a bachelor, and went to England with William Penn in 1701, and never returned to this country. This is entirely erroneous; he was married and had children, to whom he gave all his property,. dividing it equally between them; his wife died before him, and probably very early in life, which may have led others to suppose that he was never married at all. His will was dated at Shrewsbury, N. J., May 24, 1703, and proved February 18, 1705." S.

Francis Borden married February 12, 1677 Jane Vicars of Yorkshire, England, is the second marriage record in the Friend's Record.

6. MATTHEW, born May 16, 1638; died July 5, 1708.

"Matthew was born in Portsmouth, May, 1638, soon after the arrival of the first company of settlers upon the island, and it is written of him in the record of the Friends' Monthly Meeting at Newport that he was the first child born of English parents after the settlement It is to this record that I am indebted for nearly all my early dates of Richard Borden's family, and immediate descendants. Matthew must have been born at the place where the first location was made by the settlers, as they remained there only a year, and his birth occurred soon after their first arrival. In the course of time his father changed his location for a farm, which has since been owned by Andrew MacCorrie, near the east shore. Here Matthew grew up, and, at the death of his father, inherited the homestead, with all its privileges. He appears to have had the care of the place and devoted himself to the support of his parents in their declining years. After the death of his father he married Sarah Clayton, March 4, 1674. Matthew seems to have been a man of good understanding, and to have pursued the even tenor of his way through life. He was from time to time appointed to minor offices in the town, but never held any position under the colonial government.

SECOND GENERATION.

Still he seems to have possessed the most unbounded confidence of the citizens generally, as well as of the Society of Friends. If he possessed the ambition to be popular or to be wealthy and influential in the community that most men do, he must have put a strong curb upon his natural inclinations. I have met with but one instance where he engaged in the purchase of real estate in the neighborhood, and that was half a share of the Pocasset purchase of the Indians, bought of William Manchester; and as he conveyed this away before the division of the town lands occurred, I suppose that he took it as security for money loaned, and not as a purchase for use or profit. In fact he did not possess the strong desire to acquire landed property like his brother John, but seems to have pursued an entirely different course from him through life. But still Mr. Borden was an active and energetic member of the Society of Friends, and was generally known as such among all the societies of this order of people throughout this country.

The traveling Friends were generally preachers, and they always found in him a kind host to welcome them, and freely to provide for all their necessities, and his house was to them a pleasant home. In those early days, and for many years afterwards, these traveling Friends supplied the place of newspapers and private correspondence by letters; and their visits, annually, were anticipated with eagerness by the people, as they expected to receive intelligence from their friends who were settled in various remote states and territories. On their arrival at any principal town or settlement, the people flocked together to hear the news from their distant friends. And some of these preachers are said to have used great diligence in collecting information for the benefit and satisfaction of those living all along the line of their travel. Among these visiting Friends were two who traveled in company who were familiarly called the "two Dicks." These were Richard Gove and Richard Townsend from North Carolina. They used to visit Portsmouth and when going away applied for a certificate to present to their own meeting, which was granted by the monthly men and women's meeting at Newport, September, 1703.

For many years the house of Mr. Borden became the meeting place of the Friends. A note to this effect was passed at a regular meeting March 5, 1677: "Resolved, that hereafter the meetings be held at Matthew Borden's house." This resolution was passed just three years after Matthew and Sarah were married, and thus a permanent location of the Friends' meeting in Portsmouth was obtained through their liberality, which probably continued until the erection of the present house in 1706, less than two years previous to the death of Matthew Borden. On the 7th of January, 1706, the meeting at Portsmouth chose a committee of four to act in conjunction with a similar committee chosen by the Friends at Newport, to procure subscriptions, and to sell the small house and lot which John Staunton donated to the Freinds, and reimburse the same in building a meeting house at each town for the public service of truth. The committee from Portsmouth were Matthew and John Borden, Gideon Freeman and Abraham Anthony. Those of Newport were Walter Clark, John Easton, William Barker and Samuel Thurston. This committee consisted of eight of the most influential and active business men on the island. They commenced their work in earnest and before the close of the year each society was provided with comfortable accommodations for the worship of God. Both of these houses have been enlarged, and still remain as a sterling memorial to the pious zeal and labor of these early Friends, and may with proper care, stand to witness another centennial period.

Mr. Borden continued to live less than two years after the close of this year. Indeed he had already made his will, and seemed to be waiting for the signal for his departure.

Of his ten children only six are mentioned in his will, which was dated March 23, 1705. The date of his death was March 5, 1708, aged 70 years. He was on a visit to Boston when he was attacked by a malignant fever, which terminated his life in a few days. He was interred in the burial grounds of the Friends, at Lynn. His widow survived him twenty-seven years, and died April 15, 1735, aged 82 years.

And of the six sons named only three have left any male descendants. These are Joseph, Richard and Abraham. To Richard Matthew gave his land at Cooper's Neck, New Jersey, who emigrated thither, and has left male descendants there. These are now the only male descendants of Matthew living. Richard located in the township of Chester, Burlington county, New Jersey, and his heirs are known as "the Chester Bordens." The female line through the three is continued to the present time." S.

1705, March 23, will proved. Exx wife Sarah. To son Joseph, all my dwelling house and lands belonging to it in Portsmouth, he paying my wife Sarah £20 yearly for life. To wife, the use of little chamber with chimney in it, porch chamber, half of great hall, half of cellar, the garret, half of porch, liberty to keep twenty fowls, and use of ten apple trees; and son Joseph to keep a horse for her, and supply firewood. To son Joseph, two oxen, two cows, ten sheep, mare, carts, etc. To son Thomas, half a share in Tiverton, £30, silver tankard, mare, ten sheep, two cows, silver spoons, and feather bed. To son Richard land at Cooper's Creek, West Jersey, £40, a mare, ten sheep, cow and silver spoons. To son John £140 and silver spoons. To son Benjamin £140 and silver spoons. To daughter Sarah Hodson, £30. To daughter Ann Slocum, £30. To granddaughters Sarah and Ann Stodder each £10. To men's meeting of Friends on Rhode Island, £5. To wife Sarah, rest of movables.

7. JOHN, born September, 1640; died June 4, 1716.

"John Borden was born in Portsmouth, September, 1640, and married Mary Earl, the daughter of William, of the same place, December 25, 1670, less than a year prior to the death of his father. John Borden was frequently associated with his brother, Matthew, in the performance of various duties assigned by the town and religious community, of which they constituted two of the main pillars. Some of which have already been noticed in the account concerning Matthew, and I feel happy in saying that they uniformly conducted themselves and their business affairs in such a manner as to secure the entire confidence and respect of their neighbors and Friends at home, and gained for themselves among the Friends throughout every state in the union a good reputation. In fact, the name of John Borden of Quaker Hill, on Rhode Island, has been so universally spread over the country as to completely cast into the shade those of his father and brothers; so that for many years even their names had passed out of the recollection of the Borden descendants on Rhode Island; and John was supposed to have been the original emigrant from England, and the father of all that now bear this name in the country. So generally received and firmly established was this conviction that twenty-five years ago, when the Rev. Orrin Fowler published his lectures on the history of Fall River, he appended a genealogical chart of the Borden and Durfee families, in which he places John Borden at the head of the Borden family, as the original emigrant; and in a note he says: "John Borden, the first of the name in this region, and, as is believed, the father of all of the name in the United States, lived and died in Portsmouth." Nor was this impression peculiar to Mr. Fowler or the people of this vicinity. Go wherever you might, whenever you met a person of the name of Borden or inquired about the origin of his ancestors, he was sure to refer to John Borden of Rhode Island as the original emigrant. And they will invariably say "our family has been long separated from the parent stock, but we are sure we are not mistaken, for we have never heard of any other person named as our ancestor but this John Borden."

This impression has originated from the fact that for the last century his descendants have greatly outnumbered those of all his brothers. Theirs have been diminishing, while his have been increasing, until they may be found in almost every state of the union. This fact must excite the surprise of all concerned. But to account for this anomaly is beyond the power of man. Neither Thomas Borden, who settled in Providence, nor Matthew, who settled in Portsmouth, near his brother John, have now any living person bearing the Borden name to represent them here. But descendants they both have through their granddaughters, who are highly respectable and do honor to the memories of their ancestors.

The descendants of John, finding themselves thus alone in the race of life, had very naturally concluded that they had never had any competitors or companions of the Borden name, and that they had all derived their existence from John Borden. And this was in effect true, but their conclusion was too broad as this investigation has shown. John settled two of his sons near the Fall River stream; Richard and Joseph. This was the nucleus around which their descendants have rallied until Fall River has become the great Borden center in this section of the country, and no other place is known in the United

States which contains so large a number. And yet, previous to the establishment of cotton factories here it was very fashionable for the young people to remove westward to seek their fortunes among the fertile lands of New York, and subsequently of the far, far west. But time has wrought great changes in society and in the pursuits and employments of men. The introduction of manufactures in the present century has given a new impulse and a new direction to human energy. The people have concentrated, forming factory villages, towns and cities, and the country towns have been depleted of their redundant population. Even Portsmouth, the center of the early Bordens, cannot show more than four or five families of the name, but in Fall River they are very numerous, as the directory will testify, and are on the increase.

John Borden was left by his father with a good substantial estate, which gave him a fair start in the world. In addition to this he possessed, by nature, a shrewd business tact and an excellent judgment, which, with his activity of mind and untiring energy of character, insured success in whatever scheme he engaged. Accordingly, in a few years he became the owner of large tracts of land in the colonies of Rhode Island, New Jersey, Pennsylvania and Delaware, near Cape Henlopen, Lewees, and it was said by the old people that he was so eager to acquire landed property that he publicly proclaimed: "If any man has land to sell at a fair price 1 am ready to buy, and have the money ready at my house to pay for it." He also bought Hog Island and lands at Bristol Ferry, Swansea, Tiverton and Freetown. He was associated with John Tripp, and after his death, with his son, Bikill Tripp, in leasing and managing Bristol Ferry. He first commenced this connection with the ferry in 1660, when he was but twenty years of age. At that time the ferry was in the hands of Portsmouth, which established it in 1640, and the Pocasset Ferry, was established the same year, and placed in care of Thomas Gorton. Just previous to the Indian war, the general court of Plymouth, knowing that there was an intimate acquaintance existing between King Philip and John Borden, for they had several times requested John Borden to use his influence with Philip to restrain and quiet him; so now that an Indian war was imminent and the fears of the people were highly excited on account of it, they sent to him again to ask his intercession. To this request he gave heed at once, for he saw clearly that a dark cloud hung over the white settlement, still in their infancy, and poorly prepared to withstand the horrors of an Indian war. He also felt that the danger and distress which such a war would entail upon the Indians themselves would be disastrous in the extreme, and wishing to save the lives of both races, he hastened to Philip and faithfully portrayed to him the horrors and vicissitudes of such a bloody and exterminating war, the final termination of which could not be estimated or foreseen; and contrasting these with the blessings which would certainly flow from continued peace. This coming from one in whom he had implicit confidence, must have made a

SECOND GENERATION.   51

deep impression on the mind of Philip; he had said that John Borden was the only honest white man he had ever seen; he could not, even now, doubt the honesty of his intentions, nor the truth which he had spoken. But the memory of his wrongs stung him to the soul, and steeled his heart and nerved his arm for the dreadful strife; and thus he answered one whom he knew and recognized as the best friend he had on earth: "The English who first came to this country were but a handful of people, forlorn, poor and distressed. My father was then Sachem. He received them and relieved their distress in the most kind and hospitable manner. He gave them land to build and plant upon. He did all in his power to serve them. Others of their companions came and joined them. Their numbers rapidly increased. My father's counsellors became uneasy and alarmed lest, as they were provided with firearms, which was not the case with the Indians, they should finally undertake to give law to the Indian and take from them their country. They therefore advised him to destroy them before they became too strong and it would be too late. My father was also the father of the English. He represented to his counsellors and warriors that the English knew many things which the Indians did not; that they improved and cultivated the land and raised cattle and fruits, and that there was sufficient room in the country for the English and the Indians. His advice prevailed. It was concluded to give victuals to the English. They flourished and increased. Experience has taught that the advice of my father's counsellors was right. By various means they got possession of a great part of his territory. But he still remained their friend until he died. My older brother became Sachem. They pretended to suspect him of evil designs against them. He was seized and confined and thrown into sickness and died. Soon after I became Sachem, they disarmed my people. They tried my people by their own laws, assessed damages against them which they could not pay. Their land was taken. At length a line of division was agreed upon between the English and my people, and I myself was to be responsible. Sometimes the cattle of the English would come into the cornfields of my people, for they did not make fences like the English. I must then be seized and confined till I sold another tract of my country for satisfaction of all damages and costs. Thus tract after tract has gone. But a small part of the domain of my ancestors remains, I am determined I will not live till I have no country." (Arnold's History of Rhode Island, p. 394, vol. 1).

Such was the answer of Philip to the urgent appeal of his friend, John Borden, for the preservation of peace. It is a plain statement of historic facts as they had occurred from the first landing of the English at Plymouth. His sentiments were expressed with much coolness and deliberation; the deep anguish of his soul on account of the wrongs inflicted upon him and his people by those whom they had received as friends, sheltered and nourished when in a desperate and forlorn condition, nay, almost starving—shines forth in every sentence which he uttered, and he concludes by announcing to his friend the fixed determination of his mind that he would sooner perish than survive the loss of his country. How could he longer listen to proposals of peace from those whose sole object was to rob him of his country and drive him and his people far back into the wilderness to procure a precarious living among wild beasts and under their more savage Indian enemies? This he would not do, but he would die in the defence of his country. Philip was an untutored Indian, but he seems to rise to the full stature of a true patriot when he exclaimed: "I am determined not to live till I have no country." He possessed indeed strong powers of mind and a high moral sense which raised him far above the level of his treacherous neighbors and persecutors, whose miserable pettyfogging schemes to filch from him under the shadow of law, every foot of his territory, deserves universal execration.

There can be no doubt that John Borden did all in his power to dissuade Philip from engaging in a war with the English at this time. At an earlier period the chances of success would have been entirely on the side of the Indians, but now, the English had become too strong for them, and the inevitable result would be great destruction of life and property to the English, and the utter ruin of Philip and his people. And such it proved to be, and detailed accounts of this horrid tragedy had very little influence over the provincial government. Here self-manner as to exculpate the guilty party and charge the blame to those who were comparatively innocent. There was some genuine piety among the first settlers at Plymouth without doubt, but it seems to have have very little influence over the provincial government. Here self-interests and the baser passions seems to have predominated. Philip being disposed of, his Indians scattered to the four winds of heaven, the Plymouth government next attacked John Borden. It would have been supposed that this man who had hitherto been regarded by them as "the peacemaker" between them and Philip, and had served them faithfully in this capacity on several important occasions, would have been at least secure from any disturbance from them. But it seems that gratitude for any service he could render could not be reasonably expected from such men. But a few years after the conclusion of the Indian war, Mr. Borden was arrested in Bristol, in the matter of Hog Island, which the Plymouth government claimed as a part of their territory, although it had been always considered as a part of Portsmouth, and paid taxes as such. The island belonged to John Borden, who refused to pay any tax to Bristol or Plymouth. After his treacherous arrest at Bristol, Mr. Borden entered a complaint to the Legislature of Rhode Island, in 1684. In this complaint he states, as the cause of his arrest "his maintaining the true right of His Majesty's colony of Rhode Island against the intrusions of the Plymouth government."

This difficulty arose from an attempt to extend the jurisdiction of the Plymouth government over all the islands of Narragansett Bay. It does not appear whether they made their claim as a part of King

SECOND GENERATION. 53

Philip's territory or not; but no doubt that insatiable thirst for more land which they had manifested from an early day, had much to do with it. They arrested Mr. Borden's tenant first for the same purpose; but to arrest him they had to resort to strategy to get him within their jurisdiction. He was invited to come over to Bristol to receive compensation for the many services he had rendered the Plymouth government, and the ruse succeeded. The equity of their courts in such cases may be duly estimated by that of King Philip some years before. The Englishmen's cattle destroyed the Indian's cornfields, and complaint being entered, Philip was forced to sell more land to pay the damages and costs of court to the Plymouth government.

But Mr. Borden did not appear before the Plymouth court—his case took a different direction. His complaint was received by the general assembly of the colony, and his cause was adopted as their own. They immediately addressed the following communication to the Plymouth government, which states the case more circumstantially:

"General Assembly of Colony of Rhode Island:
To the Governor and Council of Colony of New Plymouth:

COMPLAINT OF RHODE ISLAND.

1. For a warrant granted by James Brown, Esq., against Morris Freelove, for possession of Hog Island.
2. We also have information that Nathaniel Byfield, in an unmanlike and deceitful manner, invited John Borden over to Bristol, pretending to requite him for former kindnesses received, and immediately caused the constable to arrest him to your court, to the intent that he might answer by virtue of a warrant granted by Daniel Smith of Rehoboth for detaining lands at Hog Island which he presumes to assert is in your colony, as by the warrant, a copy of which we have seen, is more largely demonstrated.

"Honored Gentlemen:

"We did expect that there would have been a cessation of these interruptions; forasmuch as the Honored Gov. Hinckly, Esq., and his accociates did declare at the meeting in Bristol, that, although they were not come to a final decision, yet they would live as loving neighbors until another meeting, and did hope that we should meet nearer them next time. But forasmuch as, notwithstanding we have used all fair means for a peaceable and neighborly compliance, you still persist to violate said agreement, we will, by all lawfull ways, uphold our patent right to the extent thereof. We have ordered John Borden not to give answer in your court to the matter he arrested in concerning Hog Island, forasmuch as said Island has been possessed by the town of Portmouth, in our jurisdiction, more than forty years, and none is mentioned in yours.

"Gentlemen: Thus much you have extorted from us contrary to our desires; nevertheless, we take leave, and shall be ready to serve you, wherein you serve His Majesty; and remain your loving neighbors.

"Signed by order of His Majesty's General Assembly, held for the Colony of Rhode Island and Providence Plantations June 24, 1684.

"JOHN SANFORD, Recorder

About this time the little state of Rhode Island was assailed on the west by Connecticut, and on the east by Plymouth—each endeavoring to appropriate to themselves all they could wrest from her. Connecticut claimed the jurisdiction over the western part of the state of Rhode Island, the waters of Narragansett Bay, and, of course, to all the islands those waters contained; but Plymouth claimed on the east side to tide water, and had they got possession of Hog Island, would have claimed all the remainder. There is no stopping in such a race until the object is won or lost. But the attempts on both sides failed, and the evils intended for Rhode Island recoiled upon her enemies. Capt. Christopher Almy was sent to England to lay the situation of affairs here before their Majesties, and, in 1688, four years after the arrest of John Borden, the boundaries of Rhode Island were permanently established as claimed by her citizens, though Massachusetts held on to several towns until 1746, before they relinquished their hold upon them. Capt. Almy, the agent of Rhode Island, was permitted to enter the royal palace and deliver the petition of his fellow-citizens to the Queen herself, which was regarded at the time as an act of especial favor, highly honorable to the petitioners and their little state. But their troubles did not cease. Although by an act of William and Mary the line of Rhode Island on the east was extended so as to include the five towns of Little Compton, Tiverton, Bristol, Warren, and Barrington within the jurisdiction of Rhode Island; Massachusetts claimed them as a part of the province of New Plymouth, which was quietly annexed to that state. This created a new difficulty and fresh bickerings between the new government and the people of those towns, which lasted more than half a century, very much to the annoyance of the people, and no benefit to any one interest. But in 1746, Massachusetts relinquished her claim upon the five towns, and they passed under the jurisdiction of Rhode Island. This was one of the most impudent claims that was ever set up to rob a people of their rights, for the same authority, a statute of William and Mary in 1688, that gave the province of New Plymouth to Massachusetts, had also ordered the five towns enumerated above to be annexed to Rhode Island.

So far as the documents presented for our inspection testify concerning John Borden, he was in no way concerned in raising this difficulty unless it was a sin in him to own Hog Island, which Nat Byfield wished to wrest from him for his own purposes and benefit. He was basely decoyed and betrayed within the Plymouth lines, ostensibly for state purposes, if we can credit tradition, by the meanest man that could have been found in Bristol, where he lived. In so far as he was the agent of the Plymouth government in this affair, it shows that they had sunk so low that they were no longer capable of performing the duties

SECOND GENERATION. 55

of civil magistrates, and fully justifies the decision of William and Mary to place a guardian over them. No descendant of John Borden will find in this or any other act of his life anything to lessen the respect and esteem in which his memory has always been held. For his character was always above reproach; his standing in society fully equal to that of the most elevated of his associates, and his influence over those who knew him best, was paramount to that of all others. Mr. Borden was much before the public, though not strictly speaking a public man; that is, he did not depend on the public for business nor for salary, which are now the necessary attendants upon public life. Early in life he appears as a ferryman at Bristol Ferry. From 1680 to 1708 he frequently represented the town in the general assembly. In 1706 he was associated with seven other persons in the erection of two meeting houses for the Friends, one each for the towns of Newport and Portsmouth, and often times he was engaged in minor affairs assigned him by the town or religious society to which he belonged.

Mr. Borden became very extensively known throughout the country as a Friend. To account for this we must suppose that there was something peculiarly attractive in his manner or conversation which arrested the attention of those that came in contact with him, and fixed him and his sayings in their memories. For we hear more of the sayings and doings of John Borden than of all the family of his father beside." S.

1716. February 24, will probated. Exx. wife Mary. Overseers, son Richard and Friend William Anthony. To eldest son, Richard, land in Tiverton. To son John, farm at Touisset Neck, Swanzey, half at my decease and half at death or marriage of wife, he paying my daughters Hope and Mary Borden £50 each, and to children of daughter Amey Chase deceased (late wife to Benjamin Chase of Tiverton) £15. To grandson Stephen Borden, eldest son of Joseph, my son, deceased, land in Freetown, where son Joseph built a sawmill; said Stephen paying his three brothers, William, George and Joseph, £100 each as they come of age. To grandson Joseph Borden, a half share at head of Freetown in Tiverton. To son Thomas Borden all housings and land in Portsmouth, he keeping for his mother a horse and two cows, giving her two fat swine yearly, allowing her sufficient houseroom while widow, and the keep of half a doz. fowls. To son Thomas also, rights at Hog Island. To son William, one-half of 1000 acres of land in Pennsylvania. To son Benjamin, the other half. To daughters Hope and Mary Borden all lands in Shrewsbury, N. J., and certain lands in Pennsylvania. To wife, Mary, all movables and wearing apparel, spectacles, feather bed, 2 Bibles and several other books, silver, pewter, five spinning wheels, three and a quarter years service of Indian girl, £130, cider £1, four cows, 2 two-years, 2 yearlings, calf, 40 sheep, 20 lambs, 3 swine, some pigs, etc.

1721. August, his widow Mary, declared herself to be aged sixty-six, having been married at sixteen years of age.

8. JOSEPH, born July 3, 1643; died 16—

"Joseph was born in Portsmouth January 3, 1643, and was 28 years old at the death of his father. His legacy from his father was to him a mere trifle, and one-half of that was not payable until after the death of his mother. He was married and had two children at the death of his father, and another soon after, which died in 1676. The christian name of his wife was Hope, but her surname is lost. The record of the births and deaths in his family was kept by the Friends at Newport, and he was styled John Borden of Barmadoes. From this circumstance I conclude that he emigrated to that island at an early period, and had established himself in business there some time prior to his father's death, and probably received from him some pecuniary assistance; his family residing for a time at Newport until he became permanently settled in business at Barbadoes, when they joined him. And from this period they seem to have been lost to the family circle here, as the death of his daughter, Hope, March 25, 1676, is the last reference I have been able to find of him or his family. He had a son William, who may have left some descendants in Barbadoes." S.

9. SARAH, born May, 1644; died 1705. She married Jonathan Holmes, son of Obediah and Catherine Holmes. Little is known of her life after her marriage. She had nine children.

10. SAMUEL, born 1645; died 1716, Monmouth, N. J. He married Elizabeth Crosse.

"Samuel was born July, 1645, and was 26 years old at the death of his father. The attention of the family having been drawn to the Jerseys at an early period as a place for settlement, it seems most probable that Samuel went there in company of his brother Francis as an assistant, and completed his education in the field. But however this may have been, it is certain that either before or after the death of his father, he emigrated from Rhode Island and located himself in Burlington county, near the Delaware River, where he spent the remainder of his days. This statement is strongly confirmed by the following: "When Samuel Jennings was sent over to England by the purchasers of West Jersey of Lord Berkly to organize a government for the province, Gov. Billings remaining in England, he called upon the eight counties to elect each three delegates to meet him in a general assembly to assist him in the performance of this duty. This call was made late in the fall of 1681, and the assembly convened and was dissolved, and another one assembled in 1682, of which Samuel Borden was a member." I have found in a note to Proud's early history of Pennsylvania a list of these delegates by name, and among them I notice that of Samuel Borden. This fact establishes his residence in West Jersey. Under ordinary circumstances he must have been several years among this people to have gained so much upon their confidence and to make himself sufficiently known as a man whom they could safely trust to establish their constitution and make their laws, to which they looked for the protection of their lives, their liberty and their happiness." S.

SECOND GENERATION.

11. BENJAMIN, born May, 1649; died 1718, in Burlington county, New Jersey. He settled in Middletown, N. J., in 1672, and married Abigail, the daughter to James Grover, surveyor and secretary of the Gravesend Land Co. The family of Benjamin drifted toward West Jersey.

"Benjamin was the youngest son of Richard, and at the death of his father was 21 years of age. He likewise received a legacy of only £40 on the same conditions as Joseph and Samuel did; it would, therefore, seem that his only reliance for support was in his own efforts. The will of Richard furnishes the latest date I have found concerning Benjamin. As he does not appear in the Portsmouth records I have concluded to add his name to the emigrants to New Jersey, where his two brothers, Francis and Samuel, had preceded him. The practice of surveying land in former times was a constant employment of many of the old stock of Bordens, and as they found it advantageous to themselves, they taught it to their sons as a source of revenue, for the purchase of land and all the necessaries of life. Especially was this true in the time of Samuel and Benjamin, when New Jersey and Pennsylvania first came into the market." S.

12. AMEY, born Feb. 1654; died February 5, 1684. She married William Richardson of Flushing, Long Island. They had three sons.

13. MARY, born April, 1606; died 1691. She married John Cook, son of Thomas Cook of Providence.

THIRD GENERATION.

THIRD GENERATION.

4. THOMAS, Providence, R. I.

14. RICHARD, born December, 1663; died Sept. 27, 1724. The name of his wife is unknown.

"There is but little on the records concerning him, but there is enough to show that he was the eldest son of Thomas Borden of Providence. His name is not recorded with those of his brothers and sisters, neither does it occur in the record of his father's will, though the gift was made by Thomas to his three sons, Richard, Joseph and Mercy.

In searching the records of Portsmouth I found a deed from this Richard to his three uncles, Matthew, John and Francis. The old English law at that time gave all the real estate to the eldest son, and the father's will could not convey a legal title to real estate to any of his younger children without a violation of this law. This had been done by Richard Borden of Portsmouth, and by Thomas Borden of Providence. But Thomas of Providence was the oldest son of Richard of Portsmouth, and by the law of primogeniture had a legal claim to all the real estate of his father, Richard; and Richard, the son of Thomas, being the eldest son inherited all the rights and claims to real estate which belonged to his father. So that according to the old English law he could recover all the real estate of his father and grandfather. This state of things created uneasiness among the heirs of both, particularly his uncles Francis, Matthew and John. And to remove all difficulties and restore peace and harmony in the family circle, Richard made a quitclaim to his three uncles, beginning as follows: "I, Richard Borden, son of Thomas of Providence, and grandson of Richard of Portsmouth, freely give, grant, bargain, sell and convey and forever quitclaim to my uncles, Matthew and John of Portsmouth, and Francis of Shrewsbury, in East Jersey, all my right, title and interest in all lands given them by my grandfather, Richard Borden of Portsmouth," etc.; date of deed, 1692.

I have also seen the original deed which this Richard gave to his brother Mercy for the purpose of quieting his title to the real estate given by his father Thomas. But Joseph having died in 1713, there seems to have been no quitclaim made to him.

In this deed Richard makes an allusion to the death-bed of their father, and the counsel which they heard from his dying lips, and his own intention of acting in all things according to his expressed wishes.

There appears to have been something truly noble and generous in the conduct of Richard in these transactions, and sufficiently distinguish him from the mercenary crowd who, if such an opportunity of gain should present to them, would turn out of doors their nearest relatives." S

15. MARY, born Oct. 8, 1664. The date of her death is unknown, as is that of her sister Dinah.

16. DINAH, born 1665.

17. WILLIAM, born January 10, 1667; died young.

18. JOSEPH, born November 25, 1669; died 1713. He married Margery Whipple September 12,1712. His body was found in the Nanasquatucket River December 30, 1713. His widow presented an inventory of his property to the Town Council, and was appointed administratrix upon his estate, and guardian of his child.

19. MERCY, born November 3, 1672; died April 12, 1753. He married Meribah ———, a widow, 1717.

"Mercy Borden was the youngest son of Thomas, and outlived his brothers many years. He was a very active and enterprising man, with a strong propensity for acquiring real estate. In this respect he seemed to have come nearer to the standard of John Borden of Portsmouth, his uncle, of whom it used to be said that he proclaimed publicly that "if any one had land to sell at a fair price, he was ready to purchase, and had the money in his house to pay for it." Mercy Borden must have acquired at an early day a large tract of land, as, after the death of his wife in 1723, he made twenty deeds of landed property which were recorded in Providence, and still others in Johnson during the same time, between the years 1724 and 1747. And still he had land enough to supply all his descendants and theirs with homesteads for a century. This large landed estate has passed into the hands of the Kings and Thorntons through Josiah King, who married Mary Borden, and Meribah Borden, her sister, who married Elihu Thornton, and some legacies made by John Borden, the son of Mercy, to the Thorntons. Also Richard Thornton married Meribah Borden, the daughter of William, and grand-daughter of Mercy. Her family register shows the birth of thirteen children. The Borden name, as derived from Thomas Borden is now extinct. It continued down to about 1825, in the family of Richard, the brother of Mercy. But the descendants of Mercy, through his two daughters, and his grand-daughter Meribah, bid fair to continue from generation to generation as long as the world shall endure. But the children of his daughter Mary are but few compared with the other two; she was the second wife of Josiah King, and had but two children, William Borden King and Hannah, who married Caleb Alverson. William B. King married Welthan Walton November 7, 1774. Their children were William B. King, John King and Samuel W. King, who was chosen Governor of Rhode Island about the time of the Dorr war.

It will no doubt appear strange to others as it has done to me that the name Mercy, which is so often applied to females with singular propriety, should have been ever bestowed upon a son. I have often expressed my surprise at this, and others to whom I have mentioned this case have done the same; and until recently I have written the

THIRD GENERATION.

name Marcy, supposing it to have been a surname which had been adopted as a Christian name. But recently I called upon Judge Potter of Jonston, who is a descendant of Mercy Borden, and formerly lived in his old mansion, which is still standing and occupied. He informed me that the name was Mercy, and that it was bestowed upon him by his mother under peculiar circumstances. Judge Potter said that when Mrs. Borden was in an advanced stage of pregnancy, she was sitting in her chair as usual, and her cat was lying asleep on the floor directly under her chair; a heavy shower of rain was falling at the time, accompanied with loud thunder and vivid lightning; one flash more vivid than the rest struck the house, and entering the room, killed the cat under her chair, but left her unharmed. Under the excitement produced by this sudden and terrific shock she arose quickly, and looking upon the destruction around her, she exclaimed: "The mercy of God has spared my life and that of my child; and for this great mercy my child shall be called Mercy when born." In one hour from that time the child was born, and proved to be a son; but true to her vow, she immediately bestowed the name of Mercy upon him, which he bore honorably through a long, prosperous and useful life.

Mercy Borden was born November 3, 1672, and died April 21, 1753. The death of his wife occurred September, 1723.

Mrs. Borden, the mother of Mercy, was the daughter of William Harris, a companion of Roger Williams. He had three children, Andrew, Mary and Howlong Harris. The last name reminds us of the age of Oliver Cromwell, and greatly modifies our surprise on learning that his daughter Mary had named her son Mercy. Mr. Harris was a strange man, and nothing but strange things could be expected from him. He was taken captive on a voyage to England by an Algerian pirate. He was ransomed after a year and went to England, where he died.

The will of Mercy Borden was proved May 3, 1753. His inventry amounted to 5210 pounds. After the death of his wife he sold numerous parcels of land, and at his death his real estate was very extensive for this section of country. He was a wealthy man for the times in which he lived, but by the failure of male heirs his property rapidly passed into the hands of descendants of his daughters and granddaughters and their children." S.

20. EXPERIENCE, born June 5, 1675.
21. MERIBAH, born December 19, 1667; died October 13, 1760. She was married to Phoenix Crawford February 12, 1717.

5. FRANCIS, Shrewsbury, N. J.

22. RICHARD, born Nov. 2, 1668; died 1751, at Eversham, Burlington county, New Jersey. He married Mary; her maiden name is unknown.
23. FRANCIS, born January 9, 1680; died April 6, 1759, in Shrewsbury, Monmouth county, New Jersey. He married Mary ——; maiden name unknown.

24. JOYCE, born April 4, 1682; date of her death unknown. She married John Isaac Hance; their son John Hance married Ann Borden, daughter of James Borden and wife Susan Robins of West Jersey.

25. THOMAS, born April 12, 1684.   He married Margaret ——; maiden name unknown.

6. MATTHEW, Providence, R. I.

26. MARY, born September 20, 1674.

27. MATTHEW, born August 14, 1676; died June 22, 1700. He married Peace Briggs, and left one son, Thomas, who died July 28, 1710.

28. JOSEPH, born July 18, 1678; died October 20, 1729.

"Joseph was born in Portsmouth July 18, 1678, and married Elizabeth Bryer April 8, 1708. Previous to his marriage and during the lifetime of his father, he made a journey among the Friends of New Jersey and Pennsylvania, as young men sometimes do at the present time, who are on the lookout for a partner for life, but ostensibly to enlarge the circle of his acquaintance and improve his mind by the various incidents of travel; and by contact with genteel and refined society to rub off the crude excrescences which are wont to adhere to one who has been brought up a farmer's boy.   And wishing to take with him clean papers to be used as occasion might require, he laid his intentions before the Friends' Meeting in his native town, where he was best known. After a due time spent in careful inquiry and earnest deliberation, for the Friends of the olden time did nothing by halves nor ever suffered themselves to act in a hurry, they approved, at length, of his undertaking and presented him the desired communication, which was so cautiously written as to hint the supposed object of his mission without disturbing the modesty of the adventurous traveler.

LETTER.

"To our Friends and brethren in Pennsylvania and New Jersey, wherever this may come:

"Whereas, our friend Joseph Borden, son of Matthew Borden of Portsmouth, appeared at our last monthly meeting and laid his intentions of traveling toward your parts and desired of us a certificate in order thereto; two friends were appointed to inquire concerning his clearness of entanglements as to marriage; the friends here have made report to this meeting that they do not find anything to the contrary; but that he is clear in that respect as to any woman here; and his connexions here among friends, as becometh truth and soberness as far as we know.

"Signed at our monthly men's and women's meeting at Portsmouth the 20th day of 5th month, 1703."

Witnesses:

| JOHN BORDEN | MARY BORDEN, | HOPE BORDEN, |
| MATTHEW BORDEN, | SARAH BORDEN, | and other friends. |

Joseph was 25 years of age before he started out on this very inter-

THIRD GENERATION.

esting journey, and I would gratify my readers with his notes of travel, and inklings by the wayside" if he had ever jotted them down, but if he did they are surely not extant now, and we have to regret the loss of many sage remarks, wise reflections, relusive dreams, and fond anticipations which may have fired his imagination as he passed through state after state, observing the beauties of nature, the fertility of the soil, the mildness of the weather and the numerous herds of cattle that roamed over the wide expanse of creation, unrestrained by fences or stone walls, so unlike anything he had ever seen upon Rhode Island. After a few weeks spent in travel and much cogitation, our hero turned his face homeward, where he soon arrived safe and sound. All his friends and acquaintances crowded the meeting every Sabbath hoping to hear something of the young man's visit to Pennsylvania. The girls expected to hear the old folks drop something significant about this important affair, and so they kept their eyes peeled and their ears open, and were sure to be found near every group, if no more than two persons were together. Thus they did until continued disappointment made their hearts sick, and they concluded to give it up and patiently wait for "more light,". The most censorious of his neighbors said he had beeen rejected, they knew he had, and that made him keep so dark about the matter. But the more aged men approved his course, for they had often watched him at meeting and saw that his mind was deeply engrossed by some all-absorbing subject and they did not hesitate to say that Joseph would come out all right in due time. No one told anything, because no one knew anything; for Joseph knew the inquisitiveness of his neighbors, and therefore very wisely and prudently he had kept his own secret locked up in his own heart. But after five years had expired Joseph made another trip westward, and after a few weeks' absence returned, bringing his blooming bride with him, very much to the surprise of his neighbors.

Joseph Borden was a man of good understanding and possessed of good business qualifications, and spent much of his life in the service of the public in various capacities. Of the sons of Joseph one only lived to have a family, Matthew.

Joseph Borden settled in Newport. He was annually, for ten years, chosen as treasurer of the colony, and engaged in other employments until his health failing, he declined a reëlection as treasurer, and after lingering nearly two years, he died October 20, 1729." S.

29. SARAH, born December 29, 1680. She married Josiah Hodgson.

30. ANN, born January 5, 1682. She married Giles Slocum.

31. THOMAS, born April 10, 1685; died young.

32. RICHARD, born October 10, 1687.

33. ABRAHAM born March 29, 1690. Married Elisabeth Wanton, daughter of Joseph Wanton, a Friend preacher in Tiverton, December 1, 1713. They had a numerous family of children, several of whom died in their infancy, and one, John, died at the age of 16 years.

"Abraham became a citizen and merchant of Newport, and was highly respected for his honesty and integrity. He was admitted a freeman May 3, 1715, and after the death of his brother, Joseph, Abraham was chosen to fill his place as treasurer of the Colony of Rhode Island. He did not serve, however, more than two or three years, when his earthly career was terminated by sickness and death. His life seems to have passed away like a dream when one awaketh. He died December 30, 1732, aged 42 years. His death appears to have been sudden and unexpected, as he left no will, which is much to be regretted, as that might have thrown some light upon the disposition of the land in West Jersey, bought of Arthur Cook by Matthew Borden, and given to his son Abraham, and believed to contain the site of Bordentown. Abraham's oldest son was named Joseph, but as he was lost at sea about eighteen months after his father's death, he could not have been the Joseph Borden who founded Bordentown, and figured so largely and successfully during the Revolution, as the historians relate. They say he was the son of Abraham Borden, but the same record which speaks of the birth also records the death at sea of Abraham's son Joseph, November 17, 1734.

Abraham Borden left only two sons who lived to have families, Abraham and Matthew. Abraham married and removed to Westerly, where he lived in 1771. In 1783 his residence was again at Newport. In both places he made deeds of land which he owned in Tiverton, being part of his grandfather's, Joseph Wanton's, estate. His sister, Ann, also deeded her interest in the house and lot of her grandfather, Joseph Wanton, January 11 1783.

Abraham had a son Abraham, and a daughter Mary or Molly as she was usually called. In his youth Abraham Jr., shipped on board of a privateer during the Revolution and was driven on shore by a terrible gale of wind, the weather being so cold that nearly all of the hands perished, but he was among the survivors. He was so erratic in his movements that his friends considered him idiotic. Later in life he came to New Bedford to visit his cousins, and when he left a William Rotch pleasantly remarked that "my cousin Abraham Borden came to spend a week with us, and stayed forty years." S.

34. JOHN, born August 29, 1693.

35. BENJAMIN, born April 5, 1696; died at Barbadoes, 1718.

7. JOHN, Portsmouth, R. I.

36. RICHARD, born October 25, 1671.

"Richard was born in Portsmouth October 25, 1671, four months after the death of his grandfather, and bears his name. The record of his marriage I have not found. It probably occurred about 1692, as his eldest child was born in 1694. The name of his wife was Innocent Wardell. The names of Richard and Innocent I have met with very often as witnesses to marriage certificates of their acquaintances and relatives, both in Portsmouth and Newport, and also to other documents. Richard was a man of good understanding, fair abilities, and had an eye which looked to the future. He located his homestead on the fourteen or fifteen great lots on the main road about a mile from the east shore of Mount Hope Bay, and two and a half miles south of the City Hall in Fall River. These two lots contained two hundred acres of land, and extended one mile from the shore. Here on the east side of the main road Mr. Borden erected his dwelling, small at first, but by additions made from time to time, it was made large and commodious, but exhibited not the least sign of any attempt, at any period of its erection, to imitate any particular style of architecture.

His attention at first was given principally to the clearing and cultivating his farm, procuring ship timber and plank for which there was a ready market at Newport. He was also deeply interested in the Tiverton lands, as one of the proprietors of the town, having bought one or more of the original shares. He bought and he sold land; his farm yielded abundantly and added to his means, so that in a few years he became one of the wealthiest men in the town. At this juncture, 1714, Capt. Benjamin Church and his son, Constant, who owned twenty-six and one-half shares in the mill lot and the Fall River stream, bought of the different proprietors of the Focasset purchase, proposed to Richard and Joseph Borden to sell out their interest to them for £1000. After due deliberation and consultation with their father, the purchase was made and the business was closed. But Joseph becoming sick soon after the purchase, arranged with Richard to assume the whole purchase money, and directed his wife after his death to convey his interest in this property to his brother Richard. So that he now became sole owner of the whole stream, for they had already secured the remaining shares which had not fallen into the hands of Capt. Church. And John Borden, being a part owner, arranged with Richard so as to give his grandson, Stephen Borden, the son of Joseph, deceased, the saw mill and privileges on the north side of the stream, which his son, Joseph, had erected the fall before his death. He gave him also a strip of land on the north side of the Fall River extending from the county road westward to the salt water on which his house was built.

Richard never deserted nor neglected his farm; he was always on the alert and let no opportunity for a good investment escape him. The management of his Fall River property he left with his two sons, Thomas and Joseph, whom he settled there. He purchased largely of the heirs of Edward Gray of Plymouth, as well as of other persons, so that at his death he was one of the largest landholders in the town. And it was principally to his exertions that his descendants were so bountifully supplied with the means of living without much exertion on their part. All that was needed was prudence and economy, and with a little labor properly bestowed, they have maintained themselves in independence.

But Mr. Borden's time was short for this world. And as he felt it to be so, he prepared to arrange his affairs to meet the crisis which no precautions can avert. He had already given deeds to each of his sons of one-half the property designed for them; it now remained to give to each of them by will the remaining half which he still held. By his will he confirmed to his eldest son, John, the full possession (after his own death), of his old homestead, being the fourteenth great lot, originally the property of Nathaniel Thomas, and conveyed by him to John Sands of Hempstead, L. I., for money loaned, and by said Sands sold to John Borden with one full share of the Pocasset purchase. To this was added numerous lots of woodland. To his youngest son, Samuel, he confirmed all that tract of land bought of Ephriam and Susannah Cole, a daughter of Ed Gray, of Plymouth; this was the fifteenth great lot, adjoining the fourteenth on the south, together with numerous lots of woodland, making another full share of the Pocasset purchase. To his second son, Thomas, he confirmed all his housing and lands, swamp, grist and saw mill, etc., at Fall River that lyeth to the westward of the county road, excepting that given in Pocasset purchase. To his third son, Joseph, he confirmed all his housing, fulling mill and shop where he then lived, and land whereon said housing stands, with all my lands on the eastern side of the county road up to the head or eastward thereof with the river. He also gave him sundry tracts of woodland in various parts of Tiverton, bought of Widow Tallman, Hazards, Ephriam Cole, Peter Talman, etc. To his daughter, Sarah Hazzard, he gave £160, and directed his son John to pay her £60 in addition, making £220 in money. To his daughter, Mary Gifford, wife of Christopher Gifford, a similar gift of £220; to his daughter Rebecca Borden he gave £250 in money, and for his wife, Innocent, he made an abundant provision. His executors were his three sons, Thomas, Joseph and Samuel, and to them he left all his money, bonds and a deed of every kind, giving to each the bonds owned by each of them. Richard Borden's will bears date February 12, 1731, was proved before the Hon. Nathaniel Blagrove, Esq., July 18, 1732. The date of his death I have not been able to find. He was probably about 60 years of age.

When we look over the transactions of Richard Borden during his life and see the multiplicity of his business engagements and the general thrift which attended all his operations we shall be constrained to admire the quickness of his apprehension, the soundness of his judgment, and the apparent ease and quietness with which he accumulated a large estate, so nicely arranged as to support three succeeding generations and lay a foundation for fortunes for the fourth; all that was required of the three intermediate ones being only to use and transmit to their successors, which would scatter the property but little, owing to the small number of sons in each generation." S.

37. JOHN, born 1675.

"John was born in Portsmouth in 1675, and was early settled by his father on a rough farm in Swansea, of eighty acres, on Towasset Neck, now in Warren, R. I. He married Sarah ———, her maiden surname is

THIRD GENERATION.   69

now lost. They had five children, three sons and two daughters, and had a comfortable living during the lifetime of his father. But he possessed a feeble constitution naturally, and when his father died, he imposed such a heavy burden upon him that he faltered under it. He was required to deliver one-half of his whole farming produce to his mother at her residence on Rhode Island, a requirement which had not been imposed on his brothers Richard, Joseph or Thomas, though each of them received a much larger share of their father's estate. But in truth, John possessed less energy of character than either of his brothers, he became discouraged and pined away and died about three years after the death of his father, leaving his widow to manage his farm and care for their children. She presented an inventory of his effects, and was appointed to administer upon his estate April 6, 1719; and December 24, of the same year, was married to John Earl of Portsmouth, who was probably a near relative of her former husband, if not also of herself. In this she manifested no want of tact or energy. The property and the children were well cared for until they became of age, when the farm was divided and some of it—about one-half—still remains in the Borden name. John, one of the sons, with his wife Mary, sold out his portion (2-5 of 80 acres, his father's estate) to James Mason of Warren, R. I., August 15, 1747, for £450. He then lived in Scituate, R. I., where he was admitted a freeman as early as 1739, and was engaged in a forge for the manufacture of iron upon the Ponagansett River. Benjamin remained in Swansea on a part of the homestead, and Capt. Luther Borden of Warren, lately deceased, was his grandson. He also left one son, and a brother, Martin F. Borden, now living on a part of the old homestead. Joseph married Hannah Stafford, sister of David Stafford of Tiverton, and and settled at Core Sound, Carteret county, North Carolina, where he died.

The descendants of John are quite numerous in Scituate, and are active and industrious people. They are engaged in farming, milling and once in cotton spinning, using the old forge privilege." S.

38. AMEY, born 1678, May 30; date of her death unknown. She married Benjamin Chase of Tiverton.

39. JOSEPH, born December 3, 1680; died 1715.

"Joseph was born in Portsmouth December 3, 1680, and was married to Sarah Brownell, daughter of George Brownell, and Susannah Pearce. She was born in Portsmouth June 14, 1681, and married Joseph Borden February 24, 1703. Her grandfather, Joseph Brownell, and John Cook were appointed water bailiffs by the government of Rhode Island May 20, 1647, when the laws of Oleron were adopted for the regulation of marine affairs in Rhode Island. Her grandmother's name was Ann. After the death of her husband she married John Read, October 31, 1719, for many years town clerk of Freetown.

Joseph Borden settled in Freetown, and in that part now called Fall River on the north side of the stream, and on the west side of the county road. The Richardson house now covers the spot where his house formerly stood. In 1714 the same year that the purchase of the mill lot and the stream was made from Col. Benjamin Church, Joseph Borden erected a new saw mill near where the Pocasset upper factory now stands. It was afterwards moved farther down stream, to increase the head of water near to the head of the Great Falls. His possessions extended on the north side of the stream from the county road westward to the salt water, with the privilege of joining dam or dams with his brother Richard, who owned on the south side of the stream. Together with all of John Borden's half share of the first lot in the Freeman's purchase, lying next to the Fall River on the north side. But this last tract was not included in the will of Joseph, and should be considered as a free gift made by John Borden to his grandson Stephen, in response to a suggestion of his dying son, who, in closing his will, said: "As said lands have not been conveyed to myself by deed, but remain in the hands of my father, John Borden, of Portsmouth, my request and desire is that my said father will be pleased to confirm the same in the tenour above expressed, with what more he pleaseth." John Borden, his father, being present at the time did "fully and freely declare and approve of the devises in said will, promising to give such further confirmation of said lands as may be proper, agreeably to the testator's request in said will. And this agreement was endorsed upon the will of Joseph Borden. Date of will July 15, 1715; date of codicil, July 18; will approved August 1, 1715. The codicil provided for his wife and fixed the time when Stephen should take possession of the estate.

Joseph Borden disappeared so early that he has left but little to record. He seems to have commenced the improvement of his property with considerable energy and promised fair to become an active and energetic business man, but Providence ordered otherwise, and he passed away "as a dream when one awaketh." No tradition of him unfavorable to his character has come down to us." S.

His family consisted of four sons, Stephen, William, George and Joseph These all lived to have families in and around Fall River for many years. Stephen died August 1, 1738, Joseph moved to the east side of the North Watuppa Pond, 1750, which is the date of a power of attorney to his brother George. George himself removed to Tiverton, 1755, and erected a saw-mill on the Crandall road which has been owned successively by his son John and grandson Benjah Borden, who added a grist mill to the premises in 1812. But William and his descendants have always remained in and around the place of their nativity. His youngest son was the Rev. Job Borden, the blind Baptist preacher, who is favorably remembered at the present day; so that although the married life of Joseph Borden was so very short, about 12½ years, his descendants are more numerous in this region than those of any of his brothers.

40. THOMAS, born Dec. 3, 1682, died 1745.

"Thomas was born in Portsmouth Dec. 3, 1682, and married Mary Briggs of the same town Oct. 24, 1727, he being then forty-five years of age. His wife was much younger than himself. Thomas never left the house of his father, and for many years had the direction of his farming and other interests in the town. He is spoken of as an active, energetic and industrious man, economical in his expenditures, with a mind intently bent upon the accumulation of property. In his early years he was connected with the Bristol Ferry, which was leased in his father's name, but managed by himself. By this and various other means he became possessed of a good estate in his own earnings, besides what he received from his father which was in those days an ample fortune of itself. Of his private character we know but little. He was generally considered as a shrewd business man and capable of transacting anything he chose to undertake. He received of his father, for his portion, the homestead and all the lands in Portsmouth connected with it, together with the farm on Hog Island, both upland and salt meadow. He had not possessed this property long before the price of salt marshes advanced very much under the pressure of parties who wanted to buy. But Thomas took a more rational view of this matter than most of his neighbors, and as his one hundred acres of meadow on Hog Island seemed to be an eye-sore to many people, and caused them to violate the command: "Thou shalt not covet," he at length determined to give them all an opportunity to gratify their longing desires which had so long disturbed their peace and made him the object of envy to the whole town. Accordingly when the mania for salt meadows had about reached its height, he laid out one hundred acres of his Hog Island meadows into ten-acre lots and offered them for sale at five hundred pounds per lot, putting up a notice to that effect, and a few days proved sufficient to close the sale of the whole number; and a year proved sufficient to reduce the price of this species of property so much that he bought back the whole again for three hundred pounds per lot, leaving a net profit of two thousand pounds on the whole. This land operation produced quite a sensation among the Rhode Island farmers of that day, and is occasionally alluded to at the present time when the old folks get talked over. Salt meadows have continued to decline in value in that place from the period spoken of to the present time, when it would be difficult to get a purchaser at any price, so that Mr. Borden's only error consisted in buying back again at so high a rate. He should have left this arrangement for his son William who could have secured the whole for one-quarter of the amount his father paid. Nevertheless, Mr. Borden was highly successful in his operations and accumulated a good estate which he transmitted to his descendants who were always ranked among the most respectable people in the town. To his son John he gave his homestead with all his lands in Portsmouth, together with his ferryboats. To Joseph and William he gave the Hog Island farm to be divided equally between them; and to each of his daughters, Mary and

Sarah, he gave one hundred and fifty pounds; the balance of his money went to his sons. Some years later Joseph sold his interest in Hog Island farm to his brother William, and bought the homestead of his father and the whole of John's share, and John removed to Duchess County, New York. The old family seat of John Borden of Quaker Hill is a place well known. It contained, a few years ago, two houses near each other. The oldest one, which has been taken down since the death of the late Mrs. John Borden (for she would never consent to its removal while she lived), was without any doubt the new house which Richard Borden built and gave in his will to his son John." The other house, though almost in ruins now, was called the new house, but there is no tradition of the builder, or when it was built.

Thomas Borden died about 1745, leaving a widow much younger than himself. She subsequently married Christopher Turner of Dartmouth, a widower with several children. One of Turner's daughters married Thomas' son Joseph, thus forming a twofold connection between the two families." S.

41. MARY, born 1684, died April 2, 1741. She married Thomas Potts, 1698. From this marriage came the long line of the Potts family in America. Thos. Maxwell Potts of Canonsburgh, Penn., has lately published a work entitled "Historical Collections Relating to the Potts Family in Great Britain and America."

42. HOPE, born 1685, March 3, date of death not known. She married William Almy, Jr., of Tiverton.

43. WILLIAM, born in Portsmouth, R. I., Aug 15, 1689, died 1748 in North Carolina.

"He married Alice Hull, daughter of William Hull, Esq., of Jamestown, R. I., July 7, 1715, at the Friends Meeting-house in Newport. Among the witnesses who signed the marriage certificate were ten Bordens, viz: Innocent, Hope, Mary, Sarah, Benjamin, Abraham, Benjamin, Elizabeth, Richard and Christopher. There were eight Hulls: Catherine, Sarah, Hannah, Robin, John, Alice, Tedman and John. These were probably all very near relatives of the married couple and to each other, evincing a deep sympathetic interest in this matrimonial connection and the future happiness and prosperity of their friends.

William Borden early engaged in the construction of vessels at Newport and procured his lumber and plank from Tiverton, making his home for months together at the house of his brother, Richard, about two and one-half miles south of Fall River. And the tradition in his family is that his son William was born in Tiverton. And there is an entry in the old family Bible of this son, made by himself, stating that "he was born in Tiverton, R. I." The records of that day are extremely deficient in regard to births, marriages and deaths of the permanent population and would not be likely to notice those of transient persons.

After some years spent in this employment, feeling more and more the necessity of a constant supply of duck for sails, he turned his at-

THIRD GENERATION.

tention to this also as a necessary branch of his business, and having collected what information he could on the subject, he concluded that the manufacture of duck in this country was practicable as well as desirable, and determined, if the necessary funds could be procured, to carry on the work, to commense the undertaking. The novelty of the thing itself, the wants of the shipping interest, and the credit of introducing a new branch of domestic industry and thus saving at home large sums which then were sent abroad for duck, all pressed heavily upon his mind and urged him to engage in the enterprise.

He first presented the proposition to the legislature of Rhode Island who favored the object in various ways. They first passed a law granting a bounty on every pound of flax or hemp which should be raised for this object. They then gave Mr. Borden the exclusive privilege of making duck for ten years. They offered a bounty of one pound per bolt of hemp duck made by him that should equal Holland's duck.

Under such auspices the business was commenced. Mr. Borden built the duck factory at Newport in 1722, got it into operation apparently on his own responsibility and with his own resources. After carrying on the business three years, he applied to the state for a loan of five hundred pounds for three years, which was granted him on a mortgage of real estate. This loan was for the manufacture of duck and bore date May, 1725. Another loan, or rather an advance of three hundred pounds for the same purpose, was made him by the state June, 1725, on the condition that he should manufacture one hundred and fifty bolts of duck annually for the colony. After this the business seems to have progressed without any further legislative aid until the stock of flax and hemp declining yearly, an act was passed in the assembly raising the bounty on both of these articles; that on hemp was 9d, that on flax 6d per pound. This act passed May 5, 1731.

This is about all I have been able to glean of this enterprise. I judge that he was successful in producing a good article of duck, but the high price of the stock for the manufacture and the small quantity produced by the farmers, left but a small margin for profit. He entered upon his new enterprise with the zeal of an enthusiast and seems to have devoted all his energy to promote the success of the duck manufacture. He imported laborers from Europe, who had been brought up in the business and who were intimately acquainted with the machinery used there and were familiar with all the manipulations necessary for the perfection of the work; still he found himself unable to master the difficulties of his position; and after laboring hard for ten years he was forced by circumstances he could not control to abandon the undertaking.

The real difficulties seem to have been the insufficiency of good stock— the high price of labor and the scarcity of good help. In fact the movement was premature, the country was not then prepared for it. The enterprise was a patriotic one; its object being to supply a great na tional want by the labor of our own citizens. It was pursued with energy, and patronized by the State government, but the farmers found other crops more easily raised and better adapted to their soil, requiring less labor to prepare them for the market, and therefore they would not cultivate more land in hemp or flax than what was needed for the supply of their own families. The rich bottom lands of Kentucky, Tennessee and other western States which now produce hemp and flax in abundance were then one vast wilderness which had never been trod by the foot of a white man; and with all the improvements in mechanism, the manufacture of sail duck from either has not yet been begun. Mr. Borden disposed of the duck factory about 1732 or 1733 and in the fall embarked for North Carolina and settled at Core Sound on a river which he gave the name of Newport River, in remembrance of the town from which he had emigrated. Here at a point near Beaufort he formed a settlement and soon commenced building vessels for his friends and customers at the North. And he soon became extensively known, both north and south, as William Borden, the ship builder. He was the pioneer in this business at the south, and employed a large number of men from Rhode Island in the winter season, lumbering and building vessels, year after year, most of them returning before the heat of summer had become oppressive. Mr. Borden was attracted south by the excellence and cheapness of the lumber of that country and its close proximity to his ship-yard; and many others since his day have yielded to the same temptation, though few of them have become permanent citizens of the south I recently saw a letter dated Core Sound, 1752. It was signed by Hannah Borden and directed to her mother then a widow living in Tiverton, R. I. Hanah was a Stafford and married Joseph Borden, the son of John Borden, Jr., the brother of the above William Borden Of course William was uncle to Joseph, the husband of the letter writer, and this connection accounts for their being at Core Sound at the same time. In 1771 Hannah came to Tiverton, bringing her daughter Mary, who was married to Samuel Little, the son of Fobes of Little Compton. Hannah was a widow then and resided at Nixonton, Pasquatank County.

By patient, persevering industry, Mr. Borden accumulated a sufficiency for himself and family. He left a son, William, and four daughters, Alice, Catherine, Hope and Hannah, the last of whom was born in North Carolina. The record of those born in Rhode Island I have not been able to find." S.

44. BENJAMIN, born 1692, died Nov., 1743.

Benjamin was born in Portsmouth, 1692, but the precise date cannot be found. By the will of his father he came into possession of five hundred acres of land in Pennsylvania, being one-half of a lot of one thousand acres purchased by his father and divided equally between his two youngest sons, William and Benjamin. That he was a favorite of his father appears from a condition in his will in his favor; he directed "that if Thomas dies childless, the homestead and all the property given

THIRD GENERATION.

to him should descend to Benjamin." As Benjamin did not remain long in Rhode Island, it seems probable that he followed the tracks of his uncles to New Jersey and Pennsylvania which still presented a wide field for labor, which would enable him to greatly enlarge his means. He did not settle down as a cultivator of the soil as the history of his future operations will show. S.

Extract from Howe's History of Virginia:

"In the spring of the year, 1737, Lewis (John Lewis, who settled in the valley of Virginia, on Middle River), on a visit to Williamsburg met with Benjamin Burden [Borden], who had lately come over as agent for Lord Fairfax, proprietor of the Northern Neck. Burden [Borden] accepted Lewis's invitation to accompany him to his new home in the valley. He spent several months with his friend, exploring the country and hunting buffalo with Lewis and his sons, Samuel and Andrew. But he was a more prudent hunter than Mackey (companion of Lewis). The party happened once to take a young buffalo calf, which Samuel and Andrew turned and gave to Burden, to take with him to Williamsburg. This sort of animal was unknown in lower Virginia; the calf would, therefore, be an interesting object of curiosity at the seat of government. Burden presented the shaggy monster to Governor Gooch. The Governor was so delighted with this rare pet, and so pleased with the donor, that he promtly favored his views by entering an order in his official book, authorizing Benjamin Burden [Borden] to locate 500,000 acres of land, or any less quantity, on the waters of the Shenandoah and the James rivers, on the conditions that he should not interfere with any previous grants, and that within ten years he should settle at least one hundred families on the located lands. On these conditions he should be freely entitled to 1000 acres adjacent to every house, with the privilege of entering as much more of the contiguous lands at one shilling per acre.

"Burden [Borden] returned forthwith to England for emigrants, and the next year, 1737, brought over upwards of one hundred families to settle on the granted lands. At this time the spirit of emigration was particularly rife among the Presbyterians in the northern parts of Ireland and Scotland and in the adjacent parts of England. Most of Burden's colonists were Irish Presbyterians, who, being of Scotch extraction, were often called Scotch-Irish. A few of the pure Scotch and English were mixed with the early settlers, but all, or nearly all, of the same Presbyterian stamp. Among the primitive emigrants to Burden's grant we meet with the names of some who have left a numerous posterity, now dispersed far and wide from the Blue Ridge to the Mississippi—such as Ephraim McDowell, Archibald Alexander, John Patton, Andrew Moore, Hugh Telford, John Matthews, etc."

Also settled on Borden's grant, "the Prestons, the Paxtons, the Lyles, the Grigsbys, the Stewarts, the Crawfords, the Cummingses, the Browns, the Wallaces, the Wilsons, the Carutherses, the Campbells, the McCampbells, the McClungs, the McCreas, the McKees, the McCowns, etc."

Following is a copy of a copy of Benjamin Borden's will, punctuated and corrected in spelling:

"In the name of God, Amen. The third day of April, in the year of our Lord one thousand seven hundred and forty-two, I, Benjamin Borden, of Orange County, in Virginia, yeoman, being in good state of health and sound mind and memory—thanks be it to God for it—therefore, calling unto mind the mortality of my body, I do make this my last will and testament; that is to say, principally, and first of all, I give and recommend my soul unto God that gave it, and for my body I recommend it to the earth, to be buried in a Christian-like manner at the discretion of my executors, nothing doubting but at the General Resurrection I shall receive the same again by the mighty power of God, and touching such worldly estate [as] it hath pleased God to bless me with in this life, I give and dispose of the same in manner and form following: Imprimis, I will all the funeral charges and my just debts should be paid and satisfied. Item. I give and bequeath to Zeuriah Borden, my wife, all the improvements and what lands she has or shall have occasion to clear as long as she remains my widow; and if she should get married, then she shall have but half the improvements and what land she and her husband should have occasion to clear of this plantation I now live on in Orange County, in Virginia, on Spout Run, during her natural life. Item. I give and bequeath to my son Benjamin Borden and my son John Borden and my son Joseph Borden, to them and their heirs and assigns forever, this plantation, and the lot on the said Spout Run that my well stands on, and one hundred and fifty acres I have agreed to rent to my said three sons, to be equally divided between my son Benjamin and my son John and my son Joseph Borden, in quality to be divided by way of lots drawing between my sons Benjamin and John and Joseph Borden, guardians, that is, all this plantation I now live on, excepting eight hundred acres I give to Edward Rogos and his wife Hannah Rogos and the heirs of her body forever, and five hundred acres I give William Yeamley [Feamley] and my daughter Marcy, his wife, to them and their heirs forever. Item. I give to my daughter Hannah Rogos but five shillings, she having her portion before. My will is, that all my lands and estate that I have in New Jersey should be sold, and all my lands on Bullskin and my lands on South Creek and North Shenandoah, and all my entries everywhere, and all my lands on the waters of James River should be sold, excepting five thousand acres of land that in all good I give to five of my daughters, that is, Abigail Worthington and Rebecca Bronson and to Deborah Borden and Lydia Borden and to Elizabeth Borden, that is one thousand acres of good land apiece to every one of said five daughters above mentioned, to them and their heirs and assigns forever; and all the rest of my land to be sold, as aforesaid, excepting this I now live on, to be all sold and equally divided between my wife and my son Benjamin and my son John and my son Joseph and my daughter Abigail Worthington and daughter Rebecca

THIRD GENERATION. 77

Brenson and my daughter Marcy Yeamly [Feamley] and my [daughter] Deborah Borden and my daughter Elizabeth and my daughter Lydia Borden; and my movables to be divided between my said wife and sons Benjamin, John and Joseph Borden, and my aforesaid six daughters, Abigail, Rebecca, Marcy and Deborah, Lydia and Elizabeth Borden. First before my movable estate be divided, there must be taken out my great brown riding horse and my bay mare that came of my [two words here not decipherable] Mag [or May], and the best bed, with furniture, be it good, that I have in the house, that I give to my wife first of all; the rest to be equally between my wife and my aforesaid three sons and six daughters aforesaid divided. I constitute and appo.nt my wife executrix, and my son Benjamin and my son-in-law William Feamley executors of this my last will and testament, and to execute deeds for the lands I have sold and ordered to be sold. This will I publish to be my last will and testament, and all other wills made by me void. [Here follows mention of some interlinings and erasures, made before sealing and signing.].

[Signed.]   BENJN. BORDEN. (Seal.)
Here follows the names of five witnesses.

Proved and admitted to record "at court held for Frederick County on Friday, the 9th day of December, 1743."

The name written "Yeamley" several times in the will and only once "Feamley" should perhaps be the latter. It is written William Feamley in some of the court records. The first copyist likely mistook the "F" for a "Y.".

It .s worthy of note that Benjamin Borden, the elder makes in his will mention of no negroes or slaves. The inference then is that he owned none.

As stated in his will, his wife's given name was Zeuriah, which would indicate that she was a New England woman.

It is said that he first went to Virginia in 1732. By a grant dated June 12, 1734, 1122 acres of land, lying "on the western side of the Sherandoah River, in the county to be called the county of Orange." were patented to Benjamin Borden, Andrew Hampton and David Griffith. It was in 1736 that he obtained his large grant from Governor Gooch, and in 1737 that he settled on it ten first colonists.

Benjamin Borden died in 1743, between October 21st and December 9th, most likely in November of that year. Evidently he was a man of good address and of great courage, energy and enterprise. It would appear that he crossed the ocean at least twice, to visit England on matters of business. As an explorer he must have ridden hundreds of miles in a wild, unsettled country through trackless forests, doubtless often in peril by lurking savage foes. He seems to have had the English love of broad estates, and at the time of his death, only some seven years after he obtained his conditional grant from Governor Gooch, he owned. It has been stated, 130,000 acres of land in Virginia and New Jersey. According to court records, he "was possessed of 92,000 acres of land on the waters of the James River." Because of his associations with Lord Fairfax, he was known as "Fairfax Ben." The alliterative saying, "As good as Ben Borden's bill" (pronounced Burden), handed down by those who had dealings with him, and not only common in the valley of Virginia, but often heard in other parts of the country, is a monument to the integrity of his character "more lasting than brass."

Wm. A. Obenchain, A. M., president of Ogden College, Bowling Green, Ky., gives the following account of the litigation concerning the distribution of the estate of Benjamin Borden:

"Not only according to family tradition, but as is shown by the records of the Chancery Court, in Staunton, Virginia, the history of the administration of the large estate left by Benjamin Borden, the elder, plain "yeoman," as he styled himself in his will, is a history of reckless management, injustice and wrong. Benjamin Borden, Jr., "gent," the eldest son of the testator, and one of the three executors named in the will, seems to have taken upon himself the exclusive management of the estate. He is said to have become intemperate and to have squandered a large part of it before his death. Certain it is, he was in no hurry to make a division among the heirs. On Feb. 19, 1750, more than six years after her father's death, Deborah wrote her brother a brief letter in which, after addressing him as "Loving Brother," sending love and greetings to him and his family, and entreating him to "come down" by March, she said: "Brother Joseph has chosen him a guardian, and I and my brothers are great sufferers by its [the estate's] not being divided sooner. Therefore I desire you would make no longer delay."

In 1753, three years after this letter was written, Benjamin Borden, Jr., died of smallpox, and still no division of the estate had been made. His mother had died a few days before, of the same disease.

The marriage in the following year, 1754, of Benjamin Borden's widow, Magdalen, to John Bowyer, and the marriage some years later of his widowed daughter—his only surviving child, Martha Hawkins, to Robert Harvey, complicated matters all the more. Archibald Alexander and Magdalen Borden were the executors of Benjamin Borden, Jr.'s will, John Bowyer, after his marriage, appears to have acted for his wife and sold some of the land. As shown by the records, a good deal of it was sold also by Benjamin Hawkins and Martha, his wife, and afterwards by Robert Harvey and Martha.

According to testimony now on record, Benjamin Borden, Jr., started out soon after his father's death to buy up, "at his own price," the interests of his sisters in the estate, sometimes coupling his efforts with terrible threats. With some of them he had in a manner succeeded, owing doubtless to the dire extremity to which they appear to have been reduced by the delay in the division of the estate. One of these deeds bears date as early as Sept. 17, 1745.

Robert Harvey and his wife claimed a good deal of the estate under these alleged sales to Benjamin Borden, but they were defeated in

THIRD GENERATION. 79

every case when tried by law. It is curious to note that Harvey, in some of his suits at least, styled Benjamin Borden, Jr., "eldest son and hier" (1st Leigh, Virginia, Reports; Harvey vs. Bronson. Bronson had married Rebecca Borden). Harvey had perhaps persuaded himself that, by the law of primogeniture, his wife's father was entitled to all the landed estate, in spite of Benjamin Borden, Sr.'s will.

Defeated in his efforts to establish these claims, Harvey appears to have resorted to less creditable methods to secure the coveted land. In 1st Munford, Virginia Reports, in Peck (or Beck) vs. Harvey, it is shown that in 1797 Robert Harvey inveigled Jacob Beck and Lydia, his wife, into signing a deed already prepared, making the consideration four hundred pounds, and paying cash eighteen pounds. It is also show that the land was worth between two thousand five hundred and three thousand pounds; that in 1797 Jacob Beck was about 100 years of age, and that his wife Lydia was about 80; and that they were then poor and in their dotage. Their marks were made to the deeds, but it was proved in court that they could write. Deed set aside in the lower court and decision affirmed But, owing to the delay of justice, the final decision in this case was not rendered until long after the original plaintiffs were both dead.

After long years of weary waiting, the other heirs, or their children in order to recover their interests in the estate of Benjamin Borden, the elder was compelled to file suits in the Court of Chancery, in Staunton, Va., some of which have exceeded in duration Dickin's famous case of Jarndyce vs. Jarndyce, if they have not equaled it in results.

In these suits it was shown that a great deal of the land had been sold by the "executors" at less than half its value at the time of sale According to the report returned by commissioners in 1803, in the suit of Joseph Borden vs. John Bowyer, Robert Harvey and others, 36,930 acres of the estate had, up to that time, been sold.

```
Price at which it was sold...........................  £3,570- 9-10
Value when it was sold...............................   7,825-12- 2
Value of improvements since it was sold..............  45,080-13- 0
Rents and profits ...................................  75,139- 4- 2
```

It is not unlikely that some of the land belonging to the estate was never sold at all, that is, that some of it is held today by tiller that cannot be traced back to the original grantee.

Then, as if to cap the climax, after some of the heirs had won their cases, in long and tedious suits, thousands upon thousands of dollars, some of the proceeds of sales ordered by the court, were lost in the hands of commissioners, as I was informed a few years ago by the Clerk of the Court in Staunton, Va., because of the breaking out of the Civil War before a division was made among the heirs, and the destruction of property and values, and the consequent failure of the securities, brought about by that terrible conflict.

If some of these suits, begun over a hundred years ago, are not still pending in the Chancery Court, at Staunton, Va., it is because they have been dismissed from the docket within the last few years."

The early historians of Virginia speak of the emigrants brought to Virginia by Benjamin Borden as "moral in their deportment, industrious and frugal in their habits, intelligent in their conceptions, and fixed and determined in their religious principles. Most of Mr. Borden's emigrants were Irish Presbyterians from the north of Ireland, who, being of Scottish extraction, were denominated Scotch-Irish, but are now known in this country as Protestant-Irish, widely differing in character from the Catholic-Irish of the southern counties. The remainder of them were from Scotland and the north of England. The wisdom of this selection is apparent on the face of the narrative, for though they came from places remote from each other and represent three nationalities, Irish, Scotch and English, the same caste of religion pervaded each district from which they all came. And therefore they could unite in the same church, attend upon the ministry of the same person, and commune at the same table. This was therefore, no element of discord among them either civil or religious. They met as brethren of the same church in England, and on their arrival here, they settled down into a quiet and peaceful community."

And the historian adds concerning them: "No one acquainted with the indomitable spirit of John Knox, the Scotch reformer, can fail to recognize their relationship." They no sooner found a home in America than they betook themselves to building houses, planting orchards and cultivating their lands, like men determined to provide for their families all that was necessary for their comfort and happiness. They showed less disposition than the English to hunt game or engage in traffic and speculative enterprises." Without being dull and phlegmatic, they were sober and thoughtful, keeping their native energy of feeling under restraint, and therefore capable, when exigencies occurred, of calling forth exertions as strenuous as the occasion required. As Presbyterians, neither they nor their fathers would submit to an ecclesiastical hierarchy, and their detestation of civil tyranny descended to them from the covenanters of Scotland. Another characteristic of these people was their rigid Calvanistic, or as some would call it, their Puritanical morality. Some of the churches built by this first emigrating party still remain as standing monuments of their pious zeal and persevering labors. In erecting some of these houses for religious worship, it is said that the females of the parish volunteered their services to collect the sand needed to make the mortar and to transport it in sacks upon horses several miles; while the men gathered the stones and laid up the walls of the edifice; thus securing for themselves permanent accommodations for religious worship, instead of building with logs which would soon rot down.

"Benjamin Borden left the family here (in New England) at an early period, and does not appear again on the records of his native town, and seems to have been lost to the descendants of his brothers and

THIRD GENERATION. 81

sisters in the State of Rhode Island; but his brother William removed from Newport to Core Sound in North Carolina, and the two brothers corresponded together and probably conferred with each other, and it was through a descendant of William that the position and history of Benajmin was first made known. But though nothing is known of him in New England beside his birth and parentage, I find that he was well and favorably known to all the writers of Virginia's history, to whom I am indebted for all that I have written concerning him. And if any praise is due to those who led the Pilgrims from England to Holland and afterwards from Holland to America, this action of Mr. Borden deserves a prominent place in our admiration; and his memory and high position in our recollections of the descendants of Richard Borden of Portsmouth, the grandfather of Benjamin Borden of Virginia. For when the government of Virginia joined the cotton States of the south and organized armies to establish a despotic government on the basis of slavery, the inhabitants of Western Virginia, true to the principles of their fathers and their own native instincts, refused to take part with the rebels. They at once organized their counties into an independent State and sought to be recognized as such. They were joyfully received by Congress and granted all the immunities and privileges of the other free States. Now if the question be asked, why did these counties remain faithful to the Union when their own State government and so large a portion of the south and western States cast off their allegiance, or only gave a meagre support to the government—the only true answer which can be given is this: Their ancestors were an intelligent, pious and ardent Christian people, who feared God and respected the rights of others. They were ever opposed to negro slavery and all its attending evils. Their children had been religiously educated by their parents and were deeply imbued with their sentiments. And when some of their citizens began to introduce slavery among them they tried to separate from the slave portion of the State and to organize as a free State. They did not succeed at the time. But when the day of deliverance came, they rushed to join the liberating forces, and aided in extinguishing the fires which slave holders had kindled. The judicious course of Mr. Borden in selecting such materials as he did for the basis of his colony gave a decided character to Western Virginia. The first emigrants were true men; those who followed them in long succession were their former neighbors, acquaintances, relatives and friends of the same church, and these spread over these counties gave as decided character to these counties as the Pilgrims did to New England. Their isolation secured them very much from the introduction of vicious people, and their strict morality disgusted and drove away those who may have ventured among them.

The grant to Mr. Borden was between the Blue Ridge Mountains and the Alleghenies on the headwaters of the Shenandoah and James rivers. It comprised Rockbridge County and parts of some others probably. The celebrated Natural Bridge of Vriginia is in this county, and gives to it the name. This bridge of course was once his property.

There is one fact more concerning Mr. Borden which deserves a notice as it will illustrate his character and position in Virginia, as well as the great respect shown him by the people. In his time there was very little money in the State, so that all trade was carried on by barter and the price of tobacco was the standard of valuation for all merchandise; the Legislature in imposing fines and penalties recognized this as the legal currency by ordering them to be collected in tobacco. All fines imposed called for a certain quantity of tobacco, more or less, according to the magnitude of the offense. It is said that Mr. Borden was a purchaser and shipper of tobacco which, in his time, was the principal article of export of that State. He often purchased and paid in notes of hand on time which were payable to bearer which made them negotiable and good in any person's hands, and were used oftentimes in the purchase of goods or in the settlement of debts. This was found to be a great convenience and in process of time, he was importuned by the people to increase these issues in small amounts for the accommodation of the public. These notes went into the hands of the people, and the tobacco to England and the remittances for the sales of tobacco usually preceded the presentation of the notes. Such was the popularity and acknowledged ability of Mr. Borden that the people held on to his notes as long as they held together, and only then would they part with them. The circulation of these bills, as they were called, extended throughout Virginia and North Carolina and even to this day may be heard there the oft-repeated declaration "this bill or this man's note, is just as good as Ben Borden's money," which was the highest assurance that could be given." S.

[1740.]   FAC SIMILE OF ONE OF BEN BORDENS NOTES.

THIRD GENERATION.

8. JOSEPH, Barbadoes.

45. SARAH, born April 17, 1664.
46. WILLIAM, born December 31, 1667.
47. HOPE, born December 26, 1673, died March 25, 1676.

9. SARAH, Portsmouth.

48. OBADIAH, HOLMES.
49. JONATHAN, HOLMES.
50. SAMUEL, HOLMES.
51. SARAH, HOLMES.
52. MARY, HOLMES.
53. CATHERINE HOLMES.
54. MARTHA HOLMES.
55. LYDIA HOLMES, born 1675.
56. JOSEPH HOLMES.

10. SAMUEL

57. DINAH, born November 9, 1681.
58. FRANCIS, born 1685.
59. JAMES, born 1687.
60. JOHN, born 1690.
61. JOSEPH, born 1693.
62. BENJAMIN, born 1698.

11. BENJAMIN.

63. RICHARD, born January 9, 1672, died July 9, 1744. He married Mary Worthley April 7, 1695.

64. BENJAMIN, born April 6, 1675, died 1728. He married Susannah—maiden name lost; no children are recorded of this marriage.

Benjamin Borden is styled "a weaver" in a deed which he received from John Fenwick for 200 acres of land at Middleton. The deed was dated February 28, 1697, and its pompous style sufficiently displays the arrogance of the times and the true character of Lord John Fenwick in particular. He begins his deed to Mr. Borden as follows: "I, John Fenwick, one of the Lords Absolute, or chief proprietor of the Province of New Cescrea or New Jersey, particular of Fenwick's colony, etc."

65. JAMES, born September 6, 1677, died December, 1727. He married Mary—(name not known) and lived in Freehold, New Jersey.

66. REBECCA, born June 8, 1680.

67. SAFETY, born September 6, 1682, died November, 1757, at Bordentown, New Jersey. His wife's name was Martha.

68. AMEY, born March 4, 1684.

69. JOSEPH, born May 12, 1687, died September 22, 1765. He married Ann Conover, 1717.

Gordon, in his "History of New Jersey," says of this Joseph Borden that he was an early emigrant to the State and by the record of deeds I find that his residence was in Nottingham, Burlington County. In a deed of land conveyed to him February 8, 1731, he was styled "Joseph Borden, Esq., of Nottingham," another deed May 24, 1746, he bears the same designation. But in 1750 he is styled Joseph Borden, Esq., of Bordentown. The brick house which he built for his own residence still bears the date, 1750, and there is a deed on the records of a house-lot in Bordentown, on Main, now Grove street, which he sold June 20, 1750. These dates seem to decide the point as to when Bordentown was founded and by whom. Mr. Borden laid out several other house-lots on Main street which he gave to his daughters in his will which was dated July 16, 1763, and proved 1768. His family then consisted of one son, since known as the Hon. Col. Joseph Borden of Bordentown, and five daughters as named by himself: Rebecca Brown, Hannah Lawrence Elizabeth Douglass, wife of John, Ann Potts and Amey Potts. To these he gave liberal legacies, including their children, and to his son Joseph, who appears to have been the the eldest, he gave the residue of his estate.

Mr. Borden died 1765, in his seventy-third year according to his gravestone; and having lived seventy-two years his birth must have occurred in 1693. He was buried in the cemetery which he had himself selected for this purpose on the banks of the Delaware River at the western terminus of the present Church street. His son Joseph lies at his right hand and two other Bordens on his left, Joseph, Jr., who died October 16, 1788, in his thirty-third year, and Ann Borden, who died March 11, 1774. It will be seen that Joseph Borden, Esq., lived to see the organization of the first American Congress in the city of New York, and witness the honors conferred upon his son for the faithful performance of his duty on that occasion.

"Bordentown was founded by Mr. Joseph Borden, an early settler here, and a distinguished citizen of the State, and has borne his name for nearly a century; the site is perhaps the most beautiful on the Delaware and the village is alike remarkable for its healthfulness, cleanliness and the neatness of its dwellings."—Thomas F. Gordon's History of New Jersey.

In Swank's "History of Iron in All Ages," 2nd., 182, p. 157, is the following communication from Austin N. Hungerford: "In 1722 the erection of a bloomary forge was undertaken by Isaac Horner, Daniel Farnsworth and Joseph Borden. On the west side of Black's Creek, which rises near Georgetown, in Burlington County, and runs a northeasterly course between Mansfield and Chesterfield, townships and empties into the Delaware at Bordentown. On February 1, 1725, the partly erected forge with the property connected with it was conveyed by the three proprietors to Thomas Potts, a son-in-law of Joseph Borden. On the same day he conveyed one undivided half of the property

THIRD GENERATION.

fourth to John Allen, Potts retaining a one-fourth interest for himself. to Col. Daniel Coxe (of whom mention is made that he had been interested in the manufacture of iron for several years), and one- The forge was completed in the summer of 1725, and was probably operated for a few years, but no account of it has been obtained. The property afterwards passed into the possession of a Mr. Lewis of Philadelphia. Thomas Potts had emigrated in the "Shield," which brought so many emigrants to West Jersey.

70. JONATHAN, born April 14, 1690.

71. DAVID, born March 8, 1692.

72. SAMUEL, born April 8, 1696.

12. AMEY.

73. WILLIAM RICHARDSON, born January 15, 1697.

74. THOMAS RICHARDSON, born September 10, 1680.

75. JOHN RICHARDSON, born February 1, 1683.

13. MARY,

76. MARY COOK

77. ELIZABETH COOK, born 1653.

79. JOHN COOK, born 1656.

78. SARAH COOK,

80. HANNAH COOK,

81. JOSEPH COOK,

82. MARTHA COOK.

83. DEBORA COOK,

84. THOMAS COOK,

85. AMEY COOK,

86. SAMUEL COOK.

FOURTH GENERATION

FOURTH GENERATION.

14. RICHARD, Providence.

87. RICHARD, born 1687; died 1716. He was a farmer at Johnston, R. I. The name of his wife is unknown.

88. JOSEPH, born in 1689; died 1745. He was made a Freeman of Providence May 6, 1735. He appears to hav been a man of some notoriety among his fellow citizens. He was appointed clerk of the General Assembly May 7, 1712, and in 1744 and 1745 represented the town in the same. A division deed, made by his two sons of a house and land "given them by their honored father, Joseph Borden," is recorded in Jonston, R. I. Joseph's wife was named Elizabeth.

89. THOMAS, born 1692; never married.

90. ELIZABETH, born 1694; married a Latham, Johnston, R. I.

18. JOSEPH, Providence.

91. JOSEPH, born June 2, 1713; died Feb. 24, 1795. He married Margery White, July 17, 1735. He was made a Freman of Providence May 4, 1736. He was mentioned in the will of his uncle, Mercy Borden, and received a legacy of £300 from him. Joseph's farm was on the north part of Nutecognate Hill. He is remembered by Borden descendants by the appellation of "Great Joe Borden."

19. MERCY, Providence.

92. WILLIAM, born 1719; died April 21, 1744. He married Elizabeth Hudson November 5, 1742.

94. MERIBAH, born 1722; married Elihu Thornton.

93. MARY, born August, 1720; married Josiah King.

95. JOHN, born 1723; died Wilmington, North Carolina, without heirs October 10, 1757.

22. RICHARD, Shrewsbury, N. J.

96. JAMES, born 1692; died 1771, Dec. He lived in Eversham, Burlington county, New Jersey. The maiden name of his wife Jane is unknown.

97. JOSEPH, born 1694. He married Sarah Baker at Mansfield, afterwards lived at Salem, New Jersey.

98. JONATHAN, born 1696; died 1769. He lived at Gloucester, New Jersey, where he married Martha Holmes, March 3, 1754. His descendants live in Monmouth county, New Jersey.

99. BENJAMIN, born ——; died June, 1758; lived at Eversham, Burlington county, New Jersey.

100. HANNAH, married —— Coxe.

THE BORDEN FAMILY.

101. MARY, married John Tindall.
102. AMY, married —— Foy.
103. ANN, probably died young.

23. FRANCIS, Imlaystown.

104. ELIZABETH, born July 5, 1707.
105. JANE, born July 7, 1708.
106. FRANCIS, born Dec. 24, 1709; died 1753 at Nottingham, Burlington county. He maried Mary Lippincott.
107. JOHN, born November 23, 1710; died 1772. He lived at Shrewsbury, Monmouth county, New Jersey. In 1736 he married Elizabeth ——, maiden name not on record.
108. AMEY, born December 6, 1714.
109. MARY, born July 24, 1717.
110. THOMAS, born April 27, 1719; date of death unknown. He married Massey Jackson, November 19, 1751.
111. JAMES, born August 4, 1722.

25. THOMAS, Manasquan, N. J.

112. JEREMIAH, born ——; married Esther Tilton.
113. JOSEPH, married Hannah ——.
114. RICHARD, married Hannah Tilton at Squan Meeting House, July 22, 1758.
115. BENJAMIN, married Rebecca Tilton at Thomas Tilton's house, Squan or Manasquan, December 29, 1757. Thomas Tilton's old house is yet standing about a mile from the Friend's Meeting House. Benjamin had no children.
116. SAMUEL.

27. MATTHEW, Johnston, R. I.

117. THOMAS, born 1700; died July 28, 1710.

28. JOSEPH, Portsmouth.

118. MATTHEW, born January 21, 1711, at Newport, Rhode Island.

Matthew was born January 21, 1711, in Newport, and married Hannah Clark, daughter of Samuel, of Jamestown, 1737. They had two sons and two daughters. Of these, his son William is the only one whose family has come down to the present century in the male line, but it still exists in the female line. His son William was formerly a Justice of the Peace and town treasurer of Newport as early as 1783, and continued until about 1820, when he died. His son William commanded the brigantine Nicholas and Felix, of Newport, on a voyage to the West Indies about 1825. William Williams of Newport being supercargo. And as they were never heard from after leaving port

FOURTH GENERATION.

It has been surmised that they were cut off by the pirates who infested the West Indies about that time. His death closed the male line of descent from Matthew Borden through his son Joseph. S

119. MARY, born June 3, 1710.

120. SARAH, born Oct. 24, 1718. Married Peleg Thurston of Newport, Rhode Island.

121. ELIZABETH, born August 25, 1720.

33. ABRAHAM, Portsmouth.

122. JOSEPH, born April 2, 1716, was lost at sea Nov. 17, 1734.

123. SARAH, born January 10, 1716; married Thomas Howland.

The family of Sarah was never very numerous. They lived in Tiverton, near the present stone bridge. Phebe Howland, a granddaughter of Sarah, now lives near the old homestead of Joseph Wanton.

124. ABRAHAM   Twins.

125. ELIZABETH, born September 4, 1719; died young.

126. ANN, born November 6, 1720; never married; her deed of interest in the house and lot of Joseph Wanton, her grandfather, is signed Ann Borden, January 11, 1783.

127. BENJAMIN, born July 17, 1721; died young.

128. MATTHEW, born April 2, 1723.

Matthew was the son of Abraham, Sr., born in Newport and first married Sarah Whipple, daughter of Joseph, October 15, 1749, and after her death he married Mary Borden, the widow of his cousin Capt. Thomas Borden, the son of Joseph. The history of this family is very obscure, owing to the loss of the records of Newport. But the census taken by the colonial government of Rhode Island in 1774, shows that Matthew was then dead, and that his widow, Mary Borden, then had three males and four females in her family. If all these were the children of Matthew they may have been by one or both of his wives, and the males may have been either sons or sons-in-law of Matthew. One of these daughters was named Mary Wanton Borden, and married Thomas Bush of Newport.

129. BENJAMIN, born May 9, 1726; died in the West Indies.

130. ELIZABETH.

131. MARY, born 1728.

"Mary married Thomas Rodman April 5, 1750, and was his second wife. By him she had seven children; one son, Samuel Rodman of New Bedford, deceased, and six daughters, Mary, Elizabeth, Nancy, Hannah, Sarah, Charity. Samuel Rodman marrying Elizabeth Rotch of Nantucket, daughter of William, and his sister, Elizabeth Rodman marrying William Rotch, Jr. All their descendants may be reckoned among those of Abraham Borden, Sr., of Newport. They settled in

New Bedford many years ago, and have long been ranked among its most distinguished citizens. Their enterprising zeal and personal energy and capital has done much toward building up the town and stimulating others in the pursuit of wealth. Samuel Rodman was educated in the counting room of Aaron Lopez at Newport, and for many years was his confidential clerk. After the death of Lopez, in 1779, and the seizure of his property by the British at Newport, Mr. Rodman removed to Worcester county, in Massachusetts, with his mother and sisters. After a few years they removed to Nantucket, and from thence to New Bedford. Mr. Rodman invested largely in the Pocasset Manufacturing Company of Fall River, Mass." S.

132. JOHN, born April 16, 1731; died August 28, 1747.

36. RICHARD, Fall River,

133. SARAH, born July 31, 1694, married Mr. Hazzard of Newport, R. I.

134. JOHN, born December 24, 1695; married Ann Bennett, June 20, 1768. They had no children.

135. THOMAS, born December 8, 1697; died April, 1740, at Tiverton, R. I. He married Mary Gifford August 14, 1721.

136. MARY, born January 29, 1700; married Christopher Gifford of Dartmouth, Mass.

137. "JOSEPH, born November 4, 1702. He learned the trade of a clothier, and while quite young carried on that business in the old fulling mill built by Col. Benjamin Church on the Fall River near to the head of the Great Falls. He pursued his business with diligence, and there being no competing establishment near him, soon obtained a good run of business, as it was the custom in those days for every family to manufacture their own woolen clothes, which requried fulling and dressing by those who understood the business.

Joseph Borden married Abigail Russel of Dartmouth, June 24, 1730; she was a sister of Caleb Russell, who, April 2, 1734, married his sister Rebecca, and who, in 1640, became the guardian of Joseph's children. Mr. Borden was now very pleasantly situated, his father had given him a deed of one-half of the water power on the south side of the river from the foot of the Great Falls to the main road, together with the half of all the buildings upon the adjoining lands, and in 1732, by the will of his father, proved July 25, the other half of this property was given to him, besides other landed estate. At this time he had two children, Abraham and Patience, and no doubt was looking forward with pleasing anticipation to the future. But God, without whose notice not a sparrow falleth to the ground, had marked out for him a different course. He was permitted to enjoy his prosperity but little more than two years after the death of his father, when he was suddenly struck down in a moment. He was working alone in his mill at the time, and it is supposed that in attempting to adjust some part of the machinery he received a blow which instantly killed him. He was found lying upon the floor, and the mill running at the usual speed. Nothing further than this was known by his family and friends concerning his death. He left four children, all too young to realize the magnitude of their loss, or to do anything, for many years, to repair it or manage for themselves. The death of Mr. Borden was entered on the records of the Friends' Monthly Meeting at Newport as occurring December, 1736. His widow married a young man by the name of —— Jencks, who was an apprentice to her husband, who continued the business, and greatly assisted her in managing the estate of Mr. Borden and in bringing up his children, until the eldest son, Abraham, was old enough to take charge of it. He, too, became a clothier.

Joseph Borden and his family were members of the Friends' Society in Tiverton. Caleb Russell of Dartmouth was appointed guardian of the children of Joseph Borden of Tiverton, being minors under 14 years of age, October 21, 1740. As this was done by the Town Council of Tiverton, and Mrs. Borden's house stood on the south bank of the river, it seems that the boundary line between the Freeman's and Pocasset purchases were fully recognized at that day. Under the management of Caleb Russell and Abraham Borden, the estate of Joseph was preserved intact, and descended to the two sons of Abraham, Simeon Borden and Perry Borden, who also left it to their heirs with the single exception of the upper dam, which was sold to the Troy Manufacturing Company about 1812 by the widow of Simeon.

Joseph's two daughters both married; Patience married Hon. Thomas Durfee, Esq., August 9, 1747, and had twelve children, eight sons and four daughters. Peace married Joseph Borden, son of William, and had fourteen children, nine sons and five daughters, making twenty-six children born of these two sisters, while the two brothers' families contained only three each, and six in all, and of them there were only two, Simeon and Perry, to continue the male line of Joseph Borden; and the case is still the same in the present generation, the Hon. Nathaniel B. Borden having left but two sons, and his brother Simeon dying unmarried. But the children of Joseph's two daughters all, with one exception, living to a marriageable age, and have left a numerous posterity." S.

138. SAMUEL, born Oct. 25, 1705; married Peace Mumford of Exeter, R. I., about 1735. "His opportunity for an education was the same as farmers' boys usually receive, hard work in the spring, summer and fall, and a little schooling in the winter, when it is not good logging, or the family wanted firewood drawn to the door. But amid all these hindrances, by degrees he acquired the rudiments of a common school education, to which he afterwards added a good knowledge of the art of surveying, which he might have obtained from his father, who was a practical and not a theoretical mathematician merely. Indeed, it seems to me now, after scanning the lives of so many of the old Bordens, that a knowledge of surveying was deemed by them to be

the ne plus ultra in the education of their sons, and consequently surveying seems to have become a Borden accomplishment from generation to generation from Richard Borden, the first emigrant to Rhode Island, down to Simeon Borden, Esq., his descendant of the sixth generation from him, who was the most accomplished mathematician of them all.

But to return to Samuel. He applied himself so faithfully to the study of mathematics and the practice of surveying that he soon became an accomplished surveyor, who was well-known beyond the limits of his native town or county. And when the order came from England to Gov. Shirley of Massachusetts to send a suitable person to Nova Scotia to take charge of the company of emigrants and locate them on the lands from which the neutral French had been expelled, the Governor appointed Mr. Borden.

The country to which Mr. Borden was now sent formerly belonged to France, and bore the name of Acadia, and contained a mixed population of French and Indians. By the treaty of Utretch in 1713 the Acadians were brought under the government of England, and were bound by agreement, in case of future wars between France and England, to remain neutral. Hence they were called the "Neutral French" of Nova Scotia. For forty years, says Barry in his History of Massachusetts, they were neglected by England, and in that time they prospered, and their substance increased. The crops from their fields were exceedingly rich. Flocks and herds grazed in their meadows or roamed over their hills; domestic fowls abounded; and the thickly clustered village of neat thatched roofed cottages sheltered a frugal, happy people. The spinning wheel and the loom were busily plied; and from morn till night the matrons and maidens, young men and their sires, toiled for the bread which they ate in peace. But when the French troops entered Nova Scotia, the Acadians were delighted to hear the sounds of their native tongue; and received them with great cordiality, and supplied their necessities. In all this they barely expressed their preference for the people who spoke their language, professed their religion and from whom they had derived their origin. It was perfectly natural that they should have acted thus, and perfectly consistent with their obligations to remain neutral in time of war. But Gov. Shirley, as soon as he heard of the transaction, determined to execute vengeance on these unfortunate people, and got authority for this purpose. He went to England, and by a false representation of this whole affair, procured instructions to the Lieutenant Governor of Nova Scotia sufficiently broad to annihilate the whole settlement of Acadia. Having accomplished his fiendish designs in England, Shirley hastened back to America, and sent the instructions to Gov. Lawrence of Nova Scotia, who resolved at once to carry out the instructions to the letter. In vain did they plead for mercy; in vain did they beg their oppressors to allow them to emigrate out of their jurisdiction; and finally, when they found that the Governor was determined to drive them from the homes they loved and scatter them as exiles over the whole length of the English colonies, they again besought the Governor to allow that each family might depart in the same vessel; but this he would not grant, though it was sought with tears by them all. The men were seized wherever they could be found and hurried on board of the transports which Shirley had provided, not being allowed any communication with their families. At Grand Prie they used strategy. They ordered all the males to meet in the church to hear a proclamation read from the King, and when they had obeyed the summons they were immediately surrounded by British troops and marched directly to the shore where the transports were waiting to receive them. When the males had thus been secured, the women and children who begged to go with their husbands and fathers, were coldly told that they must wait until more transports could be obtained. The horror and anguish produced by this answer cannot be described, nor even conceived. In this inhuman manner 7000 men were sent as fast as transports could be procured, and scattered along our coast from Maine to Georgia without either friends or money or even a knowledge of our language, and it is not probable that many of these husbands and brothers ever met again in this world.

This expulsion of the Acadians occurred in 1755, soon after the foundation of Halifax by Gen. Edward Cornwallis, a brother of Lord Cornwallis of Revolutionary memory. It was not long afterwards, in 1760, that Samuel Borden went to Nova Scotia, but it is not known how long he was employed there. He settled his son Perry on this tract, who married Emma Percy, the daughter of a British officer, in 1761. After his return home, Mr. Borden led a retired life, cultivating his farm. In 1768 he made his will, September 1, and it was proved in Tiverton December 7, 1778, having died probably in November. An old book of his with tables for ascertaining the latitude and longitude of places, published in London in 1720, with signature of Richard Borden, Samuel above and Capt. Samuel Borden, late of Fairhaven (the eldest son of Perry Borden of Nova Scotia), was found among the books and papers of Mrs. Betsey Hawes, the daughter of Capt. Samuel Borden." S.

139. REBECCA, born July 18, 1712; married Caleb Russell of Dartmouth.

37. JOHN, Swansea.

140. ELIZABETH, born April 7, 1708; married O. Wardwell.

141. JOHN, born June 4, 1710; died July 8, 1761. He married Mary Peters, daughter of Rev. Hugh Peters.

John Borden was made Freeman of Scituate May, 1739, and sold two-fifths of his father's homestead, consisting of 80 acres of land in Warren, R. I., formerly Swansea, Mass., to James Mason, for £450. His deed was executed at Providence August 15, 1747. He obtained a water privilege on the Ponaganset River, where he set up a forge for the manufacture of wrought iron, the ore being brought from what is now called Cranston, and the forests around him furnishing ample materials for charcoal. The old forge site is now occupied by a grist mill, and the water power carries a cotton factory, which is owned in part by John Borden, a descendant.

142. BENJAMIN, born December 2, 1716; married Elizabeth Mason.

143. JOSEPH, born November 18, 1718.

144. SARAH, born March 4, 1720.

39. JOSEPH, Fall River.

145. STEPHEN, born August 10, 1705; died August 30, 1738; he married Penlope Read February 3, 1726.

146. WILLIAM, born 1707. The date of his death and his wife's name is unknown. He lived in the vicinity of Fall River, and raised a numerous family.

147. GEORGE, born 1709; died 1767. He married Priscilla Wilcox, and lived in the villages of Tiverton and Fall River, Mass. He was appointed October 21, 1740, guardian of Mary, Stephen, Hannah, Meribah, George and Susannah, his brother Stephen's children, being minors under the age of 14 years; and on February 11, 1741, John Bowen, who married the widow of Stephen, was appointed guardian of Hannah, she being over 14 years of age.

148. JOSEPH, born 1712; died in Tiverton, 1800. He married Susannah Read January 26, 1736.

"Joseph Borden was a shoemaker, and lived in Fall River, until 1750, when he gave to his brother George a power of attorney to collect his outstanding deeds and attend to his affairs generally. He then removed to an estate given him by his grandfather, John Borden, of Portsmouth, R. I., which was on the east side of the North Watuppa Pond, near to the Indian reservation, it being then in the town of Tiverton. The births here given were taken from the Freetown records, but probably do not include all his children. There was a John Borden of Tiverton, whose wife was named Lydia, who, I think, belonged to this family. His family register I found on the records of Tiverton, and also the marriages of three of his children, viz: Keziah, Lemuel and — —. This family disappeared from the records of Tiverton about 1800." S.

40. THOMAS, Fall River.

149. JOB, born February 16, 1728; died young.

150. JOHN, born May 12, 1729, in Portsmouth, Rhode Island. He married October 16, 1746, Susannah Pearse, and removed to Delaware county, New York, from where his family scattered.

151. MARY, born January 1, 1731; married Robert Lawton, February 12, 1748.

FOURTH GENERATION.

152. JOSEPH, born August 12, 1733. He was a farmer at Portsmouth, Rhode Island; married Catherine Turner April 21, 1751.

153. SARAH, born April 29, 1738; never married.

154. WILLIAM, born January 25, 1741; died 1803. He was a farmer at Hog Island, Rhode Island, and married Sybil Smith November 25, 1766. By his will Hervey and Thomas, his sons, received Hog Island, and Smith, his youngest son, his house and buildings in Portsmouth, and the residue of his property. Hervey and Smith were to pay to each of their four sisters £400. The will was dated June 14, 1803.

43. WILLIAM, Beaufort, North Carolina.

155. ALICE,

156. CATHERINE

157. HOPE,

158. WILLIAM, born February 6, 1731; died November 2, 1799.

"He was the son of William, the shipbuilder, and was born, according to an entry made with his own hand in his family Bible, in Tiverton, R. I., February 6, 1731, and came to North Carolina with his parents. When arrived at man's estate he married Mrs. Comfort Small, a widow whose maiden name was Lovett, and settled upon the bank of Newport River, Carteret county, N. C. He became a planter, and being a Friend, as his father was before him, he was opposed to negro slavery, and therefore carried on the business of his plantation with hands hired for the purpose. He was successful in his farming operations, and gradually accumulated property. He was a warm and devoted friend to the cause of the American colonies during the Revolutionary war, and consequently suffered much from the depredations of the enemy, having his store, mills and warehouses destroyed by the British cruisers, who omitted no opportunity to harrass our citizens whenever they could gain access to their property. Mr. Borden was not a mere politician, more clamorous for an office than anxious to promote the best interests of his country; but he was one of those upright, honorable, straightforward kind of men whose intelligent look and bearing fixes the attention and secures the confidence and respect of the communities where they reside. Such men have to be sought for in times of great tribulation and actual danger, and urged to come forward and aid in supporting the cause of their country. At the outbreak of the Revolution, North Carolina had no written Constitution. It was considered by the friends of liberty and law that one should be established forthwith as a declaration of their rights, and a basis of their laws, so as to combine the whole population into one body politic, and thus to secure unity of action throughout the state at this alarming period. A convention for this purpose was called to meet at Halifax, N. C., to form a state Constitution. They met December, 1776, and after due deliberation succeeded in forming one which was finally adopted by the people. Mr. Borden represented Carteret county in this convention;

and when the convention was called to meet at Hillsborough, July 21, 1788, to ratify and adopt the Constitution of the United States, by which North Carolina would become an integral part of the American Union, or as it is otherwise expressed "a component part of the United States of America," Mr. Borden was again sent by his county to perform the pleasing duty of voting in the affirmative. In both instances, he thought, felt, and acted with the majority in adopting the two Constitutions, and his course was duly appreciated by his constituents, and proved highly beneficial to all concerned. Mr. Borden suffered from the French depredations upon our commerce prior to 1800. The government has published a list of all the petitions which have been presented to Congress since its organization. In this list I find the following petitions for the heirs of William Borden for French spoliation:

"Petition of James W. Borden, executor of Joseph Borden, to 24th Congress, second session; Also to 25th Congress." The petitions were not granted, if ever acted upon; the indemnity not having been collected from France by the government. It has since been paid over, and a court of claims instituted for the distribution of it, but all the parties originally interested have passed away; and so much of the history of these transactions forgotten and lost that many who then lost a large portion of their property and toiled through life for a subsistence and left to their children the semblance of a shadow, which a rule of court would quickly dispel, and the government will probably secure the lion's share.

The Hon. William Borden died November 2, 1799, aged 48 years, less than two months previous to the death of Gen. George Washington. He was buried in the cemetery at Core Sound, Carteret county, North Carolina. He left a family of six children, four sons, John, William, Benjamin and Joseph; and two daughters, Alice and Hope, both of whom married." S.

159. HANNAH, born in North Carolina; married a Mace.

44. BENJAMIN, Botetourte, Virginia.

160. BENJAMIN, born —; died 1753, of smallpox caught from Mary Greenlee, his mother-in-law. He married a widow, Mrs. Magdalene McDowell, her maiden name was Greenlee. Her first husband, John McDowell, who was a surveyor in the employment of Benjamin Borden, the elder, was killed by Indians in 1742, near where the North River empties into the James.

161. HANNAH, married Edward Rogos.

162. MARCY, (in one list called Mary), married William Feamley.

163. JOHN, date of birth unknown, as is the name of his wife. He removed to Kentucky, near Covington, and died there in 1785. He left several sons and daughters.

164. ABIGAIL, married (1) Jacob Worthington; (2) James Pritchard.

FOURTH GENERATION.

165. REBECCA, married Thomas Bronson; had two sons, Eli and Levi.

166. DEBORA, married Thomas Hendley; her marriage must have taken place sometime after 1750.

167. LYDIA, born about 1720; died 1801. She married, 1745, Jacob Beck, a German by birth. Tradition says he belonged to a prominent, if not a noble family in Germany, and that he left his country because of some political or personal troubles, and came to America. He was a tall, fine looking man, and was always reticent concerning his early life. He died in Botetourt county, Virginia, in 1802, at a very old age; some say he was 102 years of age at the time of his death.

168. ELIZABETH, married a Mr. Patton from whom are descended the Pattons of Tennessee.

169. JOSEPH, dates of birth and death unknown. He emigrated to North Carolina and little is known of his descendants.

63. RICHARD.

170. FRANCIS, born Aug. 5, 1717; died Sept.11, 1782. Married Mary ——, October 29, 1740. They lived at Mansfield, N. J.

65. JAMES, Freehold, N. J.

171. RICHARD.

172. RICHARD,

173. INNOCENT,

174. PHOEBE,

175. REBECCA,

176. ABIGAIL,

177. HELEN,

178. MARY,

179. ANN or Nancy married John Hance.

67. SAFETY, Bordentown, N. J.

180. RICHARD,

181. CATHERINE, married Jsoeph Britton,

69. JOSEPH, Bordentown, N. J.

182. JOSEPH, born August 1, 1719; died April 8, 1791. "He married Elizabeth Rogers, who was born July 10, 1725; died November 2, 1807, having outlived her husband sixteen years, seven months and twenty-four days. She was a woman of great ability and fine Christian character. Her will was dated September 15, 1798, and proved November 5, 1807. Among her legatees were her daughter, Ann Hopkinson, and her three daughters, Elizabeth, "daughter of my brother Isaac Rogers, lately deceased." Lastly her grandson, Joseph Hopkinson, who had the use and occupancy of the old brick family mansion, which is still retained by his descendants. It is from this will that I

have obtained the maiden name of Mrs. Borden. Mary Rogers was buried in the same tier of graves with the Bordens and Hopkinsons, and may have been a sister of Isaac and Elizabeth. She was buried in the ground now known as the Hopkinson Cemetery, at the western terminus of Church street, on the bank of the Delaware River. Here lie in the same tier of graves on the north side, Joseph Borden, Sr., on his right Col. Joseph Borden and his wife Elizabeth, Joseph Hopkinson and his wife Emily; on his left hand are Joseph Borden, Jr., died October 16, 1788, in his 33d year, and Ann Borden, died March 11, 1744. These last probably were near relatives to the family.

Joseph Borden was 30 years of age when his father commenced the settlement of Bordentown, and this was done to facilitate the design of his son to establish a new line for the transportation of the mails, merchandise and passengers between Philadelphia and New York. His new house and store house were erected in 1752, and occupied and open for the accommodation of passengers early in 1753. In evidence of this fact I will here make an extract from the journal of Conrad Weiser, a noted interpreter of several Indian languages, who resided at Heidelburgh, in Berks county, Pennsylvania, and traveled this route in 1753 on a mission from the Governor of Pennsylvania to the Mohawk Indians. He says: "I waited on His Honour, the Governor, at Philadelphia, July 26. On the 28th I took the stage boat from Bordentown and arrived in New York on the first of August." The stage boat from Philadelphia usually arrived at Bordentown in the afternoon, and the passengers the next morning were forwarded in stage coaches or covered wagons to some intermediate station, where they spent another night, and they arrived the following day at Amboy or Elizabethtown, where they remained another night, and the next day they arrived in New York by another stage boat, wind and weather permitting.

It appears from this quotation from Mr. Weiser's journal that Mr. Borden was fully established in the transport business in 1753. But he did not lack competitors. The old line through Trenton became alarmed at his success, and made great efforts to surpass him, both in speed and in the accomodation of their passengers. They built a new style of carriage which they called "flying machines," and set up flaming advertisements all over the country, promising to take passengers from city to city in three days. But, unfortunately for the public, all their extra speed laid in the brains of him who wrote the advertisements. It still required from four to five days for a person to go from New York to Philadelphia. In connection with the transportation of the mails, which had greatly increased in weight and bulk, Mr. Borden went to England to confer with the Postmaster-General on the subject, the colonial agent of the government in New York not having any authority to increase the compensation for this service. During his stay in England, it was reported that he visited Borden, the ancient home of his ancestors, but no account of what he saw and heard relating to the family connections has been preserved.

FOURTH GENERATION.   101

The next subject connected with the Colonel's history occurred in 1765, when he was chosen with two others to represent the state of New Jersey in the general convention of the colonies, or Congress, which met in New York October, 1765.

The convention met at the time appointed, organized and proceeded without delay to a full discussion of matters connected with the relation of the colonies with the government of Great Britain. They enumerated the grievances felt in each of the colonies. They asserted and unanimously agreed upon the rights of their constituents, as subjects under the government, and closed their labors by adopting an address to His Majesty, the King, a petition to the House of Lords in Parliament, and another to the House of Commons. In these documents they had plainly set forth their views of the legal power of the government over the colonies and the rights of the people in America. They were Englishmen, and as such were entitled to be represented in Parliament, and to all other privileges of citizenship of which they were deprived. Having accomplished the objects for which they had been called together, the convention was dissolved October 24, 1765.

Elliot, in his Debates in Congress, p. 326, says of this convention: "We must never forget that the egg of this republic was the Congress which met in New York in 1765." Of the thirteen colonies, nine only were represented, viz: Massachusetts, Rhode Island, Connecticut, New York, New Jersey, Pennsylvania, Delaware, Maryland and South Carolina. Elliot says again: "This convention of sages was the parent plant of our present confederacy of republics." Only two of its members refused to sign the petitions, Mr. Ogden of New Jersey, and Mr. Ruggles of Massachusetts. They contended that each colony should petition separately. Mr. Borden and his colleagues, on presenting their report to the assembly, received a vote of thanks from them for the judicious, able and faithful discharge of their duty to the state on this important occasion."

As time rolled on, the disposition of the British government to oppress and enslave the colonies became more apparent; and they began to realize the fact that resistance to these encroachments upon their rights and privileges must be made or they would be lost forever. Consequently, in 1774, each colony acting on its own responsibility, appointed a committee of correspondence, whose duty it was to communicate with the committee of other colonies on all subjects connected with the alarming state of the country's affairs. This arrangement produced an excellent effect on each of the colonies. Each of them being equally exposed to the desolating power of England, sympathized more intensely with those colonies on which this power had fallen, and stimulated them to make the greater exertions to support and encourage them in resisting. And when the first blow fell upon Boston, the sound reverbrated along the Alleghanies, until it aroused the attention of every dweller on American soil. The committee of correspondence for New Jersey consisted of nine persons, one of whom was Col. Joseph Borden. The following year, 1775 he, with nine others, were chosen to form a council of war, on which he acted during the war with untiring zeal and unflinching energy, whether in the council or in the field. He and his son were both attached to the troops of the New Jersey line, and were in most of the battles fought in that state. Indeed, so great was the influence of Col. Borden with the army and the people both, and so untiring in his efforts to oppose the British army and uphold the cause of his own country, that he was soon selected by the British commander as a mark for his special vengeance. He was determined either to detach him from the service of his country or ruin him by the destruction of his property. Lord Cornwallis was then encamped on the Delaware, near Bordentown, and established his headquarters at the house of Col. Borden, his wife and daughter being all of the family at home; the Colonel and his son being with the American army at the time. This great English general abused his position by commencing his attack on Col. Borden by assailing his good lady. As this attack and its results have been graphically described by Mrs. Ellet in her work entitled "The Women of the American Revolution," at pages 305 and 306, of the second volume, I will insert here an extract from her excellent work relating to this subject:

"At the time when New Jersey was overrun by the enemy and when the prospects of the colonies were darkest, an officer stationed at Bordentown, N. J., said by Maj. Gordon to be Lord Cornwallis, endeavored to persuade and then to intimidate the wife of the Hon. Joseph Borden of that place to use her influence with her husband and son to abandon the American cause. They were both absent in the Continental army, at the time she was visited at her residence for that purpose. The officer promised her that if she would induce them to quit the standard of their country that they followed and join the Royalists, that her husband's property should be protected, while in case of refusal, their estate would be destroyed and their elegant mansion burned to the ground. Mrs. Borden answered by bidding him defiance, saying: 'The sight of my house in flames would be a treat to me; for since you have been here I have seen enough to know that you never injure that which you have power to keep and enjoy. The application of a torch to my dwelling I should regard as a signal for your departure.' The house was burned in fulfillment of the threat, the property laid waste, the animals slaughtered or driven off, but, as the owner predicted, the retreat of the spoiler quickly followed."

"This truly noble woman was one of the most active patriots of the Revolution. She was willing to see her property sacrificed, and the lives of her gallant husband and son exposed to the dangers of battle for the good of her country; and when called upon to make personal efforts for the glorious object, which she was often called upon to do, she met the crisis with heroic firmness and exerted herself to the uttermost to urge forward the good cause. Her character and personal influence were highly appreciated by all who knew her per-

FOURTH GENERATION.

sonally, or had heard of her name." And when the ladies of New Jersey met at Trenton on July 4, 1780, and chose Mrs. Col. Joseph Borden as their agent to collect funds for the Continental army—it was no unmeaning compliment, but their choice was the result of a just appreciation of her noble character, influence and patriotism. Such an exhibition of the character and spirit of Mrs. Borden as Mrs. Ellet has given very naturally excites a strong desire to know more of her history and that of her family. But at present the subject is involved in much obscurity, both as regards Col. Borden and his lady. If Mrs. Ellet had been acquainted with more of her history she would probably have given it." S.

183. REBECCA, married a Brown.

184. HANNAH, married a Lawrence.

185. ELIZABETH, married a Douglass.

186. AMY, date of birth unknown; married in 1756 to William Potts, born in 1745; died July 25, 1783, at Bordentown, New Jersey.

FIFTH GENERATION.

FIFTH GENERATION.

87. RICHARD, Johnston, R. I.

187. RICHARD, born 1735; never married.

188. WILLIAM, born 1748; died April 18, 1824. He was a farmer at Johnston, Rhode sland; his wife's name was Huldah; maiden name unknown.

189. OLIVER, born 1750. "The name of his wife was Phoebe; they lived in Johnston, and I think he was the last of the Bordens descended from Thomas bearing the Borden name. He was a Baptist preacher or exhorter, but without much intelligence, natural ability or education. A person who had frequently heard him preach remarked to me that Elder Wilson of Providence once said of Elder Borden: "There is a minister in Jonston, I hear, who preaches for nothing, and I think that is as much as it is worth." But it is no disparagement to the reputation of anyone to bear the reproaches of Parson Wilson. The Apostles were all exhorters, and worked, not for wages, but the glory of God, and were eminently successful; and so may others do if they have the ability to sustain themselves and families. Mr. Borden had several children. In an old neglected burial ground I found a headstone with the names of Joseph, Sarah and Oliver, children of Oliver Borden; the last died December 17, 1806. He may have left others at his death, but I could not hear of any. The last entry of his name on the records was in a deed of land bought by him of Amos Collins March 22, 1822 for $220. The date of his death I could not find, nor the place of his burial." S.

88. JOSEPH, Providence, R. I.

190. ELIZABETH,

191. SARAH,

192. JOSEPH,

193. ABRAHAM,

92. WILLIAM, Providence, R. I.

194. MARY, born August, 1743; married Josiah King.

195. MERIBAH, born 1744; married Richard Thornton.

46. JAMES, Burlington, N. J.

196. JACOB,

197. ASA,

198. JAMES,

199. WILLIAM, born 1734; date of death unknown. He lived at Eversham, Burlington county, New Jersey; married a widow, Mrs.

THE BORDEN FAMILY.

Mary Baker, whose maiden name was Hazleton.

200. ABIGAIL,

201. MARY,

98. JONATHAN, Gloucester, New Jersey,

202. THOMAS, born 1755; married in Monmouth county, N. J.; the name of his wife is unknown.

203. JOSEPH,

204. SAMUEL,

205. JONATHAN,

206. BENJAMIN,

207. HANNAH,

208. MARY,

106. FRANCIS, Nottingham, N. J.

209. JOSEPH,

210. THOMAS,

211. FRANCIS, born 1743; died ——; Imlaystown, New Jersey. He married Elizabeth Parker, born November 14, 1745.

212. JAMES,

213. RHODA, married —— Robbins,

214. MARY, married a Lawrie.

215. LYDIA, married James Ford,

216. HANNAH, married John Hawkins,

107. JOHN, Shrewsbury, Monmouth county, New Jersey.

217. JOHN, born 1737; died 1833; married Lydia Worthley.

218. MARY, born January 4, 1739; married Samuel Pintard.

219. JOEL, born August 6, 1744.

220. ELIZABETH, born February 2, 1744; married John Corlies.

221. ANNE, born September 30, 1749.

222. JEAN, born March 24, 1753.

223. HANNAH, born August 15, 1755.

224. LYDIA, born November 30, 1759; died April 22, 1763.

112. JEREMIAH, Manasquan, N. J.

225. REBECCA, married John Wooley of Shrewsbury, New Jersey.

226. HESTER, married Capt. Richard Lippincott. Pensioned one-half pay for life and granted 3000 acres of land near Toronto, Canada. They had one daughter, Esther, who married George Taylor Dennison, an officer in the English army, stationed in Canada.

FIFTH GENERATION.

227. JEREMIAH, died young.
228. SAMUEL, went to New York.

113. JOSEPH, Manasquan, N. J.

229. RICHARD, married Catherine Chambers.
230. JOHN, married —— Bennett.
231. AMY, married Robert W. Morris.
232. ELIZABETH, married Jesse Chandler, had one son and two daughters; son went to Illinois; one daughter married John Pettit.

114. RICHARD, Manasquan, N. J.

233. SARAH, born April 11, 1759; she married John de Bow, near Clarksburg, Upper Freehold.
234. THOMAS, born November 6, 1760; never married. He and Benjamin were farmers together at Rumson.
235. RACHAEL, born February 16, 1763; married Thomas Cook of Point Pleasant, Manasquan.
236. BENJAMIN, born December 22, 1766; died October 10, 1839; married Mary Lloyd of Philadelphia, Penn.

134. JOHN, Fall River, Mass.

237. RICHARD, born ——; married Priscilla Westgate November 12, 1754. They had several children, but all died before their father.
"Ruth, the last of these children, died in 1790; Richard Borden's last will was dated February 1, 1791. His children having all died, he gave all his estate to his wife durnig her life, and after her death to his cousin, Abigail Westgate, who afterwards married John Durfee. This will was contested by the agents of John and Hannah, his brother and sister, and was rejected by the Town Council of Tiverton. But the widow appealed to the Supreme Court of Rhode Island, who reversed this decision, and ordered the Council to probate and establish the will, which was done. After the death of Mrs. Borden, John Durfee took possession of the estate in right of his wife, which gave rise to a long and expensive lawsuit, which was carried on by Isaac Wilbour and William Howland of Little Compton, both connected with John Borden through his sister Hannah Wilbour, the wife of Charles and mother of Isaac. After spending much time and money, as usual in suits at law, the matter was compromised, John Durfee retaining the homestead of Richard, and the contestants getting enough of the outlands to cover their costs, and perhaps a little more." S.

135. THOMAS, Fall River, Mass.

238. RICHARD, born 1722; died July 4, 1795; married Hope Cook March 12, 1747.
"Richard Borden was the son of Thomas, and grandson of Richard

Borden of Tiverton. His father, Thomas, owned that portion of the Fall River stream which lay below the great falls on the south side of the stream, and the land adjoining down to the salt water, besides other landed estate which he gave to Richard with other outside lots. This portion of the stream was the site of the saw mill first erected by Caleb Church of Watertown, near Boston, who purchased of the original proprietors of the Pocasset purchase thirteen shares of the mill lot and stream. From this it would appear that he designed to settle at Fall River, but changing his mind afterwards, he sold his thirteen shares of the mill lot, with one-half of the saw mill, to his brother, Benjamin Church, the Indian warrior, who probably owned the other half. They were each purchasing mill rights at the time, and had secured twenty-six and a half thirtieths of the mill lot and stream, which they sold to Richard and Joseph Borden in 1714, who had secured the balance. The property had been occupied by Thomas Borden during his life time, and was transmitted to his son Richard. This property has always been held with the utmost tenacity by the Bordens descended from the first Richard of Tiverton down to 1812, all feeling that the time would come when this property would become immensely valuable, and they held on to it from generation to generation as an investment for the benefit of their heirs. But some of the heirs of Joseph Borden, the brother of Richard, sold their interest in the saw-mill built by Joseph on the north side of the stream and given by him to his son Stephen who married Penelope Read, sold their interests in their father's (Stephen's)estate to John Bowen, who had married their mother. This one privilege remained mostly in the Bowen family until it was bought by the Pocasset Company about 1821.

But among all these descendants of Richard, there was none who placed the prospective value of this property so high as the subject of this article. He was a man of ordinary abilities, but when he got upon his favorite topic, "the prospective value of Fall River," he became quite excited and interesting, so much so that his friends would often oppose him just to hear the old man talk. He usually closed with: "Do ye see, neighbor, do ye see, I tell ye, the time will come when every dam on the stream will be sought after by men who have the money to pay for it at a great price, and every stone and tree around Fall River will be wanted—yes, I tell ye, the time will come when the rocks on Rattlesnake Hill will bring the gould." Those who heard him then no doubt considered him very enthusiastic, to say the least, but those of us who have survived him and witnessed the changes which only half a century has wrought upon the place must admit that Mr. Borden had formed a just appreciation of the prospective value of Fall River. This prepossession led him to hold, with a firm grasp, every foot of land, and all his water-power, and his grandchildren are now reaping a rich harvest from them. They certainly owe him a marble column or cenotaph for the deep interest which he manifested for them though yet unborn.

During the war of the revolution, the British landed a force at Fall

FIFTH GENERATION.

River, burnt the saw-mill belonging to Mr. Borden and a large quantity of lumber which was owned by his two sons, Thomas and Richard, who ran the mill on their own account. On their retiring they seized upon Mr. Borden, Capt. Benjamin Borden and John Negus as prisoners, and after setting fire to Mr. Borden's dwelling house, hastily embarked in their boats, taking their prisoners with them to Newport. On approaching Bristol Ferry they met a storm of chain shot, language and balls thrown by the fort on the British side of the ferry. Mr. Borden, not being a military man, lay down in the bottom of the boat and resisted every effort to make him stand up. The storm of shot coming more annoying as they neared the ferry, killing and wounding some of their men, two strong men laid hold on him, determined to make him stand up, swearing that the rebel should take his chance to be killed like themselves; but as the old man resisted them with all his might, they found it no easy task—a chain shot swept across the boat, killing both of the Englishmen and leaving Mr. Borden unharmed. The attempt to pass the ferry was then abandoned, and the boats beached near the town pond creek and the men reached Newport by land. Mr. Borden was detained at Newport but a short time and came home again in about a week. The general commanding on Rhode Island had given orders to the captains of these marauding parties to make prisoners of some of the principal men, hoping to get some information from them which he could use against the rebel cause. But he soon found, in the examination, that he had waked up the wrong passenger, and let Mr. B. go in disgust. It is worthy of remembrance that this general afterwards declared that though hundreds of men from Tiverton, Little Compton and vicinity had been brought before him for examination, he had never found one who would communicate any information advantageous to his enterprise, or injurious to the cause of their own country." S.

239. CHRISTOPHER, born October 10, 1726, at Cranberry Neck, Tiverton, Rhode Island, married Hannah Borden, daughter of Stephen, December 24, 1748.

240. DEBORAH, married David Brayton, November 25, 1742.

241. MARY, married Samuel Shearman March 27, 1748.

242. REBECCA, married Benjamin Borden of Warren, Rhode Island, September 8, 1759.

137. JOSEPH, Fall River, Mass.

243. PATIENCE, born August, 1731, married Hon. Thomas Durfee, August 9, 1747. The following reminiscences of Col. Joseph Durfee, a son of Patience Borden, is taken from Mr. Earl's book, Fall River and its Industries:

"There is still treasured by a very few of our oldest citizens a modest pamphlet, coverless, not exceeding twelve pages and altogether unpretentious in typographical execution, yet exceedingly valuable for the true picture of the settlement as it was about the middle of the last century and for the record of local patriotism it has preserved. Its author, referred to in the early pages of our narrative, was a conspicuous citizens, identified with the original industrial enterprise of the settlement (then Tiverton, R. I.,) as the projector of the first spinning factory, and noted for his intelligent and comprehensive observation. In 1864, still possessing a vivid recollection of the incidents of his youth and maturer years, he wrote the interesting, though much too brief, record of local events, which is here reproduced in its entire volume."

REMINISCENCES OF COL. JOSEPH DURFEE,
RELATING TO THE
EARLY HISTORY OF FALL RIVER, AND OF REVOLUTIONARY SCENES.

"Joseph Durfee was the eldest son of the late Hon. Thomas Durfee. He was born in April, in the year 1750, in what is now the city of Fall River. At that time, and until within a few years, the Fall River stream was owned by the Bordens. Much of what now is the city, where are elegant buildings and a dense population, was then a wilderness, where the goats lodged in the winter seasons. The Bordens and the Durfees were then the principal proprietors of the Pocasset Purchase, and owners of the land on the south side of what is now Main street, for more than a mile in length. Thomas and Joseph Borden owned the south side of the stream, and Stephen owned the north side. Thomas Borden owned a saw-mill and a grist-mill at that time, standing where the old saw and grist mills stood near the iron works establishment.

"Thomas Borden left a widow and four children, viz.: Richard, Christopher, Rebecca and Mary. Joseph Borden, brother of Thomas, owned a fulling-mill, which stood near where the Pocasset factory now stands. He was killed by the machinery of his fulling-mill. He left four children, viz.: Abraham, Samuel, Patience and Peace. Patience was my mother. Stephen Borden who owned the north side of the stream, had a grist-mill and a saw-mill, standing near where the woolen establishment has since been erected. He left six children, viz.: Stephen, George, Mary, Hannah, Penelope and Susannah.

"The widow of Joseph Borden was afterwards married to Benjamin Jenks, by whom she had six children—John, Joseph, Hannah, Catherine, Ruth and Lydia. The widow of Stephen Borden was married to John Bowen, by whom she had two sons—Nathan and John.

"At that time, and until within a few years, there were but two saw-mills, two grist-mills, and a fulling-mill standing on th Fall River. There are now about forty different mills on the river. The stream was very small; but the falls were so great that there was little occasion for dams to raise a pond sufficient to carry the wheels then in operation. A small foot bridge, which stood near where the main street now crosses the stream, afforded the only means of passing from one side to the other of the stream, except by fording it. There was for-

FIFTH GENERATION.

merly a small dam near where the Troy factory now stands, over which the water flowed the greater part of the year. When it failed, those who owned the mills near the mouth of the stream hoisted the gates at the upper dam and drew the water down. It was no uncommon thing, twenty-five or thirty years ago, for the water to be so low and the river so narrow at the head of the stream, that a person might step across without difficuty. It was frequently not more than six inches deep. At one time there was a foot bridge of stepping-stones only across the Narrows between the North and South Ponds.

"Our country has been involved in three wars since my recollection. The first was with the French and Indians—when we fought for our lives. The French offered a bounty for every scalp which the Indians would bring them. It was therefore certain death to all who fell into the Indians' hands. I distinctly recollect the time when General Wolfe was killed—and of seeing the soldiers on their march to reinforce the army I saw many men enlist into the service, and among them Joseph Valentine, father of William Valentine of Providence. I was then about ten years of age.

"The second war was with Great Britain, during the greater part of which I was actively engaged in the service of my country. We then fought for our liberty. We were divided into two parties, called Whigs and Tories—the former, the friends of liberty and independence; the latter, the enemies of both. Before the revolution broke out the Whigs were busy in making saltpetre and gunpowder, in making and preparing small arms, in training and learning the art of war. At this time we of this State were British subjects and constituted what was then called the Colony of Massachusetts. Conventions were held in the colony to transact the business and consult upon the affairs of the colony. At one of these conventions I received a captain's commission, signed by Walter Spooner, Esq., and took command of a company of minute men.

British ships commanded by Wallace Asque and Howe, early in the revolution, were off our coast, in the river and bay, harrassing and distressing the towns of Newport, Bristol and other towns on the river. I was called upon with my company and such others as could be mustered to guard the shores and pervent the British from landing, until the colony could raise a sufficient force to protect the inhabitants from their depredations.

In 1776, after the battle of Long Island, a reinforcement was called for to cover the retreat of the American troops. I was orderd to take the command of a company of sixty men and march forthwith to the army then retreating from New York. These orders were promptly obeyed. With the company under my command I found the regiment commanded by Colonel Thomas Carpenter, and by a forced march we reached the army a few days before the battle of White Plains. In that engagement I took an active part.

Soon after my return home from the battle of White Plains the British landed at Newport, in Rhode Island. I was called upon to proceed immediately with my company to assist in covering the retreat of the small forces then commanded by Colonel John Cook from the island of Rhode Island. This was effected without loss, though attended with difficulty and delay, as there was then no bridge from the island to the mainland.

At that time the inhabitants in the south part of Massachusetts and Rhode Island were in a critical situation. They were nearly surrounded by British emissaries. A part of the English squadron lay off our coast, and their troops had possession of the south part of Rhode Island. Both were harrassing our towns, destroying our property and making prisoners of the inhabitants. In addition to this we had the Tories at home, enemies in disguise, who were aiding and abetting the British, while they professed friendship for the cause of liberty and for those who were shedding their blood to obtain it. Early in the spring of 1777, I received a majors commission, and was stationed at Little Compton in the State of Rhode Island, in the regiment under the command of Colonel John Hathaway of Berkley, Mass. At Little Compton, and in that neighborhood, I continued several months on duty spring of 1777 I received a major's commission, and was stationed at with the regiment, often changing our station to repel the invasion of the enemy and to protect the inhabitants from their frequent depredations. In the fall of 1777 I returned home to Fall River. I found the citizens, among whom were my relatives and best friends, exposed and continvally harassed by the enemy. I applied to several of the leading and influential men of this place and proposed raising a guard for the safety and protection of the inhabitants They coincided with my views, as to the necessity of a guard to protect our defenseless inhabitants. I went to Providence to consult General Sullivan, who was commander-in-chief of all the forces raised in this section of the country, and to obtain assistance from him. He approved of my plan of raising a guard, and gave me an order for two whaleboats, and an order also for rations for twenty men, drawn upon the commissary, then at Bristol. I soon raised a guard, procured the store now standing at the end of the Iron Works Company's wharf in this place for a guardhouse, where we met every day, called the roll, and stationed sentinels for the night to watch the movements of the enemy and give the alarm when approached. The orders of the sentinels were peremptory—that if a boat was seen approaching in the night, to hail them three times, and if no answer was received to fire upon them. It was not long before one of the guard, Samuel Reed, discovered boats silently and cautiously approaching the shore from the bay. The challenge was given but no answer received. He fired upon the boats. This created an alarm, and the whole neighborhood were soon in arms. I stationed the guard behind a stone wall, and kept up a constant fire upon the enemy until they brought their cannon to bear upon us, and commenced firing grapeshot amongst us—when, as we were unable to return the compliment, it was deemed advisable to retreat. Two of the guards were sent to remove all the planks which laid over the stream for foot people to cross upon, and to cut off, as far as possible, every facility for crossing the stream, except the upper bridge. We then retreated slowly until we reachd the main road, near where the bridge now crosses the stream. I then gave orders to form and give them battle. This was done, and never were soldiers more brave. So roughly were the enemy handled by your little band of Spartans, that they soon beat up a retreat, leaving behind them one dead and another bleeding to death, besides the wounded, whom they carried away.

"The wounded soldier, left by the enemy, before he expired, informed me that the number of the enemy who attacked us was about 150, commanded by Major Ayers. When the enemy landed, they set firs to a house of Thomas Borden, then nearly new. They next set fire to a gristmill and a saw-mill, belonging to Mr. Borden, standing at the mouth of the Fall River. These buildings I saw set on fire. When the British troops retreated, as they were compelled to do, from the shots of our little band of volunteers, they set fire to the house and other buildings of Richard Borden, then an aged man, and took him prisoner. We pursued them so closely in their retreat that we were enabled to save the buildings which they had last fired. The British were frequently fired upon and not a little annoyed by the musketry of our soldiers, as they passed down the bay in their boats on their retreat. Mr. Richard Borden, whom they took prisoner, was in one of their boats. Finding themselves closely pursued by a few American soldiers, who from the shore poured in their shot and balls upon them as fast as they could load and fire, and finding themselves in danger from the musketry of these few brave Whigs who pursued them, they ordered Mr. Borden, their prisoner, to stand up in the boat, hoping that his comrades on the shore would recognize him and desist from firing upon them. But this he refused to do; and threw himself into the bottom of the boat. While lying there, a shot from the Americans on shore killed one of the British soldiers standing by his side in the boat. Mr. Borden was obstinately silent to all the questions which were asked him; so that not being able to make any profitable use of him they dismissed him in a few days on parole. This engagement took place on a Sabbath morning, the 25th of May, 1778. The two British soldiers killed in the engagement were buried at twelve o'clock on the same day of the battle, near where the south end of the Massasoit Factory now stands. During a considerable part of the month of August following, we were busily engaged in procuring arms, ammunition and provisions for the soldiers, and in building flat-bottomed boats and scows for the troops to cross over the river on to Rhode Island, with a view to dislodge the British army, who then had possession of the island. A barn, now standing near the stone bridge, was occupied for a commissary store, of which I had charge until things were in readiness, and the troops prepared to cross over to the island, when I left the store in charge of my friend and relative, Walter Chaloner.

In the fore part of August, 1778, the American troops embarked in the boats and scows prepared for them and landed on Rhode Island, where I joined them, having been appointed a major in Colonel Whitney's regiment. Our troops were then marched to a spot but a short distance to the north of what is called Butt's Hill, where they encamped for the night, with but the canopy of heaven for a covering and the ground for our beds. But we were animated with the hope of liberty—with a belief that we were engaged in a righteous cause—and that He who sways the scepter of the universe would prosper our undertaking. At this time we were anxiously looking for the French fleet, from which we hoped for assistance against the enemy, whose numerous bodies of troops were before us. Soon the French fleet hove in sight, when the British set fire to the shipping in the harbor and blew up most of the vessels within their reach. Not long after the French fleet came up, the British fleet appeared in the offing. Immediately the French fleet tacked about and went out and attacked the British squadron, when broadsides were exchanged and a bloody battle ensued. A tremendous storm came on, long remembered as the August storm, in which the two fleets were separated, and many who had escaped the cannon's mouth found a watery grave. The French fleet, or as much of it as survived the storm, went into Boston to repair; the remnant of the British fleet went into New York. Soon after this storm our troops marched in three divisions towards Newport. One on the east road, so called, one on the west road and the brigade commanded by General Titcomb moved in the center, until we came in sight of Newport, when orders were given to halt, erect a marquee and pitch our tents. General orders were issued for a detachment from the army of three thousand men, our number being too small to risk a general engagement with the great body of British troops, then quartered on the south end of the island. Early in the morning a detachment of troops, of which I was one, was ordered to proceed forthwith and take possession of what was called Hunneman's Hill.

"The morning was foggy, and enabled us to advance some distance unobserved by the enemy; but the fog clearing away before we reached the hill, we were discovered by the British and Tory troops, who commenced such a heavy cannonade upon us, that it was deemed expedient by the commanding officers, to prevent the destruction of many of our brave troops, that we should fall back and advance under the cover of night. Accordingly, when night came, we marched to the hill unobserved by the enemy. We immediately commenced throwing up a breastwork and building a fort. When daylight appeared we had two cannon mounted—one twenty-four pounder, and one eighteen—and with our breastwork we had completed a covered way, to pass and repass without being seen by the enemy. The British had a small fort or redoubt directly under the muzzles of our cannon, with which we saluted them, and poured in shot so thick upon them that they were compelled to beat a retreat. But they returned again at night to repair their fort, when they commenced throwing bombshells into our fort, which, however, did but little damage. I saw several of them flying over our heads, and one bursting in the air, a fragment fell upon the shoulder of soldier and killed him.

"At this time we were anxiously waiting the return of the French fleet from Boston, where they had gone to repair. But learning that they could not then return, and knowing the situation of the British troops, that they were enlarging and strengthening their forts and redoubts, and that they had reinforcements arriving daily from New York, it was deemed expedient by our commanding officers, Lafayette, Green and Sullivan, all experienced and brave generals, that we should retreat to the north end of the island.

"Accordingly, on the 29th day of August, early in the morning, we struck our marquee and tents and commenced a retreat. The British troops followed, and soon came up with our rear-guard and commenced firing upon them. The shots were briskly returned and continued at intervals, until our troops were joined by a part of our army a short distance to the south of Quaker Hill, so called, when a general engagement ensued, in which many lives were lost on both sides. At night, we retreated from the island to Tiverton. On the following day we left Tiverton, crossed over Slade's Ferry and marched through Pawtucket and Providence to Pawtuxet, where we remained until our time of service expired.

"Some time after this, I received a lieutenant-colonel's commission, and took the command of a regiment to guard the sea-shores, and a part of the time my regiment was stationed at Providence. I soon received orders from General Gates, who at that time was principal in command, to march with my regiment to Tiverton and join General Cornell's brigade. The war now raged throughout the country. Old and young, parents and children, all, excepting the Tories, were engaged in the common cause of their country--in breaking the shackles of colonial bondage—in obtaining her liberty and achieving her independence. Old England now began to examine the prospects before her. She found after a bloody contest, what she might and ought to have known before, that her rebellious colonies, as she was pleased to term them, could be ruled, but not ridden upon; that by mild and liberal measures she might have retained a valuable part of her kingdom. She discovered her error too late to profit by it. The brave people of her colonies were resolved to throw off the yoke, and themselves be free. On the 29th day of October, 1779, the British troops left Rhode Island and the American troops under the command of Generals Gates and Cornell marched on to the island and took possession of the town of Newport. On the 20th day of December following my time of service having expired, I returned home to my family. This was the coldest winter known during the last century. The river and bay were frozen over so thick that people with loaded teams passed all the way from

Fall River to Newport on the ice. I continued in the service of my country until about the close of the revolutionary war, when I removed from Fall River to Tiverton, where I lived about thirty years. During this time I was elected by my fellow-citizens to several offices in town and was a member of the General Assembly for many years. When Thomas Jefferson was elected President of the United States in 1801 and the Democratic fever raged to the highest pitch, I was what was then called a Federalist, and having repeatedly sworn to support the Federal Constitution, could not consent to turn my coat wrong side out. I was therefore not permitted to hold any office. But in time this party fever abated and the people united in electing Mr. Monroe, under the general appellation of Federal Republicans. Attempts have since been made to alter the constitution, that noble fabric reared by the revolutoinary patriots, and should they succeed it will in my estimation be like sewing new cloth to a good old garment."

244. ABRAHAM, born 1733, died 1769, married 1756, Ann Mumford, born December 8, 1734, died October, 1808.

245. SAMUEL, born April 12, 1735, a farmer in Tiverton. He married Mary Sanford November 9, 1760.

246. PEACE, born February 13, 1736, married Joseph Borden February 19, 1758.

138.   SAMUEL, Fall River, Mass., and Nova Scotia.

247. JOSEPH, born October 14, 1736. He was a farmer at Fall River, married Ann Durfee December 6, 1764.

248. "PERRY, born in Tiverton November 9, 1793, received his education there. His early life was spent in cultivating his father's farm and aiding him in surveying, by which practice he soon acquired a good knowledge of the art, and qualified himself for the expedition on which he afterwards entered with a numerous company of adventurers. Soon after the expulsion of the neutral French from Acadia, or as now called, Nova Scotia, a name given to it by the English, great efforts were made to Governor Shirley of Massachusetts to resettle the valley once cultivated by the French people, but then lying waste and deserted. The reported richness of the soil, the labor which had been bestowed upon it to bring it under tillage, and the offer of a free grant to every actual settler of a good farm, presented an inducement sufficiently powerful to attract the attention of young men of small means who were ambitious of acquiring a competency for themselves and families if they should have any. During the winter of 1759 a company of emigrants was organized of one hundred and fifty persons from New England. Of this company Perry Borden was one, though he may not have gone there with any fixed determination to settle; he probably went as an assistant to his father who was commissioned by Governor Shirley to survey the lands and exercise a general supervision over the emigrants till they were located on their several allotments. When this duty had been performed, Samuel Borden returned home (1761) and left his son

FIFTH GENERATION.

Perry, who concluded to settle there with the company, and as a preliminary step he married Emma Percy, the daughter of an English officer, 1761, he being twenty-two years of age. By her he had two sons, Samuel and Joseph. His wife dying, he married for his second wife Mary Ells, October 22, 1767, by whom he had nine other sons, which comprised all his family. These sons are now all dead except one, Edward. David and Benjamin died 1865, and Perry died 1862. As this family lived so remote I thought a description of the country they inhabit and some general account of them would be interesting to their relatives here. Accordingly I wrote to Jonathan R. Borden, Esq., on the subject, and he has very kindly and very promptly furnished me with the following narrative, which I trust will be acceptable to others as it has been to myself:

"Skirting the southern shore of the Bay of Fundy, which forms the northwest boundary of Nova Scotia, is a range of hills, dignified by the appellation of the North Mountains. These extend from Cape Blomidon on the east, jutting out into the basin of Minas to Digby Neck on the west, a distance of more than one hundred miles. On its southern side it forms an abrupt declivity some four hundred to five hundred feet in height. About twelve miles distant from this range may be seen the sloping terrace of the South Mountains running nearly parallel, and rising gradually to about the same height. Between these chains of hills may be seen a level and beautiful valley bounded at its two extremities by the Basin of Minas and the Atlantic Ocean. This valley has been the scene of the most interesting events connected with the history of the province. In the western part is the town of Annapolis, formerly called Port Royal, founded by the French explorers, Demots and Pontrincourt, in the year 1604, being, I think, the first settlement of Europeans on this continent. This town was for nearly one hundred and fifty years the capital of all Acadia and figured considerably in the wars which, during that period, occurred between the English and French. The surrounding country was also occupied by the followers of Pontrincourt who have been since known in provincial history as the French Acadians. The eastern part of this valley, now Horton and Cornwallis, the latter comprising the eastern part of the North Mountain and extending nearly across the valley; the former lying along the base and extending up the base of the South Mountains, were also occupied at a very early period by the Acadians, a hardy and industrious race. Here they lived for most part in peace, felling the forest, reclaiming the marshes, cultivating and improving their land, and securing for themselves and their families a comfortable independence until 1755, when on account of refusing to take the oath of allegiance to Great Britain they were by an edict of Governor Lawrence expelled from their homes and scattered among the other colonies (from Maine to Georgia.) The character and mode of life of this people, their expulsion and subsequent sufferings form the subject of Longfellow's touching story of Evangeline. Whether this act of government was justifiable or not remains a question which we have little means of determining, for even we who live upon their lands and walk over their graves, know but little more of them than any one may learn from the fanciful sketch of the poet."

It was these lands, thus deserted and offered without purchase to the New England colonists that induced the company of emigrants to which Samuel Borden and his son Perry belonged, to leave their old homes and settle in the more remote province of Nova Scotia. The band of "first settlers" consisted of one hundred and fifty individuals. They landed on the 8th of June, 1760, at a place which, on account of its being first selected as the site of a future town, is still known as "the town plot," though the town had never been built. The settlers immediately obtained a grant of two hundred and fifty thousand acres or thereabouts to which they gave the name of Cornwallis, after that of General Edward Cornwallis, who had previously commanded the English troops in Nova Scotia and who in 1749 laid the foundation of Halifax by erecting government storehouses and barracks for the British army. In 1750 this collection of buildings and their surroundings received the name Halifax, a mark of respect to Lord Halifax, then president of the Board of Trade. This Edward Cornwallis was the brother of Lord Cornwallis who figured in the war of independence some years later. Cornwallis, Aylsford and Horton form Kings County which includes all, or nearly so, of the lands cultivated by the Acadians, with large tracts of woodland on the North and South Mountains. On taking possession of their lands the newly-installed owners proceeded to divide them into one hundred and fifty shares. The most valuable portion of Cornwallis is the valley which now contains the greater part of the population and of the wealth, and which has been styled the garden of the province. It is certain that few places possess greater agricultural capabilities. Though situated in 45 north latitude, owing to its sheltered position and the near proximity of large bodies of salt water, its climate is milder and less changeable than many places farther south. The soil when enriched is very productive, especially in potatoes and all the varieties of fruit and vegetables adapted to the climate. The upland hay crop is generally light, owing to the peculiar nature of the soil, but this deficiency is supplied to a great extent by the large and valuable marshes which skirt the banks of the numerous rivers running into the interior. These lands deserve more than a passing notice to give a correct idea of the agriculture of Cornwallis and Horton, for it is through these that they have acquired and maintained their high position as a farming country. This alluvial soil is a marine deposit peculiar to those parts which are washed by the rapidly flowing tides of the Bay of Fundy. Near its head this bay separates into two parts, the first includes Minas Bay and basin and Cobequid Bay, both lying in Nova Scotia; the second part in Chignecto Bay, which nearly separates Nova Scotia from the mainland, leaving an isthmus only fourteen miles wide. It is only in the first two bays that the tides attain their greatest force and height;

the extreme height being, at high spring tides, about sixty feet, while at the entrance of the bay at St. Johns it is about thirty feet. The Bay of Fundy, being forty miles wide at its entrance and one hundred and forty miles long, the ebb and flow of such a body of water twice every twenty-five hours must produce a very strong and rapid current. The retreating tide leaves some of the rivers bare, but the appearance of th first flow or bore, as it is called, up these rivers, is rather a peculiar sight. At high spring tides an almost perpendicular wall of water, from two to five feet high moves majestically along at the rate of two to five miles per hour, the velocity being in proportion to the height of the bore. Small boats stand but little chance in its way, and even large ones require to be managed with great care and skill to insure their safety under such circumstances. It is by the action of these tides that the marshes already spoken of are formed. The shores are continually wearing away by the attrition of the water and the finer particles of the debris, with materials from other sources, are carried up the estuaries of the rivers and deposited either along their banks or at their heads. This operation being repeated twice every twenty-five hours, large mounds are formed high above ordinary highwater mark, and then a dyke either across the stream or along its banks effectually excludes any further encroachment of the water. In this way lands are reclaimed which in natural fertility are excelled by none perhaps in the world. After a period of from forty to one hundred years these lands, cultivated only by occasional ploughing and reseeding, without the addition of any manure except what is dropped by the large herds of cattle which are turned upon them in the fall of the year to crop the luxuriant after-growth, now cut from two to three tons and some four tons to the acre of excellent hay, at a single cutting. These lands are formed at the heads of all the offshoots from the Bay of Fundy, but are not of equal extent nor of equal value. Cornwallis and Horton have about 7000 acres of this soil, some of them the finest quality, which when offered for sale command rising from forty to one hundred dollars per acre.

Although the country is rapidly improving, yet agriculture, which is the chief occupation of the people, cannot be said to be in a very advanced state. The necessity of a thorough cultivation of the soil or the careful husbanding of manures, the superiority of improved breeds of stock and the economy of introducing and using labor-saving implements are things which, for the most part, are but beginning to be understood. The chief exports are fat cattle and potatoes, which by raising and exporting to the American markets the farmers have, during the last fifteen years, accumulated much wealth. Wood, butter, cheese, fruit and pork are exported in considerable quantities. The fisheries in Minas Basin and the Bay of Fundy are prosecuted to some extent, but not sufficient to supply the home consumption. We import almost all our flour from the United States and Canada, and our manufactured goods we receive from England and the United

States, chiefly from the latter. Considerable attention is given to shipbuilding and commerce.

The early history of the Borden families here is similar to that of most early settlers. Privations were to be endured, obstacles to success and permanent improvement to be overcome, their families to be supported and property accumulated for their children. To accomplish these objects successfully, was the great aim of those who took up their abode here, the men of the first generation. And those of the second generation bear a strong resemblance to them in this respect. The first division of lands, though equitable, was found very inconvenient on actual occupancy, which led to frequent changes by sale and purchase among the proprietors and subsequently. The most shrewd and far-seeing, among whom was Perry Borden, taking advantage of this state of things secured for themselves some of the finest lands which in a few years advanced so much that they found themselves possessed of extensive and valuable properties. All the sons of Perry Borden, by the rise in the value of lands thus selected by their father and their own industry became men of independent means, leaving in some cases large properties to their children in whose possession they are chiefly found at the present day. Many of their descendants have changed their locations. The Bordens of the early times considered the mere elements of education amply sufficient for the wants of very-day life, physical power and skill in the various departments of manual labor being more highly prized by them than mere intellectual power. The Bordens of this period, although several of them were noted in the first respect, do not appear to have been deficient in the latter, on the contrary, they were generally men of sound judgment, good, practicable ability and extensive influence. David and William were induced by their friends to accept nominations for seats in the Provincial Parliament. They were unsuccessful, but not on account of a want of ability. Since that time Andrew Borden of the next generation offered himself as a candidate for the same position, but did not succeed; after which no other person of the name has shown any inclination to win for himself political honors. Indeed the general characteristics of the family were not such as to warrant them in aspiring to such positions. They were men of strict integrity, fearless and independent in thought and action, with qualities of mind more solid than showy, more fond of comfort than fame, and generally with little ability as public speakers; they were better qualified to give weight and influence to a community than to secure honor or distinction to themselves. They have been and still are mostly farmers, although some have turned their attention to other pursuits, almost always with success. Among the younger members many have received and are receiving a liberal education and bid fair to attain distinction for themselves in the various departments to which they have directed their attention. In social life they are to be found in every rank, some occupying high positions in society, while very few have ever

FIFTH GENERATION.

brought discredit upon an old, respected and honorable name."

Perry Borden's first wife was Amy Percy, married September 6, 1761. She died December 2, 1765. She was the daughter of an English officer. He then married Mary Ells October 22, 1767. She was born May 25, 1745, died 1831.

249. BENJAMIN, born 1740; he married Rachel Cobb June 28, 1772.

This Benjamin was distinguished from several others of the same name by the title of Christian Ben. He was a farmer in Tiverton, an honest and industrious man. His son Samuel was an officer in the army of the United States, serving in the western country till his health became so much impaired that he started for home, but died on the passage down the Mississippi River, and was buried on its banks. Catherine inherited the homestead of her father, and her heirs still retain it. Benjamin Borden was a member of the Friends' Society in Tiverton. S.

250. ANN, born March 8, 1743; married James Durfee January 3, 1765.

251. ABIGAIL, married Joseph Durfee February 4, 1770.

252. EDWARD, married Elizabeth Borden, daughter of Samuel and Mary.

141. JOHN, Scituate, R. I.

253. SAMUEL ASA, born July 12, 1738. He was a physician at Gloucester, R. I., also lived at Providence, where he married the daughter of Ezekiel Hopkins. After the death of his father he removed to Canada, near Quebec. His son Samuel continued to live there, and raised a numerous family.

254. JOHN, born March 30, 1740; died October 14, 1819. He married Elizabeth Colwell at Scituate, R. I.

255. NATHANIEL, born October 10, 1743. He died unmarried.

256. GAIL, born September 9, 1745; died 1777, in Gloucester, R. I. His wife was Mary Knowlton, daughter of Thomas Knowlton and Lydia Ballard, who was a direct descendant of Gabriel Bernon, and his wife Esther LeRoy. Gabriel Bernon was born in 1644, died 1736. He came to America from Rochelle, France, during the Huguenot persecution. He was a merchant at Rochelle, and early in life engaged in commercial enterprises in Canada, in which he acquired great wealth. He was thrown into prison in Amsterdam for the crime of Protestantism. He fled on his release from prison to London, from thence to Boston in 1688. From which place he removed to Providence. Several interesting memorials of his are still preserved, as carved chairs, a gold rattle, a sword (with date 1414), a psalm book, etc. He was buried beneath St. John's Church, Providence, and a great bronze tablet was placed there to his memory. The wife of Gail, Mary Knowlton, after his death, married Ezekiel Hopkins of Providence, and had two daughters, Elizabeth and Amy Hopkins.

257. ABIGAIL, born February 3, 1747.

258. EBENEZER, born November 23, 1750; he married Lydia Knowlton; left one daughter, Cyrena.

259. REBECCA, born April 10, 1755.

142. BENJAMIN, Newport, R. I.

260. BENJAMIN, born 1744; married Elizabeth Mason. "He lived at Newport. He was engaged in the African slave trade, and contracted a disease from long exposure to that climate, which tormented his life. He left a widow and small children, who soon passed away, so suddenly that suspicions were excited among the neighbors that they had been poisoned. The widow of Benjamin dying, his property fell into the hands of his legal heirs, and his brother Thomas sold his house in Newport for $550. In making his deed to it Thomas Borden inserted the name of his wife, Ruth, and his two sisters, Elizabeth, the wife of William Owens of Blooming Grove, in the county of Orange, N. Y., and Rosanah Borden of the same place; the deed bears the date September 21, 1875." S .

261. THOMAS, born 1747; married (1) Ruth Mason; (2) Sarah Easterbrooks.

262. JOSEPH, born November 18, 1748; married Hannah Stafford. "He removed at an early day to North Carolina, probably with his uncle, William Borden, who established himself as a ship builder on Core Sound or Newport River, near to the present site of Beaufort, N. C. He probably went out unmarried, and after some years returned North and married Hannah Stafford, the sister of David Stafford of Tiverton, R. I. Their family residence was on the Stafford road and near to the South Watuppa pond. The Stafford family of that day were connected with the Friends' Society in Tiverton, and were highly respected, not only in their own town, but also among the Friends in the neighboring towns, where the meetings of Friends were held. Most of the sisters of David Stafford, on marrying, removed with their husbands to remote places. A letter from Samuel and Sarah Stafford addressed to their honored mother, then a widow in Tiverton, was dated February 17, 1751, came from the Nine Partners, N. Y. Another from Hannah Borden to the same, was dated at Core Sound, N. C., July 27, 1752. A third letter is from Patience, who had married a man by the name of Earl, was dated at Nine Partners, and was signed by Patience Earl October 21, 1752. Another from Patience Earl to her mother was dated North Carolina, the 19th of the 3d mo, 1773. Their sister Priscilla married a Lowden of Newport. These letters all show a respectful and kind regard for all their friends, and a tender solicitude for the comfort and welfare of their mother, which does honor to their feelings, and contrasts strongly with the conduct of families whom we have known in later times." S.

FIFTH GENERATION.

145. STEPHEN, Fall River, Mass.

263. STEPHEN, born October 28, 1728; died August 15, 1802. He married Mary Gray, daughter of Thomas Gray, of Tiverton, Oct. 8, 1848.

264. HANNAH, born November 10, 1730; married Christopher Borden, son of Thomas.

265. MERIBAH, born February 7, 1732; married Jabez Barker October 21, 1747.

266. GEORGE, born May 2, 1735; died June 2, 1810. He married at Tiverton, Rebecca Church.

267. SUSANNAH, born May 19, 1757; married John Brownell, November 15, 1753.

146. WILLIAM, Fall River, Mass.

268. SARAH, born 1732; married John Francis, Feb. 17, 1751.

269. JOSEPH, born August 12, 1733; died 1809. He married Peace Borden, daughter of Joseph, February 19, 1758. He was a farmer at Tiverton.

270. WILLIAM, born February 26, 1736. He was a mariner of Fall River. Married Ruhama Jennings July 5, 1761. He was lost overboard off Point Judith from a vessel in which he was sailing, during a violent storm.

271. BENJAMIN, born 1738, at Tiverton, Rhode Island. He married Patience Cobb.

272. RUTH, born 1740, married Nathan Durfee January 30, 1762.

273. STEPHEN, born —. He married Mary Church, daughter or Joseph Church of Fall River, November 3, 1763.

274. ANNE, married William Jameson February 1, 1764.

275. PARKER, married Susannah Jennings February 19, 1769. He lived in Fall River.

276. THOMAS, born 1751; died 1845, in Nova Scotia. He married there (1) Susanna Cox ,born 1761, died June 27, 1826. (2) Louis Lanford, born July 2, 1805, died 1876. Thomas Borden, alone of all his family, sympathized with the mother country and joined the British army before their raid on Fall River in 1777. He continued in the service of the King as corporal in the British army and assisted in gaining possession of Canada for the English. The very property that Thomas Borden drew from the government for his services is at Grand Prie, and still in possession of his descendants. The family name seems to have been spelled Bardain in Nova Scotia, but in our mind there can be no mistake in taking this Thomas Bardain to be the Thomas Borden who, as history tells us, joined the British army and was not again heard from by his American relatives.

THE BORDEN FAMILY.

277. GEORGE, born at Fall River, Mass., married Susannah Church there.

278. GIDEON, married (1) Joanna Barlow, September 24, 1774, no children. (2) Mary Pettice, December 24, 1779.

279. JOB, born 1756, died December 31, 1832. He was a Congregational minister, commenced preaching in 1792 and was pastor of his church for forty years. He married Lois Tilton; had no children.

147. GEORGE, Tiverton, R. I.

280. JOHN, born July 6, 1736, died November 23, 1816, at Tiverton, Rhode Island. He married Patience Willcox, December, 1760.

281. RUTH, born March 19, 1738, married Mr. Hart.

282. PATIENCE, born June 13, 1739.

283. WAIT, born August 13, 1744, married Moses Simmons, October 13, 1765.

284. PRISCILLA, born September 24, 1753.

148. JOSEPH, Fall River, Mass.

285. ELIJAH, born May 29, 1737, died May 25, 1822. He married Sarah Baker December 9, 1759, and removed to Pompey, New York.

286. MARTHA, born December 1, 1739, married Amos Dresser August 15, 1771.

287. PEACE, born December 18, 1741.

288. PHOEBE, born April 26, 1744.

289. STEPHEN, born December, 1745, married Lydia — in Newport, Rhode Island. They removed with his father, Joseph, to New York State in 1809.

150. JOHN, Fall River, Mass.

290. HOPE, born October 28, 1747, married Ezra Brownell October 20, 1768.

291. THOMAS, born January 7, 1748.

292. RICHARD, born Novmber 9, 1750.

293. BENJAMIN, born July 26, 1752.

294. ANN, born April 20, 1754.

295. JOSHUA, born April 22, 1756. He lived in Delaware County, New York, and married Elizabeth Pierce, a relative of President Pierce, at Pomfret, Conn., January, 1780.

He was a soldier in the revolution and belonged to the State troops of Rhode Island. After his death his widow petitioned the Twenty-seventh Congress at its third session for a pension, but her request was not granted. He made or helped to make the chain which was stretched across the Hudson River from West Point to Constitution

FIFTH GENERATION.

Island, during the revolutionary war, to prevent the British ships from attacking the fort. A part of this chain is now preserved at West Point.

296. MARY, born March 28, 1758.

297. SUSANNAH, born September 17, 1762.

298. JOHN, born September 26, 1764.

299. RUTH, born July 24, 1766.

152. JOSEPH, Fall River, Mass.

300. JOHN, born February 16, 1752, died May 28, 1828. He married (1) Eleanor Durfee at Portsmouth, Rhode Island. (2) Sarah Shearman, December 8, 1784.

301. STEPHEN, born April 1, 1754, died October 10, 1794 at Portsmouth. He married Prudence Earl.

302. MARY, born January 28, 1756, married Alexander Thomas May 25, 1820.

303. ELIZABETH, born July 27, 1758, married Pardon Cook of Tiverton.

304. WILLIAM, born April 14, 1760; died September 22, 1798; married Elizabeth Corey February 11, 1788.

305. JOSEPH, born October 14, 1763.

306. RUTH, born ——; married Gideon Luke.

154. WILLIAM, Portsmouth, R. I.

307. HERVEY, born March 13, 1761, at Portsmouth, R. I. He married Ann Barrington July 6, 1789.

308. THOMAS, born ——; married Lydia Anthony. He was a shoemaker, and farmer at Portsmouth.

309. SMITH, married Lucy Shaw of Portsmouth December 12, 1802. He was a farmer at Easton, N. J.

310. ABRAHAM, born ——; died at Havana.

311. SYBIL, married Thomas Monroe of Bristol.

312. PHOEBE, married Pearse Anthony of Portsmouth.

313. ELIZABETH, married Capt. George Monroe of Bristol.

314. MARY, married Nathaniel Gladding of Bristol.

158. WILLIAM, Beaufort, North Carolina.

315. JOHN, died aged 18 years.

316. WILLIAM, born in Beaufort, North Carolina; married Ann Delany.

317. ALICE, born in Beaufort, North Carolina; married Col. David Ward.

318. BENJAMIN, born in Beaufort, North Carolina; married (1) Nancy Wallace, daughter of David Wallace of Orachoke; (2), Rebecca Staunton.

Benjamin died before 1825; just how long I do not know. It was this Benjamin to whom his half-free slave (Hycen) wished to sell back the free half. Hycen was a good deal of a sailor, and hired his time from his old master, Ben, for some years, and ran on his own account a small schooner up the Neuse River as far as New Berne, then the principal town of eastern North Carolina. From the profits of this schooner venture, Hycen bought a half interest in himself, still hiring of his master the other half. Not long after Hycen became half owner of himself, his schooner was caught in a violent squall at the mouth of Neuse River, which is there a wide estuary of the ocean, and he and his assistant (who constituted his crew) came very near going to the bottom.

Hycen was profoundly affected, and when he reached port tied up his schooner, and went off to see his old master. After telling of his great peril and narrow escape, he said: "Marse Ben, I want to sell my half back—nigger property is poor property."

319. HOPE, born in Beaufort, North Carolina; married Asa Hatch of Jones county, N. C.

320. JOSEPH. "He was born August, 1769, and died in 1825. He was a Quaker, and late in life set free all of his slaves except such as were entailed to his children. These he felt he had no right to set free. So though he divided a large landed estate to his heirs, he gave them no slaves. He made his will April 4, 1823, and that will shows the most conscientious effort to divide his property fairly and equally among his children. By said will he divided his large landed estate to his seven sons for life, specifying the tracts each son was to take—remainder over to the children of each son in equal parts.

To his own daughter he gave money and bank stock; but gave her no land. He married in 1795 Esther, widow of Capt. John Easton. Her maiden name was Esther Wallace, and she was a daughter of David Wallace and Mary Willis, his wife. David Wallace was the son of Robert Wallace and Esther West, his wife, whom he married in the Island of Guernsey about 1700. Robert Wallace was a son of a brother of Sir James Wallace, a colonel in the British army, whose estate lay in Argyleshire, Scotland, and was confiscated by the English government on account of his opinions, and his devotion to the Presbyterian side about the reign of James II.

Robert Wallace was a surveyor, and came to Virginia with Sir George Pollock about 1700.

Joseph was born at the residence of his father on Newport River, May 5, 1769, and married Esther Wallace, daughter of David Wallace, Esq., of Portland, Carteret county, June 16, 1796. He had been deprived of the advantages of a good education, owing to the disturbed state of the country during the Revolutionary war. He was, how-

FIFTH GENERATION.

ever, endowed by nature with good natural abilities, and a discriminating mind, which well qualified him for the performance of the various duites of life; and his diligent attention to business and persevering energy in whatever he undertook supplied in a great measure the deficiency of his early education. He settled upon the estate his father had occupied. But in consequence of the destruction of his father's improvements by the British, he was compelled to begin life with very limited means, and, necessarily, had to endure the hardships and privations incident to a country comparatively new, and then just recovering from the ravages of a civil war.

He and his wife were frugal and preserveringly industrious, and, with the blessing of God upon their labors, they soon acquired a competency of this world's goods, and reared up a numerous family, who were permitted "to rise up and call them blessed." Joseph and his wife were firm believers in the truth of the Christian religion, and were active members of the Society of Friends, as their parents and grandparents were before them; and it may in truth be said of him that he possessed a spirit devoid of guile, and was truly "an honest man, the noblest work of God." As such he was highly esteemed by the community in which he lived, and dying he left a large circle of friends and acquaintances to mourn his early departure. He died January 6, 1825, aged 55 years, 8 months.

Joseph Borden and his ancestors as far back as Richard Borden, the emigrant who settled in Portsmouth, R. I., in 1638, belonged to the religious society of Friends, and were from principle opposed to negro slavery. They neither owned slaves themselves, nor would they countenance it in others. On the contrary, they tilled their fields by hands who received a just remuneration for their labor, and it was a part of their religion to see that every man in their employ received his just due; in this way recognizing the authority of the divine precept, "the laborer is worthy of his hire." For a time the Friends in the South got along in their quiet peaceful way, managing their affairs according to their own sense of propriety and the just rights of others. They possessed the confidence of the slave owners; they were respected by them for their strict adherence to the principles of their religion, and were generally considered by the people as good neighbors and worthy members of society. But when the anti-slavery excitement commenced, a great change took place in the South towards them on a sudden. The kind feelings of many were alienated from them, and the bitter hostility of the multitude soon became so great as to endanger the peace of the community and even the lives of the Friends. This state of feeling caused a thrill of horror to run through every community of Friends. They were denounced as a great anti-slavery society, not fit to be tolerated in a slave state; and for a time it seemed as though they must be sacrificed to the demon of slavery. They were surrounded on all sides by the myrmidons of slavery, and their only chance for safety seemed to be in the forebearance of madmen who thirsted for their blood. At this time Mr.

THE BORDEN FAMILY.

Borden had a wife and nine children for whom he felt the deepest solicitude, and for whom he felt it his duty to provide some way of escape from the impending danger. After surveying calmly the prospects around him, he concluded that prudence as well as natural affection demanded of him the prompt removal of his family to one of the free states; as the first outbreak of popular fury would fall upon the Friends located in the midst of slavery, and, of course, he and his family and all their substance would perish together. Having resolved fully on a removal, he immediately commenced to put it into execution, quietly, and so cautiously that his design was not suspected by his nearest neighbors. He collected what was due him and sold his real and personal estate, and was nearly ready to remove when he was attacked by sickness, under which he soon passed away. He died at the old mansion house January 6, 1825, and was interred by the side of his father, near the old Friends' Meeting House. His family no longer having any home in the South, removed to the free states. Some of the sons and daughters who married slaveholders, have remained during the late rebellion and have participated in the trials and hardships usually attendant upon a state of civil war." S. 7

160. BENJAMIN, Botetourte county, Va.

321. MARTHA, married (1) Benjamin Hawkins; (2) Robert Harvey, son of Col. Mat Harvey.

322. HANNAH, died in infancy.

163. JOHN, Knoxville, Tenn.

323. JOSEPH, date of birth unknown; died in Knoxville, Tenn. He married Mary Echols.

324. WILLIAM, settled in White county, Tenn., and afterwards removed to Washington county, Arkansas; he died there. The name of his wife is unknown.

325. BENJAMIN, married; had several daughters; no sons.

326. JOHN, settled in Hardeman county, Tenn., and died there, leaving several sons and daughters; names not known.

327. REBECCA, married —— Overstreet of Overton county, Tenn.

328. MARGARET, married Judge Keith of Knoxville, Tenn.

329. NANCY, married John McWilliams of Bledsoe county, Tenn.

330. SARAH, married Alexander McCoy of Knoxville, Tenn.

167. LYDIA, Botetourte county, Va.

331. BENJAMIN BECK, born 1746; settled in Giles county, Va.

332. JACOB BECK, born 1748; settled in Augusta, Va.

333. JOHN BECK, born 1750.

334. ADAM BECK, born 1752; went to Tennessee.

FIFTH GENERATION.

335. MARY BECK, born 1755; married Jacob Carper.
336. JOSEPH BECK, born 1757; settled in Botetourte county, Va.
337. HESTER BECK, born 1760; married Isaac VanMeter in 1775.
338. HANNAH BECK, born 1762; married —— Hohn.

170. FRANCIS, Mansfield, N. J.

339. JOSEPH, born January 17, 1741; died October 7, 1810, in Mansfield, New Jersey. He married Sarah Baker February 13, 1762. There were perhaps other children.

182. JOSEPH, Bordentown, N. J.

340. ANN; date of her birth unknown. "She married Judge Francis Hopkinson. He was born in Philadelphia in 1737, and his death occurred in 1791. His parents emigrated from England, and his father was an intimate friend of Benjamin Franklin, and he is said to have been the first person to whom Franklin exhibited the experiment of silently drawing the electric fluid from the clouds by a pointed, instead of a blunt instrument. Francis was the first student that entered the College of Philadelphia after its organization, and completed his course there. After graduating he studied law and in 1765 he visited England, where he remained two years. On his return home he fixed his residence at Bordentown, having married Miss Ann Borden, the only daughter of the Hon. Joseph Borden. At the first breaking out of the Revolution Mr. Borden and Mr. Hopkinson both took strong ground in favor of the liberties of the colonies, and devoted all their energies to the advancement of the good cause. In 1776 Judge Hopkins was delegated to the Continental Congress, which met in Philadelphia, and on the 4th of July they declared these colonies to be free and independent states, and affixed their names to this celebrated declaration. What he had thus avowed in connection with this, he labored to sustain by the power of his pen. He was, by nature, a man of a versatile genius, and by his education he became a man of varied accomplishments. To the knowledge of the law he added those of painting, poetry and music, and was considered quite proficient in them all. He commenced his warfare upon British rule as early as 1774, by the publication of several pamphlets designed to awaken the attention of the community to their true condition, and to arouse them to stand forth in defence of their rights. Among others were the following satrical compositions, as "The Admirable Political Catechism," "Letters of Tories and British Travelers," "Answers to British Proclamations," etc., and so well did he sustain his position throughout the war that a writer remarks of him that "during the Revolution he distinguished himself by satrical and political writings which attained such popularity that it has been truly said that few persons effected more than Hopkinson in educating the American people for political independence." He also exercised his raillery in prose and verse at most of the social follies of his time. In 1779 he was made judge of the

Admiralty Court for Pennsylvania, which office he held for ten years, until the organization of the Federal government, when his commission expired. As soon, however, as Gen. Washington entered upon the duties of his office as President of the United States, he addressed to Judge Hopkinson a highly complimentary letter, enclosing a commission of United States District Judge for Pennsylvania, a position he held during life.

He was not only familiar with the sciences, but skilled in painting and musical composition, often arranging the music of his own songs. The best of his poems were "The Battle of the Kegs," a humorous ballad, and "The New Roof," a song for Federal mechanics. The miscellaneous essays and occassional writings of Judge Hopkinson were collected and published by Dobson of Philadelphia in 1792 in four volumes.

Judge Hopkinson left one son, Joseph Hopkinson, and three daughters, Elizabeth, who married a Petit; Mary Letitia, married a Buchanan, and Ann married a Buchanan. They were all named in their grandmother's will, dated September 15, 1798, and proved November 5, 1807. Francis Hopkinson and wife both died and were interred at Philadelphia." S.

Copy of clipping taken from a Burlington county, New Jersey, paper of recent date:

"Sunset Cox, a descendant of Joseph Borden, showed me the other day, as a curiosity, the announcement of the marriage of a daughter of a parental great, great grandfather, as it was published in the newspapers of New Jersey 125 years ago. It reads as follows:

" 'On Thursday last Francis Hopkinson, Esq., was joined in the velvet bonds of hymen to Nancy Borden of this place, a lady noted both for her internal, as well as external accomplishments. In the words of the celebrated poet:

'Without all shining, and within all white,
Pure to the sense, and pleasing to the sight.'

"Nancy Borden was a daughter of a Gen. Borden, who laid out the city of Bordentown, N. J., and he was also the grandfather of Sunset Cox's grandfather. Francis Hopkinson was one of the scholars of Revolutionary days. He wrote a number of humorous and patriotic pieces, among which were the "Battle of the Kegs," and others. He was a graduate of Princeton, a signer of the Declaration of Independence, and a member of Congress. His son, Joseph, was one of the ablest lawyers of his time, and was the author of 'Hail Columbia.' Speaking of 'Hail Columbia,' the music of this song was at first known as the 'President's March,' and it used to be played while Washington was preceded by an orchestra as he came into the theatre. Its music was composed by a fellow named Phyles, and it was played for the first time on Trenton Bridge as Washington rode over it on his way to be inaugurated. It became popular at once, and Hopkinson wrote and adopted these lines to it, beginning 'Hail Columbia.' During the political campaign, while John Quincy Adams was President, the name

FIFTH GENERATION.   133

'President's March' was dropped, and it has since been known as 'Hail Columbia.' "

341. JOSEPH, born 1755; died October 16, 1788. He married Elizabeth Biles, daughter of Langhorn Biles of Bucks county, Penn. He was a gallant officer in the war of the Revolution, and commanded a troop of light horse of Burlington county. He was wounded by a musket ball at the battle of Germantown, while acting as aide-de-camp to Gen. Forman. He never fully recovered from the effect of the wound, and died October 16, 1788, at the age of 33. His wife was daughter of Langhorn Biles, of Bucks county, Penn., and was one of the matrons who assisted at Washington's reception by the people of New Jersey in 1789, as did also his only daughter Elizabeth, then a girl 13 years of age. Elizabeth Borden married Azariah Hunt. Mrs. Borden's mother was a sister of Col. Joseph Kirkbridge of the patriot forces. (Woodward's History of Bordentown).

342. MARY, married Thomas McKean, member of Congress from Delaware in 1776, signer of the Declaration of Independence, and Governor of Pennsylvania. He is described as "a violent Democrat in profession, but exceedingly aristocratic in disposition and practice." Their son, Judge Joseph Borden McKean, was a distinguished lawyer of Pennsylvania. A daughter of Governor McKean became the wife of the Senor Don Carlos Fernando Martinez de Yrugo, Marquis de Yrugo, Minister from Spain, Prime Minister, etc. There is a genealogy of the McKean family edited by Mr. Roberdeau Buchanan of No. 2015 Q street, Philadelphia, Pa.

186. AMY, Bordentown ,N. J.

343. ANN POTTS, born February 13, 1757, died March 26, 1818, married February 29, 1776, General James Cox, born October 16, 1753, died September 12, 1810.

SIXTH GENERATION.

SIXTH GENERATION.

188. WILLIAM, Johnston, R. I.

344. RICHARD, born March 8, 1780, died September 5, 1804.
345. WILLIAM, born April 6, 1783, died December 24, 1800.
346. RUTH, married William Fraley of Johnston, R. I.
347. HULDAH, married Solomon Sprague of Johnston, R. I.

189. OLIVER, Johnston, R. I.

348. JOSEPH.
349. SARAH.
350. OLIVER, died December 17, 1806.

199. WILLIAM, Burlington Co., N. J.

351 WILLIAM HAZLETON, born 1759, died March, 1834. Married Lucia Reynolds, born 1761, died April 23, 1845. She was the daughter of Rev. Chichester Reynolds, an Episcopal clergyman who established and taught the first academy at Columbus, N. J., who married the daughter of Thomas and Joanna Woodward.

William Hazleton Borden was a trooper in the war of the revolution. He joined the New Jersey militia when the Britih invaded that State in 1776 and again in 1778. Historical records prove how vallently the New Jersey Bordens rallied around the cause of Liberty and with their strong efforts or their lives helped to make America free.

352. PETER, born in Burlington County, New Jersey, was killed in the revolutionary war.
353. THOMAS, born 1766, in Burlington County, New Jersew, married Charlotte Gibbs, daughter of Joseph and Mary Gibbs.

202. THOMAS, Monmouth Co., N. J.

354. THOMAS, born September 26, 1779.
355. AMOS, born September 26, 1779, died October 10, 1854. He married Jemima Schenck of Middletown, New Jersey, who was born January 31, 1783, died January, 1823, and was a sister of Mr. Schenck, who emigrated to Ohio from New Jersey and settled near Columbus. This Mr. Schenck was the immediate ancestor of General Robert Schenck whose services during the civil war won for him many honors.

211. FRANCIS, Imlaystown, N. J.

356. MARGARET, born November 9, 1763, married Samuel Allen.
357. LYDIA, born January 7, 1765.
358. WILLIAM of Evesham, born August 15, 1767. He married Lucy Harrison. Was High Sheriff of Burlington Co., New Jersey.
359. JOSIAH, born November 10, 1769. Married Mary Robbins, daughter of Aaron Robbins. She was born February 13, 1779.
360. DANIEL, born November 17, 1771, of Emlytown, N. J. He married Rhoda Stout.

THE BORDEN FAMILY.

361. ASHER, born September 7, 1773; lived in Allentown, N. J. He married Mrs. Ann Mellon, a widow.

362. SARAH, born November 4, 1775; married Thomas Black.

363. ELIZABETH, born February 2, 1778; married Thomas Tooley.

364. FRANCIS, born March 20, 1780. He lived at Imlaystown, New Jersey; married (1) Mary Erwin; (2) Letitia Erwin, her sister.

365. EDWARD, born March 23, 1782; died unmarried.

366. SAMUEL, born April 28, 1785.

367. MORRIS JOHNSON WOOLEY, born August 20, 1794; by a second marriage.

217. JOHN, Shrewsbury, N. J.

368. JAMES, born March 1, 1768.

369. ZILPAH, born December 14, 1796.

370. ELIZABETH, born April 15, 1771.

371. JOHN, born February 8, 1773.

372. LYDIA, twins.

373. RICHARD, born February 16, 1775.

374. FRANCIS, born April 20, 1777; married Margaret Parker.

375. MARY, born August 12, 1778.

376. JEREMIAH, born February 17, 1781.

377. SAMUEL, twins.

378. ANN, born February 20, 1781.

379. TYLEE, born February 20, 1787; died September 15, 1854. He married Hannah Chambers, born June 7, 1792; died July 14, 1832.

236. BENJAMIN, Manasquan, N. J.

380. JOHN L., born November 30, 1794; died April 3, 1877. He married Miriam Allen March 4, 1832. She was born ——; died April 25, 1871. They had no children.

381. RICHARD, born November 27, 1796; died April 4, 1863; married Catherine T. Williams, born October 30, 1808; died Dec. 27, 1867.

382. WILLIAM LLOYD, born March 29, 1798; died June 2, 1875; married Jane Ann De Grauw of New York.

383. THOMAS T., born June 21, 1800; died January 29, 1862; married Susan Corlies April 16, 1822. Susan Corlies was born June 20, 1794; died August 24, 1880.

384. HANNAH T., born November 9, 1801; married John B. Hartshorn.

385. JOSEPH L., born November 27, 1804; died April 10, 1879; married Huldah Combs, born ——; died 1883.

SIXTH GENERATION.

386. BENJAMIN, born November 26, 1806; died May 17, 1889; married Deborah Woolley.

387. ANN LEVIS, born November 21, 1808; died April 15, 1892; married Asher Hance. She was a member of the Society of Friends.

388. SARAH TABER, born August 8, 1811; died December 25, 1898. She married James A. DeGrauw of New York.

237. RICHARD, Fall River, Mass.

389. ELIZABETH, born October 5, 1756.

390. RUTH, born August 18, 1759; married George Harris.

391. ISAAC, born October 20, 1761.

238. RICHARD, Fall River, Mass.

392. THOMAS, born 1750; died Noveber 29, 1831; married Mary Hathaway, born 1757; died 1824.

393. RICHARD, born 1722; died July 4, 1795; married Patty Bowen,

394. PATIENCE, born August 9, 1747; married Abner Buttler.

395. HOPE, married James Graves April 29, 1769.

396. BETSY, married William Valentine September 7, 1785.

397. MARY, married Edward Bailey October 25, 1783.

239. CHRISTOPHER, Tiverton, R. I.

398. JONATHAN, born May 3, 1761; died May 9, 1848. He married Elizabeth Bowen February 21, 1790.

399. ABRAHAM, born May 1, 1770.

244. ABRAHAM, Fall River, Mass.

400. SIMEON was born in Fall River in 1759. "His father having died at the age of 40, and he being the eldest son, he was forced to engage in active business at an early date. Like his father, he worked in the fulling mill till he had acquired a full knowledge of the business, aiding his mother at the same time in the general management of the estate left by his father, which comprised all that his father and grandfather Joseph had left. At the age of 27 he married Amy Briggs, the daughter of Capt. Nathaniel Briggs of Tiverton, June 15, 1786. Some time previous to this he was engaged in lumbering, furnishing white oak timber and plank for the building of ships, and other vessels for the coasting trade, to replace those destroyed by the English during the war. He also took up a new privilege on the stream, now owned by the Troy Company, where he erected a saw mill to facilitate his operations in lumber. This, of course, created some sensation, and a lawsuit; but he maintained his position, the suit terminating in his favor. This added materially to the value of his estate. About

THE BORDEN FAMILY.

1790 Mr. Borden, Seth Russell of Dartmouth, and Capt. Samuel Borden, late of Fairhaven, fitted two or three small vessels from Fall River for whaling on our coast, to cruise from George's Banks to Cape May. I can find no records of these transactions, and cannot speak of any great success. They followed it several seasons, until Capt. Borden settled in Fairhaven, and Seth Russel in New Bedford, and Mr. Borden abandoned the business, and the only article left to remind us of this enterprise is an old try pot which came out of these whalers, which Mr. Borden brought to his farm in Tiverton to boil potatoes for his hogs. It is now in the possession of his daughter, the widow of the Hon. Job Durfee, late Chief Justice of Rhode Island. Mr. Borden removed to Tiverton in 1806. He was an amiable, kind-hearted man, animated in conversation, and a pleasant companion; but in the latter part of his life he became negligent, and, consequently he left his estate in an embarrassed condition, and exposed to great impositions from dishonest persons. He died November 27, 1811, aged 52 years. Mrs. Borden died May 26, 1817, aged 52 years. Her name was Amy Briggs. They were married June 15, 1786." S.

401. PERRY, born 1761; married Phoebe Sisson May 20, 1785; lived at Fall River.

402. JUDITH, born 1763; died unmarried.

245. SAMUEL, Tiverton, R. I.

403. PATIENCE, born August 11, 1762; married Jonathan Slade.

404. RESCOME, born June 29, 1766; died young.

405. ELIZABETH, born August 23; married Edward Borden.

247. JOSEPH, Fall River, Mass.

406. ELIZABETH, born July 11, 1765; married Samuel Wales of Boston.

407. BENJAMIN, born July 12, 1766; married Priscilla Westgate, daughter of John Westgate of Swansea, April, 1795.

408. AMEY, born October 24, 1767; married William Westgate of Tiverton, R. I.

409. RHODA, born April 24, 1771; married George Westgate.

410. HOPE, born February 26, 1776; married Nicholas Durfee.

248. PERRY, Cornwallis, Nova Scotia.

411. SAMUEL, "so well known in Fall River as a successful business man, was born in Cornwallis, in the county of Kings, Nova Scotia, September 1, 1762. When about eighteen months old he was brought to Tiverton, R. I., the residence of his grandsire, Samuel Borden, Esq., and when his parents returned home, Samuel was left in the charge of his grandparents, at their urgent solicitation. This they did out of regard to Mrs. Borden's health, who was of slender constitution, and already far advanced in pregnancy again. He remained with them

SIXTH GENERATION.

several years, until he had acquired the rudiments of knowledge as taught in the country schools at that day. In 1778, when he was 16 years old, and the year in which his grandfather died, Perry Borden sent a request to his father to send his son Samuel to him, as he greatly needed his services. The request was complied with, and Samuel returned to Cornwallis as great a stranger as though he had not been born there. But his own mother he did not find; she had already rested in her grave twelve years, and another had taken her place, who had sons of her own who engrossed all the maternal affection she had to bestow. Of course, between his step-mother and him there could be no feeling of sympathy; on the contrary, a kind of dog-war soon commenced, which was evidenced by a long series of low mutterings, growlings, and occasional sharp barkings, until 1783, when Samuel Borden became legally a freeman, and his first act was to assert his independence by returning to Tiverton again, declaring "he would not live in Nova Scotia if the government would give him the whole province." After the death of his grandmother, he came to Fall River and boarded in the family of Simeon Borden, Esq., for several years. Among other enterprises started by them was that of whaling, Seth Russell, late of New Bedford, being one of the concern. After much inquiry, I have not been able to obtain any reliable account of this transaction. They fitted one or two sloops to cruise between Cape Henlopen and Nantucket Shoals, which made several trips in a season with varied success, until Samuel Borden went to Fairhaven, and Seth Russell to New Bedford, which broke up the concern, and the remaining partner abandoned the pursuit. The course taken by the two first-named partners afterwards seems to indicate that the business had been profitable to some extent, as they both engaged in it again, though under more favorable circumstances, and followed it during their lives. But the removal of Mr. Borden at that time was occasioned by the death of a young lady to whom he was engaged to be married, Miss Judith Borden, the sister of Simeon Borden, Esq., his late partner. He afterwards married Elizabeth ——, and had two children. He died in 1850." S.

412. JOSEPH, born June 3, 1764; died 1841; married in 1793 Elizabeth Cogswell.

413. LEMUEL, born September 26, 1768; died 1834; married 1795 Esther Pineo, Horton, Nova Scotia.

414. DAVID, born January 28, 1768; died 1864; married in 1793 Elizabeth Kinsman, Horton Nova Scotia.

415. JONATHAN, born July 29, 1771; died October 18, 1835; he married Mary Miner February 17, 1814, in Horton.

416. PERRY, born February 17, 1770; died 1862; married Lavina Fuller 1809, in Horton, Nova Scotia.

417. JOSHUA, born December 3, 1774; died March 10, 1857; married

THE BORDEN FAMILY.

Charlotte Fuller in 1809. She was born January 22, 1788; died March 31, 1872.

418. WILLIAM ,born January 13, 1777; died November 4, 1853; married Margaret Rand, August 2, 1804, in Cornwallis, Nova Scotia.

419. BENJAMIN, born April 28, 1779; died 1865; married (1) Martha Wells March 23, 1802 in Cornwallis, Nova Scotia; (2) Lavinia Pineo, November 13, 1823; no children by last marriage.

420. EDWARD, born August 9, 1781; married November, 1814, to Abigail Eaton in Cornwallis.

421. ABRAHAM, born January 18, 1787; married December 3, 1817 Martha McGowan Dickey in Cornwallis, Nova Scotia.

249. BENJAMIN, Tiverton, R. I.

422. CATHERINE, born June 15, 1773; married Samuel Gardiner of Tiverton.

423. SAMUEL, born February 17, 1780. He was an officer in the U. S. army.

252. EDWARD, Westport, R. I.

424. RESCOME, farmer at Westport, R. I.; married Henrietta Sanford, daughter of David Sanford of Tiverton.

425. SAMUEL, married Eliza Borden, daughter of Joseph, September 29, 1816.

426. EDWARD, farmer at Westport; married Patience Crapo, January 13, 1824.

253. SAMUEL ASA, Pro. Quebec, Canada.

427. JOHN,

428. SAMUEL, born 1776; name of his wife was —— Hopkins, of Gloucester, R. I. He removed to Canada in 1812, and raised a large family there.

429. EBENEZER,

430. ASA,

431. WILLIAM,

432. BENJAMIN,

254. JOHN, Scituate, R. I.

433. NANCY born March, 1782.

434. RHODA, Born 1784.

435. JOHN, born November 25, 1786; died March 20, 1861; married Priscilla Hill November 6, 1811. She died August 19, 1848.

436. ROBERT, born February 6, 1789; never married.

437. ISAAC, born January 27, 1792; died March 29, 1856; married Susan Eldred July 4, 1818.

GAIL BORDEN.
439.

SIXTH GENERATION. 143

256. GAIL, Gloucester, R. I.

438. MARY, born October 28, 1775; died 1809; married in Gloucester, R. I., William Bradford of Providence.

439. GAIL, born August 23, 1777; died 1863 in Texas; he married in 1800 Philadelphia Wheeler, born in 1780; died 1828. She was the daughter of Henry Wheeler and Esther Williams, his wife, who was a lineal descendant of Roger Williams, through her father, Silas Williams, son of Peleg, son of Daniel, son of Roger Williams.

261. THOMAS, Newport, R. I.

440. LUTHER, born April 1Y, 1797; died ——. He was an owner of a vessel, and sailed for years as her captain; he married Elizabeth Pierce June 4, 1820, in Warren, R. I.

441. MARTIN, was born 1800; married Sarah Gardner, daughter of William Gardner of Gardner's Neck, November 6, 1825.

263. STEPHEN, Fall River.

442. DANIEL, born 1749; lived in Fall River, and married July, 1769, Anna Brightman.

443. BENJAMIN, born 1750, at Tiverton, R. I.; marrier Lovice Cook, daughter of Oliver Cook, April 15, 1775.

"He commanded a company of militia in Tiverton when the English troops landed at Fall River. They were led by Thomas Borden, formerly a citizen of the place, who was acquainted with all the inhabitants and their location, and improved this opportunity to vent his spleen upon one of his associates who was probably obnoxious to him. The house of Capt. Borden was surrounded and he was taken from his bed before daylight and hurried down to the English boats before the alarm was given or the presence of the enemy was generally known in the place. After completing their work of destruction, they departed, carrying with them Capt. Borden and Richard Borden. The latter was released and returned home in the course of a week, but the former was confined in Newport jail without any prospect of a release. For a while Capt. Borden bore his confinement quite stoically, but toward the close of the first year he became very impatient, and greatly excited at times, and let no opportunity pass unimproved of abusing those who came once a week to inspect the prison and see if the prisoners were all in their rooms. On one accasion he became greatly excited and played the madman to perfection. He cursed the prison, the fare, the officers before him; the commanding general and the whole British nation generally, and concluded that he would not remain there much longer. The officer in charge asked how he could help it; how he expected to get out of prison. To which Borden replied, at the same time pointing to the iron gratings of his cell: "Do you see those bars?" The officer replied: "Yes, I see them; but how do you expect to overcome them?" Borden replied hastily: "I'm going to eat them up." This conversation, far from exciting any fears for the security of the prisoner, only confirmed them in the opinion that his confinement had unhinged his mind, and that he was insane. On their turning to leave the room, Borden once more informed them that some morning they would come and find his room vacant. Capt. Borden was allowed to go into the basement of the jail through a trap door in the floor of his cell, and had discovered that the bars of the window there were of wood painted the color of iron. All that he needed to recover his liberty was a prudent, trusty companion and a good jackknife. This boon was soon granted in the person of Capt. William Taggart of Middletown, R. I., formerly, but then of Little Compton, where he and his father, Maj. William Taggert, resided at this time. As Capt. Taggart has published a statement of their escape from Newport jail, and return home, I will here quote from his narrative: "Towards the latter part of July, 1779, a large party of refugees from Newport came to Little Compton for the express purpose of making prisoners of my father and his two sons. The party landed unobserved, though there were sentinels at the shore and at the house. Two of the sentinels discovered a boat, and hailed and fired at the same time, but were immediately seized by the enemy, then at their backs, with threats of immediate death for daring to fire. We were alarmed at the house by the report of the muskets, and my unfortunate brother and myself, having armed ourselves, were the first to reach the shore, and were instantly made prisoners by the enemy. In the confusion of the moment my brother attempted to escape by leaping over a stone wall, when he was fired at and wounded through the thigh. One of the merciless desperadoes pursued and ran him through with a bayonet. They then took four of our party on board their schooner and lodged us in Newport jail. I there remained a prisoner about a fortnight, when, with Capt. Benjamin Borden of Fall River, we made our escape in the following manner: The prisoners were occasionally allowed to go into the cellar; when we discovered that instead of iron the windows were furnished with wooden bars, which might easily be removed with a good knife. But even then there were difficulties to be surmounted, which, to persons less determined than ourselves, would doubtless have appeared insuperable. Sentinels were placed in front and rear of the prison, and were continually patroling. From the window by which we escaped a few steps brought us into the street in front and in view of the sentry, who, fortunately for us, was in the sentry box at that time on account of the rain which was falling. We had agreed to walk deliberately and without betraying any signs of fear, and were providentially enabled to pass in the twilight safely through the compact part of the town near the hay scales in Broad street. We went into the fields to the southeast of that street, and at a short distance from this we crossed the line which enclosed the town, although they were strictly guarded. We then attempted to cross the road and steer our course between the

DR. JOSEPH BORDEN.
604.

SIXTH GENERATION. 145

forts on Irish's Hill and Tammany Hill in order to avoid the regiment of Anspack, which was encamped near by. The darkness protected us. We came out into West Road, and having proceeded eight or nine miles toward Bristol Ferry, we halted at the house of Nathan Brownell, who received us with great kindness. As the troops at that season were encamped in the fields, it was extremely hazardous for us to visit at seasonable hours those inhabitants who were friendly to the American cause; but still greater, and apparently insurmountable obstacles opposed any attempt to leave the island unobserved. As the shores were closely guarded we could not possibly obtain a boat, and our only alternative was to procure rails and construct a raft. This was truly hazardous; for we were compelled to launch our unseaworthy craft between the nightly guards upon the shore. But the same kind Providence still shielded and protected us. We left the shore with out raft unperceived. A thick fog soon came up, and we were all right upon or rather in the water, for our frail bark was not sufficiently buoyant to keep us above the surface. At daybreak we were so near the Rhode Island shore that we could see the sentinels upon the beach. At an hour after sunrise we landed on the south end of Prudence; from there we were taken in a boat to Bristol, and from thence we went to our respective homes." S.

444. JOHN, born 1751; was a farmer at Tiverton and New Braintree; married Peace Cook, daughter of Oliver Cook, March 10, 1774.
445. SARAH, born 1753; married Benjamin Durfee June 10, 1780.
446. NATHAN, was a blacksmith at Tiverton and Fall River; he married Phoebe Earl March 15, 1783.
447. PATIENCE, married Deacon Richard Durfee June 10, 1780.
448. HANNAH, married Thomas Earl, Octoboer 25, 1788.
449. LUCY, married J. Rogers. 1786.
450. MERIBAH, married Abner Borden December 29, 1791.
451 MARY, married John Cook.
452. LYDIA, married Arnold Borden.

266. GEORGE, Fall River, Mass.

453. ISAIAH, born June 1, 1700; married (1) Abigail Snell, December 25, 1782; (2) Mercy Read March 15, 1789.
454. THOMAS, born May 15, 1763; Universalist minister.
455. LYDIA, born April 2, 1766; married Wanton Hathaway February 11, 1808.
456. SYLVIA, born 1768; married Joshuna Weeks, January 23, 1793.
457. GEORGE, born Oct. 4, 1770; died Dec. 3, 1806; married Phoebe Borden, daughter of Thomas and Mary.
458. STEPHEN, born October 22, 1772.
459. PENELOPE, born February 15, 1775; married Ezekiel Brownell of Westport, 1797.

THE BORDEN FAMILY.

460. MARY, born September 21, 1778; married Joshua Weeks after her sister Silvia died.

461. PELEG, born February 27, 1780; married Sarah Cole June 12, 1802.

462. PATIENCE, born February 17, 1782; married Thomas Hazzard January, 1809.

463. ADAMS, born June 11, 1784; died 1864; married Lucy Borden, daughter of Parker, November 17, 1806.

269. JOSEPH, Fall River, Mass.

464. AARON, born October 5, 1758; mariner; he married Mercy Durfee, daughter of Job Durfee of Portsmouth.

465. SUSANNAH, born Jan. 2, 1760; married Noah Hart May 9, 1793.

466. ELIZABETH, born July 13, 1761; never married.

467. JOSEPH, born October 20, 1762; died September 3, 1845; married Susan Church.

468. PARKER, born June 15, 1764; died ——; he married (1) Susannah Borden, no children; (2) Dolly Church, no children; (3) Eliza Diamond.

469. ABEL, born Feb. 16, 1766; married Ann Church.

470. ABNER, (twins), married Meribah Borden, daughter of Stephen, December 29, 1791; he died in 1839.

471. DAVID, born April 5, 1768; killed in a mutiny on a slave voyage.

472. ABIGAIL, born January 6, 1770; married Christopher Wordell.

473. WILLIAM, born February 9, 1772; died May 2, 1834; he married Rebecca Church September 22, 1796.

474. PEACE, born November 9, 1773; married Abraham Warren.

475. RHODA, born March 21, 1776; married Joseph Warren.

476. ABRAHAM, born March 23, 1778; married Lucy Borden.

477. ISRAEL, born December 5, 1782.

270. WILLIAM, Fall River, Mass.

478. AVIS, born March 25, 1736; married Shubael Hutchins of Killingly, Conn.

479. ANNA, born December 4, 1764; married Ashaiel Fisher November 1, 1789.

480. RHUHAMA, born March 4, 1767; married Samuel Sprague.

481. RUTH, born March 17, 1771; married Nathan Durfee.

482. SUSANNAH, born February 3, 1769; married Rev. James Boomer October 12, 1792.

THOMAS RICHARDSON BORDEN.
605.

SIXTH GENERATION.

483. ROSANNAH, born May 26, 1773; married Dyer Ames of Sterling, Conn.

484. ROBY,

    271. BENJAMIN, Tiverton, R. I.

485. SARAH, married Henry Negus.

486. LEMUEL, born ——; died August 12, 1841.   He married Abigail Evans August 20, 1790.

487. JOANNA, married Israel Orswell.

488. WILLIAM,

    273. STEPHEN, Fall River, Mass.

489. ELIZABETH, born 1764.

490. SETH, born 1766; died 1812; married November 1, 1794, Ruth Brown, daughter of Gideon Brown of Middleton, R. I.

491. ARNOLD, born October 2, 1770; married Lydia Borden, daughter of Stephen and Mary.

    275. PARKER, Fall River, Mass.

492. SARAH, married Oliver Cook of Tiverton.

493. ANNA, married Peter Cook of Tiverton,

494. LUCY, married Adams Borden November 17, 1806.

495. DURFEE died young.

496. RUTH, married (1) William Davis, November 6, 1806; (2) Lemuel Borden.

    276. THOMAS, Grand Prie, Nova Scotia.

497. THOMAS COX, born December 17, 1761; died December 23, 1880; married Elizabeth Wood July 13, 1820; born 1790; died May 23, 1856.

498. HENRY, born 1793; died 1889; married in Kings county, Nova Scotia, Sarah Sanford.

499. JOSIAH, died 1876.

500. JOHN,

501. SARAH,

502. JULIA,

503. BYARD, died 1883.

504. DANIEL SANFORD, born November 11, 1831; married Annie Maria Borden in Cornwallis.

505. SUSAN MARIA, born December, 1833; married —— Kennard.

506. FREEDOM ADELIA, born April, 1835; married —— Brompton.

THE BORDEN FAMILY.

277. GEORGE, Fall River, Mass.

507. EARL, born May 24, 1775; married Hannah Borden, daughter of Daniel, April 24, 1798.

508. ABRAHAM, born November 29, 1777; married Sarah Brown of Portsmouth, October 14, 1806.

509. THOMAS, born January 2, 1779; married Ruth Borden, daughter of Daniel.

510. AVIS, born September 7, 1781; married Silas Bessey November 4, 1809.

511. LYDIA, born March 4, 1783; married Elisha Hathaway.

512. RUTH, born April 26, 1790; married John Stillwell.

513. ISAAC, born April 24, 1786; died May 15, 1861; he married Elizabeth Durfee.

514. ABEL, born February, 1788; died in Georgia.

515. GEORGE G., born Feb., 1793; married Peace Cook April 7, 1816.

278. GIDEON, Fall River, Mass.

516. JOANNA, born April 22, 1780; married Abner Hoskins of Newport.

517. PATIENCE, married William Briggs February 21, 1806.

518. JOB, married Rachael Brownell August 28, 1822.

519. RUTH, married William Sabin December 15, 1811.

520. ELIZABETH, married G. T. Gifford of Westport.

521. STEPHEN, married Catherine Hart November 22, 1819; lived at Portsmouth; (2) Phoebe Tripp November 14, 1822.

522. ELIHU, married Deborah B. Sowle, November 10, 1824.

280. JOHN, Tiverton, R. I.

523. SARAH, born September 28, 1761; married Ivory Simmons of Little Compton.

524. GEORGE, born April 25, 1763.

525. PRISCILLA, born January 28, 1765; married Lawton Hicks May, 1787.

526. RUTH, born March 19, 1768; married Samuel Earl Nov. 1789.

527. AMAZIAH, born July 23, 1771; died young.

528. ALICE, born August 13, 1773; married Adam Manchester.

529. MARY, born November 12, 1775; married Stephen Crandall.

530. BENJAH, born March 8, 1778.

531. JOHN, born August 17, 1781; married Elizabeth Cook October 26, 1801.

JOHN BORDEN.
555.

SIXTH GENERATION. 149

532. ABRAHAM, born December 6, 1785; died December 5, 1860; married Malinda Hart 1811.

285. ELIJAH, Pompey, N. Y.

533. SUSANNAH, born August 7, 1761; married John Boomer.

534. PHOEBE, born January 8, 1863; married Thomas Wordell.

535. NATHAN, born March 4, 1765; he was a farmer at Pompey, N. Y. His wife's name was Sarah.

536. MARY, born March 21, 1768; married Clothier Hathaway.

537. SARAH, born Feb. 14, 1770; married Deacon Nathaniel Boomer.

538. PATIENCE, born April 18, 1774; married William Cook.

539. PEACE, (twins), she married Gershom Wordell July 27, 1794.

540. JOSHUA, born July 4, 1777; died April 27, 1849. Married Elizabeth Parker, February 11, 1807, in Pompey, N. Y.

541. BETSEY, born July 19, 1779.

542. LYDIA, born January 4, 1782; married Elisha Shearman.

543. SYLVIA, born October 9, 1785.

289. STEPHEN, Newport, R. I., and New York State.

544. TIMOTHY, born 1768; name of his wife unknown.

545. JOHN, born 1772; married Lydia.

546. JOSEPH, born 1774; died 1812; removed from Rhode Island to Madison county, New York, where he died soon after. His wife's name was Sarah.

295. JOSHUA, Delaware county, N. Y.

547. JOSEPH H., born 1781; died September 14, 1806. He married Philura Beckwith, born April 12, 1788.

548. RICHARD,

549. JOHN,

550. SARAH, married a Mr. Kelley.

551. SUSAN, married George Meade.

300. JOHN, Fall River and Portsmouth.

552. WAIT, born August 8, 1776; married Peter Lawton.

553. ELIZABETH, born January 2, 1778; married Isaac Cook of Tiverton.

554. RUTH, born November 27, 1781; married Daniel Greene of Warwick.

555. JOHN, born November 19, 1785; died November 7, 1824; he mar-

ried (1) Comfort Hicks, daughter of Capt. Gabriel Hicks of Tiverton, R. I.; (2) Lydia Bellows of Preston, Conn. In 1816 he removed to Indiana, where he died, leaving three sons.

556. ISAAC, born September 9, 1787; died February 9, 1870; he married Finis G. Maynard May, 1824.

557. STEPHEN, born May 3, 1789; died March 21, 1869; he married Ann Bronen.

558. ELEANOR, born June 26, 1791; died February 12, 1846; married Gardener Thomas.

559. ASA, born January 24, 1793; married Lydia Patch April 8, 1821.

560. ANN, born February 13, 1795.

561. WILLIAM, born November 21, 1805; died March 5, 1885; married Mahalah Hadley of Portsmouth.

562. MARY, born February 28, 1797; married John Hamblin.

563. SARAH, born November 9, 1799; died December 3, 1881; married Peleg Shearman.

564. LEVI, born July 5,1801; died August 30, 1823.

565. CYRUS, born April 2, 1803; died young.

301. STEPHEN, Portsmouth, R. I.

566. JOSEPH, born May 27, 1890. He was appointed ensign in 1775 in the colonial army. He died soon after his marriage, leaving no children.

304. WILLIAM, Fall River, Mass.

567. JOSEPH, born February 11, 1788; died at Havana August 19, 1809.

568. SUSANNAH, born August 8, 1790.

569. JOHN C., born March 24, 1793; died March 26, 1833. He married (1) Eliza B. Slade; (2) Amanda Bowen, January 2, 1823; (3) Mary Ann Manchester, April 28, 1828.

570. THOMAS, born August 11, 1795.

307. HERVEY, Portsmouth, R. I.

571. JOHN, never married.

572. THOMAS, born about 1800; married Caroline Wellington; removed to Lowell Mass., from Tiverton.

573. WILLIAM, removed to Warwick, R. I.

574. SARAH,

308. THOMAS, Portsmouth, R. I

575. SUSAN, married Andrew McCorrie.

309. SMITH, Portsmouth, R. I.

JUDGE JAMES W. BORDEN.
606.

SIXTH GENERATION.

576. Russel, married Jane Brown,

577. HENRY, married Caroline Taber,

578. ANNE,

579. RUTH,

316. WILLIAM, Beaufort, N. C.

580. BARCLAY, married ——; left one son, B. F. G. C. Borden.

581. ANN TOOKS,

317. ALICE, Beaufort, N. C.

582. ABIGAIL WARD,

583. WILLIAM WARD,

584. SHEPPARD WARD,

585. COMFORT WARD,

586. HOPE WARD, married David Wallace Borden.

587. HANNAH WARD,

588. ALICE WARD,

589. ELIZABETH WARD,

590. RUFUS WARD,

318. BENJAMIN, Beaufort, N. C.

591. JOHN,

592. HOPE,

593. BENJAMIN FRANKLIN,

319. HOPE, Jones county, N. C.

594. ALICE HATCH,

595. WILLIAM HATCH,

596. ASA HATCH,

597. BETTY HATCH,

598. HOPE HATCH,

599. GEORGE HATCH,

600. MARY HATCH,

320. JOSEPH, Carteret county, N. C.

601. WILLIAM, born August 5, 1800, in Beaufort, North Carolina; died 1853, in Alabama. He married Elizabeth Dickson in 1808 in New Bern, Alabama.

602. BENJAMIN, born December 11, 1801; died November, 1886; married (1) Margaret Hill of Onslow county, North Carolina; (2) Mrs. Martha Gray of Green county, Alabama. She was a Miss Cocke of Lynchburg, Virginia; born March 27, 1808; she died in Eutaw, Ala. June 13, 1897, aged 81 years.

603. DAVID WALLACE, born August 19, 1803; died 1853; he married his cousin Hope Ward, born March 10, 1804; died November 1, 1883, in New Berne, Alabama.

604. JOSEPH, born June 8, 1806, in Beaufort, N. C; died March, 1875, in Fresno county, California, whither he had emigrated just after the Civil War (in 1868) with a colony of other Alabamans, and settled in Fresno county, at what was afterwards called Borden. He was married twice; the first time to Sarah Margaret Bryan of Marengo county, Alabama, by whom he had two children, Thomas and Mary Esther, both of whom died, leaving no children. His second marriage, on June 15, 1848, was to Juliet Elizabeth Rhodes, daughter of James Rhodes of Alabama, and granddaughter of Gen. James Rhodes of Wayne county, North Carolina, who commanded the State troops when they were ordered into the field when war was declared against France in 1797, and Washington was made commander in chief.

605. THOMAS RICHARDSON, born January 24, 1808; died 1857. He married Ann Maria Jones of New Berne, Alabama, June 8, 1830; she was born November 8, 1809; was the daughter to Frederick Jones, a Welch family, who came to America in early times.

606. JUDGE JAMES WALLACE BORDEN was born near Beaufort, N. C., February 5, 1810, and was the son of Joseph and Esther Borden. The young Borden received a superior classical education at Fairfield Academy, Herkimer, N. Y., and at Windsor, Conn. After leaving school he entered the law office of the Hon. Abjah Mann, Jr., member of Congress from the Herkimer district. At the age of twenty-one he was admited to practice at the bar of the Supreme Court of New York. At the age of twenty-two he was married to Miss Emeline Griswold of Middleville, N. Y., and in 1835 removed with his wife to Richmond, Ind., where he practiced law successfully, and was elected Mayor of the city. In 1839 he removed to Fort Wayne, Ind., to take charge of the United States Land Office, then located there.

By his first wife Judge Borden had five children, Esther Anna, Rebecca, K., both dead; William James, a merchant in New York; George Pennington, Captain Fifth United States Infantry, now stationed at Fort McPherson (1896), Georgia; and Emeline, wife of Capt. Hargous of the Fifth United States Infantry, and who is now residing in New York. August 15, 1848, Judge Borden was married to Miss Jane Conkling of Buel, Montgomery county, New York, who survives him.

In 1841, Mr. Borden was elected judge of the Twelfth Judicial Circuit, then composed of nine counties. He was a delegate to the State convention in 1850, to revise and amend the State Constitution. After the adoption of the new constitution, in the formation of which Judge Borden played an important part, he made the race for Congress in 1851, but was defeated by a small Whig majority. In 1852 he was elected Judge of the Court of Common Pleas, District of Allen, Adams, Huntingdon and Wells. In 1857 he was appointed by President Bu-

DAVID WALLACE BORDEN.
603.

SIXTH GENERATION. 153

chanan resident minister to the Sandwich Islands. In 1863 he relinquished this diplomatic position and made an extensive tour of China, Japan, Asia and Europe, acquiring a vist fund of information regarding the political and social condition of these countries. In 1864 Judge Borden was reëlected to the Common Pleas bench. In 1867, he was elected Judge of the Criminal Court, which office he held at the time of his death.

Judge Borden was one of the ablest men in his section of the country. No man was better informed on general topics and on questions of finance, he possessed the most minute information. His memory was singularly retentive, and his mind was as active and vigorous as in his younger days. He was an able lawyer, and his decisions were just, and evidenced a thorough knowledge of the law and its bearing.

607. MARY WALLACE, born June 28, 1811; died 1840; she married Israel Sheldon of Orange, N. J.

608. ISAAC PENNINGTON, born July 26, 1816; died about 1875; he married Elizabeth Marest of Tuscaloosa, Alabama.

609. HANNAH G., born 1815; died young.

321. MARTHA, Botetourte county, Va.

610. SALLIE HAWKINS, married William Mitchell.

611. MAGDALEN, married Col. Matthew Harvey.

612. WILLIAM.

613. JAMES.

323. JOSEPH, Knoxville, Tenn.

614. ELI, married in Bledsoe county, Tenn., Martha Wheeler.

615. HAWKINS, born ——; died 1842; settled in Walker county, Alabama; wife's name not known.

616. JOHN, born September 16, 1795; died May 7, 1875; he married Catharine Matlock, daughter of William Matlock, and Catherine Sevier, daughter of Gov. John Sevier of Tennessee, the pioneer Governor of that State.

617. JOEL, removed from Tennessee to Calhoun county, Alabama in 1818.

618. ANNIE, married Moses Johnson; had one son, Allan Johnson.

324. WILLIAM, Washington county, Ark.

619. JOHN.

620. JOSEPH.

621. WILLIAM.

337. HESTER PECK, Boutetourte county, Va.

622. HANNAH VAN METER, born 1776; married —— McFarran.

623. MARY VAN METER, married Charles Hedrich.

624. ELIZABETH VAN METER, married Benjamin Carper.

625. PLACENTIA VAN METER, married —— McFarran.

626. JACOB VAN METER, born January 24, 1788, at White Stone Tavern, near the dividing line of Botetourte and Rockbridge counties, Va., while his father was moving his family from Berkeley county to Botetourte. At the close of the war of 1812 he went to Charleston, Va., (now West Virginia), where, on January 24, 1816, he married Patsy Usher Shrewsbury, born September 7, 1792, a granddaughter of Col. John Dickinson, whose wife, Mary Usher, was a granddaughter of Counsellor Perry of Dublin, Ireland. In 1818 he moved to Bowling Green, Ky., where he engaged in the mercantile business, and spent the remainder of his long, active and useful life. In connection with James Rumsey Skyler he obtained a charter and built at considerable cost the Portage Road, leading from Bowling Green to the boat landing below the Double Springs. This was the first railroad constructed in Kentucky. In 1833 he was elected "against his consent" to the State Legislature, receiving more votes than the other five candidates combined. While in the Legislature he was instrumental in obtaining an appropriation for the improvement of the Green and Bowen rivers, which gave to Bowling Green and to the Green River country their first commercial advantages. Jacob Van Meter was a man of many excellent traits of character. His success in business attests his energy, enterprise and fine judgment. He was kind-hearted and charitable, upright and sincere. His motto, in a long, busy life, extending over half a century, was "Never give up." He died February 27, 1874, in the 87th year of his age. His wife, though apparently in her usual health, passed away calmly on the following day. They were buried at the same time and in the same grave, side by side. For more than fifty-eight years they had journeyed together through life; they were not to be separated by death.

627. JOSEPH VAN METER, was born near Fincastle, Botetourte county, Virginia, September 7, 1790. He was a soldier of the war of 1812. He married, September 15, 1815, Damaris Lockland, who was born near Hagerstown, Md., July 29, 1795. They moved from Botetourte county to Marion, Smythe county, Virginia, in 1855, where they both died and were buried. Joseph Van Meter, November 8, 1873; Damaris, his wife, November 8, 1879.

628. SALLIE HAWKINS VAN METER, born August 4, 1794; married Dr. Eleazer Sweetland, December, 20, 1814; she died in Greenup county, Kentucky, April 13, 1881.

A descendant of William Sweetland, whose name appears in the "Original List of Persons of Quality, etc who went from Great Britain to the American Plantation from 1600 to 1700," as commander of the ship James, trading between London and New York, in 1678-79. He and his wife, Agnes, were residing in Salem, New London county, Ct.,

BENJAMIN BORDEN.
602.

SIXTH GENERATION.

In 1703. Eleazer (5) Sweetland, son of the Rev. Eleazer (4) Sweetland of East Haddam, Ct., (Joseph, 3; Joseph, 2; William, 1) was born in East Hadam September 23, 1782. He was commissioned by Gov. George Clinton April 14, 1804, captain of a company of light infantry in the regimental militia of the county of Chenango, N. Y., whereof Nathaniel King, Esq., is lieutenant colonel commandant. He was a Mason of the highest degree then known, and was for many years an honorable Justice of the Peace of Botetourt county, Va. He died in Pattonsburg, Botetuort county, Va., October 28, 1838.

The father of these Van Meter children, Isaac Van Meter, was a descendant of Jan Gysbertsen Van Meter, who emigrated from Bommel, in South Holland, and settled in New Utrecht, on Long Island, in 1663, and was appointed magistrate in 1673. There is good reason to believe that Jan Gysbertsen was a descendant of Emanuel Van Meter, the Dutch historian, who was born in Antwerp, June 9, 1535, and died while Dutch Consul in London, in 1612. Isaac Van Meter was born about 1752 or 53, and died in 1798. Hester, his wife, died in 1835.

339. JOSEPH, Mansfield, N. J.

629. PELETIAH, born January 24, 1764; died December 17, 1820; married Eleanor Gardiner November 1, 1785. She was born Aug. 28, 1768.

630. DANIEL,

631. AMOS,

340. ANN, Bordentown, N. J.

632. JOSEPH HOPKINSON, "was born in the city of Philadelphia, November 12, 1770, and died January 15, 1842. He was educated at the University of Pennsylvania. He studied law and commenced practice at Easton in 1791, from whence he returned to Philadelphia. He soon became quite noted as a jurist and poet. In the latter character he is best known as the author of the national song of "Hail Columbia," which he composed in the winter of 1798, for the benefit of an actor by the name of Fox. The words were adapted to a tune which had been composed in 1789 by a German named Feyles, and was then called "The President's March," and was first played publicly on the occasion of Gen. Washington's visit to the New York theatre. This production of young Hopkinson is of a higher order than those which have been left us by his father, and exhibits more care in the selection and arrangement of the ideas and greater attention to harmony in the versification. It is emphatically a national song, and being so finely adapted to a truly national tune, it will be admired for ages yet to come. It does not appear that Mr. Hopkinson produced many poetical effusions, and it is certainly to be regretted that one who wrote so well, wrote so sparingly. His profession was the law, and to this he devoted all his energies. If the character of the cases committed to his charge may be considered as an indication of his standing as a lawyer, we shall be forced to regard him as one of the most dis-

tinguished in our country. In the celebrated case of Dr. Rush against William Cobbet, he was leading counsel for the plaintiff in 1799. He was also employed in the insurgent trials before Judge Chace in 1800. And subsequently when Judge Chace was impeached before the United States Senate he chose Mr. Hopkinson to defend him, and he was successful. He was afterwards twice chosen a representative to Congress from Philadelphia, and, after serving four years in Congress, from 1815 to 1819 inclusive, he declined a reëlection. Having returned to Bordentown, he was elected to the Legislature of New Jersey. In 1823 he returned to Philadelphia to resume the practice of his profession. In 1828 he was appointed Judge of the United States Court for the eastern district of Pennsylvania, a position which had been honorably filled by his father under the administration of Gen. Washington. In 1837 Judge Hopkinson was chairman of the Judiciary Committee of the convention to revise the Constitution of Pennsylvania. He was also employed by Joseph Bonaparte for many years as his legal advisor and confidential friend, and during his absence from America Mr. Hopkinson was entrusted with the sole management of his estate. Such were some of the important trusts committed to his care from time to time, and such the weighty responsibilities which he had to bear. Yet he seems to have passed through them without flinching or stumbling, but gathering confidence and strength at every step of his progress; he achieved a reputation for himself which may be envied by all, but seldom attained by any of his competitors. His ode of "Hail Columbia" should alone endear his memory to every American heart.

The family of Joseph Hopkinson retained the homestead of their great grandfather, Joseph Borden, Esq., who erected the substantial brick edifice in 1750, soon after the name of Bordentown was given to the place. In a deed to him, dated May 24, 1746, he is described as Joseph Borden, Esq., of Nottingham, which name is still retained by Nottingham Square, a place distant about one mile from Bordentown. The old mansion is still occupied in the summer by the descendants of Joseph Hopkinson, and closed the remainder of the year." S.

341. JOSEPH, Bordentown, N. J.

633. ELIZABETH, born 1776; married **Azariah Hunt**.

343. ANN POTTS, Zanesville, O.

634. RICHARD COX.

635. WILLIAM COX, born March 29, 1777; died July 28, 1803.

636. JOSEPH COX, born September 28, 1778; died September 28, 1830; married Deborah Kimen January 9, 1800.

637. LEWIS COX, born April 27, 1780; died April 3, 1835; married Elizabeth, daughter of James and Mary Lawrence; she was born December 29, 1785; died June 3, 1861.

ISAAC PENNINGTON BORDEN.
608.

SIXTH GENERATION.

638. JAMES COX, born January 17, 1782; died January 22, 1808.

639. MARY COX, born Dec. 19, 1783; died Sept. 3, 1834; married David Bateman.

640. AMY COX, born December 19, 1783; died March 19, 1820; married John Vandevoie.

641. THOMAS COX, born July 8, 1787; died December 21, 1813.

642. JONATHAN COX, born July 28, 1792; died September, 1818.

643. REV. SAMUEL COX, born November 2, 1789; died August 23, 1870. Married Hannah Lodge December 30, 1822.

644. DAVID JONES COX, born July 28 1792; died August 20, 1827.

645. HON. EZEKIEL TAYLOR COX, born May 30, 1795; died May 18, 1873, at Zanesville, Ohio. Married Maria, daughter of Samuel and Mary Sullivan, born March 16, 1801; died April 3, 1885.

646. MORGAN RHEES COX, born November 27, 1798; died January 9, 1881; married Mary Rittenhouse.

647. HON. HORATIO J. COX, born August 14, 1801; died March 6, 1883; married Ann Chambers.

JOHN ALLEN BORDEN.
648.

SEVENTH GENERATION.

351. WILLIAM, Burlington county, N. J.

648. JOHN ALLEN, born in Burlington county, New Jersey, November 20, 1797; died May 26, 1873. He married Sarah Armstrong Davison December 8, 1825. She was born July 16, 1803; died March 18, 1881.

649. THOMAS JEFFERSON, born April 26, 1800; died May 2, 1845. He married Martha De Camp, who died in 1868.

650. PETER, born 1803; married Martha Borden, widow of his brother Thomas.

353. THOMAS, Burlington county, N. J.

651. JOSEPH, born March 21, 1791.

652. WILLIAM, born August 13, 1793.

653. SARAH, born February 9, 1797; married David Allen.

654. COLLAN, born April 15, 1799.

655. GEORGE, born December 3, 1802.

656. MARY ANN, born January 1, 1809.

358. WILLIAM, Burlington, N. J.

657. CHARLES, married Effie Earl in Burlington county, N. J.

658. WILLIAM, married, name of his wife unknown.

659. THOMAS,

660. TACY, born January 17, 1804; married (1) Charles Kennedy; (2) John Valentine.

355. AMOS, Monmouth county.

661. AARON, born in Monmouth county, New Jersey; married Sarah A. Emmons, daughter of David Emmons, of an old New Jersey family.

662. DANIEL SCENCK, born October 26, 1819; died January 14, 1898; married Mary Morris December 2, 1845; born January 22, 1820.

359. JOSIAH, Emleytown, N. J.

663. AARON ROBBINS, born July 2, 1804; died August 22, 1873; married Eliza Ann Emely, born November 20, 1809; died September 1, 1861.

664. MARY R., born December 15, 1805; died September 14, 1867; married Samuel Tilton, born June 4, 1797; died July 27, 1882.

665. APOLLO W., born December 16, 1807; died March 11, 1874; married Hannah Kerlin January 26, 1831; she was born November 26, 1812; died June 2, 1882.

666. EDWARD, born August 28, 1810; died February 18, 1859; married Rebecca Penn.

THE BORDEN FAMILY.

667. CHARLES S., born February 28, 1813; died January 7, 1876; married (1) Amy Fowler, 1835; she was born March 25, 1814; died March, 1847; (2) Clarissa Fowler, in Jacobstown, N. J.

668. GEORGE W., born May 11, 1815; died March 22, 1862; unmarried.

669. SAMUEL W., born July 13, 1818; died April 17, 1857; married April 20, 1843, Julia Elizabeth Strawbridge of Philadelphia, Pa.; she was born December 20, 1818; died September 9, 1887; was the daughter of John Strawbridge and Frances Taylor, his wife.

670. ELIZABETH A., born April 14, 1822; married Joseph W. Emley November 14, 1847.

671. JOHN E., born July 13, 1819; died July 5, 1892; married April 2, 1843, Sarah Ann Emley; she died December 15, 1888.

360. DANIEL, Emleytown, N. J.

672. WILLIAM D., married Ann Stout.

673. TENBROCK W., married Sarah Cook of Ocean county, N. J.

674. EDWARD T., born December 22, ——; married in Burlington county, N. J., Maria Silpth.

675. RICHARD P., never, married.

676. BETSEY, married Thomas Hayes; went to Ohio.

677. ANN, married Richard Potts.

678. LUCY, married Joseph Coward.

679. RACHAEL, married Joseph Lawrence.

361. ASHER, Allentown, N. J.

680. SIDNEY PARKER, born November 15, 1798; died 1860; married Maria Hughes of Allentown, N. J.; she died in 1895.

681. GEORGE, born July 26, 1800; wife's name unknown.

682. ANN ELIZA, born March 3, 1807; married William Kirby.

683. FRANCIS, born June 7, 1809.

364. FRANCIS, Philadelphia, Pa.

684. LEWIS, ERWIN, born 1781; died June 10, 1821; drowned while bathing.

685. HENRY MORRIS, born November 16, 1817; died September 17, 1856.

686. EDWARD PARKER, born in Philadelphia, Pa., April 20, 1814; died November 12, 1871; married in Philadelphia Mary Jane Hopkins, daughter of Capt. C. B. Hopkins, U.S.A.; she was born July 14, 1814; after her death he married Cecilia Irvin, June 22, 1862.

SAMUEL W. BORDEN.
669.

366. SAMUEL, New Jersey.

687. JOHN, born 1800,
688. JAMES,
689. ASHER,
690. THOMAS, married Ann ——.

373. RICHARD, Monmouth county, N. J.

691. JOHN, born November 29, 1801; died June 23, 1884.
WILLIAM,
692. RICHARD,
SARAH,
CAROLINE,
JOSEPH,

374. FRANCIS,

FRANCIS married Hannah Lambert Holmes; the mother of Francis was Margaret Parker, daughter of Joseph Parker and Mary Woolley.

379. TYLEE, Shrewsbury, New Jersey.

693. HARRIET, born December 1, 1817; died 1892.
694. ISAAC PINTARD, born August 20, 1819; died October, 1865.
695. EDMUND WOODMANSE was born March 30, 1820; died 1893. His father was a carpenter. His mother a woman of great force of character, died suddenly with cholera in New York city in 1832. Edmund was bound out to a market gardener on her death, bought up his time, learned tailoring; schooled himself and studied and recited at night schools in New York city; he became a licensed Methodist Episcopal exhorter in 1842, when he married and moved at once to Michigan, 1843, settling at Battle Creek, which he reached by ox cart. He was a Methodist Episcopal circuit rider for fifteen years, and then joined the Congregationalist body and preached fifteen years. He then went over to the Presbyterian order and preached twenty-one years, a total of over fifty-one years' continuous service in the ministry, with no vacations. He gave all of his children a college training, and gained a comfortable property. He took a short course at the University of Mihcigan. He did much to plant and develop the church. His wife was Margaret Hopper Borden of Holland, Dutch, ancestry. Her father served in the war of 1812. On the death of her parents she sold her share in the homestead in New Jersey to her brother Abraham. The place is still in possession of his family.

696. THOMAS HENRY, born June 22, 1826, near Red Bank, N. J.; died May 20, 1897, in New York city; losing his mother in early life, he was placed on a farm in Shrewsbury until he was 14 years of age, then he went to New York and learned the tailor's trade. He joined the Fifty-sixth Regiment of New York Volunteers in 1864, and served until the

THE BORDEN FAMILY.

close of the war as commander of Co. D; received his commission from Gov. Seymour of New York; also held one from Gov. Holley of Connecticut as commander of Co I, Eighth Regiment of Volunteers. He was a charter member of Samuel T. Ferris' Post, G.A.R., of New Canaan, Conn.; served as Justice of the Peace in that place from 1873 to 1877. He joined the Old Bedford Street Methodist Church in New York in 1846, and has held every office in the Church from Sunday-school teacher, superintendent, class leader, steward and trustee. He married Henrietta VanKirk in New York November 26, 1846.

697. HANNAH, born June 27, 1828; died at Methodist Home for the Aged in New York September, 1895.

698. MARY, died in infancy.

381. RICHARD, Manasquan, N. J.

699. RICHARD ALBERT, born November 12, 1842; married Ann Davis.

700. DANIEL WILLIAMS, born February 16, 1845; died Jan. 6, 1881. Married Katherine Hutchinson of Trenton, N. J.

701. CHARLES E., born December 20, 1853; resides at Brooklyn; married Sarah E. Jordan of Jersey City Heights, July 18, 1877.

382. WILLIAM LLOYD, Red Bank, N. J.

702. WALTER HENRY, born at Red Bank, N. J., married Jane E. Hance of New York; no children.

703. GEORGE F., married Julia Hitchcock; he has a farm and summer hotel near Red Bank.

704. ABRAHAM, died while a boy.

383. THOMAS, Eatontown, N. J.

705. ELIZABETH C., born April 11, 1831; married January 24, 1853 Henry Wardell; married after his death John R. Berger.

706. JAMES EDWARD, born May 4, 1836; married Julia Harvard, born July 30, 1843. He resides in Eatontown, N. J., and has rendered valuable assistance in this work on Borden Genealogy.

707. HANNAH TABER, married J. B. Hartshorn.

385. JOSEPH L., Manasquan, N. J.

708. MARY ELIZABETH, born March 25, 1836; died 1861.

709. ANNA L., born April 16, 1836; married Thomas Montgomery; resides in Philadelphia, Pa.

710. EMILY T., born July 10, 1840; married Lewis Monell, Newburgh, New York.

711. SALLIE D., born December 7, 1843; died 1880; married Sylvester Lawson of Newburgh, N. Y.

EDWARD PARKER BORDEN.
686.

SEVENTH GENERATION.

712. JOSEPHINE, born January 7, 1845; died July 2, 1846.

713. HARRIET, born June 7, 1847; married Sylvester Lawson.

714. JOSEPH E., born June 11, 1851; lives in Chicago.

386. BENJAMIN, Manasquan, N. J.

715. JOHN W., born February 4, 1839; died October 7, 1866; never married.

387. ANN LEVIS, Shrewsbury, Monmouth county, N. J.

716. BENJAMIN HANCE, born February 8, 1833, in Shrewsbury Township, Monmouth county, N. J. He married Louverina Wooley January, 1860, in Philadelphia, Pa. His wife was a member of the Society of Friends.

388. SARAH T., New York City.

717. WALTER NELSON DE GRAUW, born September 28, 1841; died November 23, 1894; he married Hester Morehouse; left one daughter, Ella.

718. CHARLES ELWOOD DE GRAUW, born March 15, 1845; married in Orange county, N. J., had two daughters, Fannie and Mable.

719. MARY LOUISE DE GRAUW, married Charles Sherman of New York; has one daughter, Cora.

392. THOMAS, Fall River.

720. JOSEPH, born November 16, 1777; died March 16, 1842; married at Fall River November 20, 1800, Hannah Borden.

721. PHOEBE, born December 22, 1779; married (1) George Borden; (2) Bradford Durfee.

722. WILLIAM, born December 28, 1781; died November 12, 1814, at Fall River. He married Sarah Durfee February, 1804; she died April 15, 1862.

723. ISAAC, born March 7, 1784; died April 28, 1828, at Fall River. He was a marine; married Susannah Durfee, daughter of Deacon Richard Durfee.

724. THOMAS, born February 6, 1786; died June 25, 1855. He was a sea captain; married (1) Lydia Durfee, daughter of Deacon Richard Durfee; (2) Susannah Borden, widow of his brother, Isaac.

725. SARAH, born March 9, 1788; married Nathaniel Luther April 29, 1810; she died November 15, 1857.

726. HOPE, born October 8, 1790; married Robert Cook, 1810.

727. IRENE, born June 4, 1793; married Capt. Joseph Butler Dec. 1821.

728. RICHARD, born April 12, 1795; married February 22, 1828, Abbey W. Durfee of Fall River.

'Fall River, in every development of its thrifty daily life, its marvellous, yet substantial, progress; its financial stability in the storm that has shaken older communities; its constant advancement in the industrial arts; its conservation and harmony of industrial forces; its industrious, law-observing population, bears the impress of the Bordens, Durfees, Anthonys and Davols, the sterling mark of honest artisans upon pure coin. As Samuel Smiles says of Josiah Wedgewood: "Men such as these are fairly entitled to their rank as the industrial heroes, of the civilized world. Their patient, self-reliance amidst trials and difficulties; their courage and perseverance in the pursuit of worthy objects are not less heroic of their kind than the bravery and devotion of the soldier and the sailor, whose duty and pride it is to heroically defend what these valiant leaders of industry have so heroically achieved."

"One of the largest debts of gratitude which Fall River owes to Col. Borden (and in this connection his brother, Jefferson Borden, still living and honored in his native city, will not be forgotten) is for the present admirable system of communication with New York and Boston. Up to 1846 there was no communication direct by steam with either city, though the traveler could, by going to Providence or to Stonington, catch a train or a boat. At this time Col. Borden projected, and mainly by his own effort, constructed a railroad from Fall River to Myrick's, to connect with the New Bedford and Taunton Railroad, and using the latter to join the Providence Railroad and complete the route by rail to Boston. This was an eccentric way of reaching the State capital, and the next advance was, consequently, made to South Braintree, striking the Old Colony Railroad of that day. A satisfactory through route was thus secured; but Col. Borden, not satisfied yet, was ambitious not only to have the communication opened for his favorite city, but to make it self-sustaining. With this view he organized the Cape Cod Railroad Company, of which he was president, and constructed a line from Middleborough down to the Cape, as a feeder for his Fall River route. The care, administrative and executive ability, and the financial involvement—for he was not only the designer, but the banker of the enterprise—were excessive demands to be made upon one man in that comparatively early day; but Col. Borden's resources in all respects were equal to the exigency. It was his good fortune soon to see his railroad enterprise at least relatively a success. His purpose in freeing Fall River from its isolation was at any rate accomplished, and in a year or two he was relieved of his new responsibility by a consolidation of the roads he had constructed with the Old Colony.

In the mean time, being the second year (1847), of the Fall River Railroad, observing the success of the two steamboat lines running between Stonington and Norwich, Ct., and New York, Col. Borden determined to inaugurate a similar water communication for Fall River. His sole associate in this enterprise was his brother Jefferson. The

RICHARD BORDEN,
728.

capital appropriated was $300,000, and the line was started in 1847 with the Bay State, a fine craft for that day, built for the company, and the old Massachusetts, chartered as an alternate boat. The following year the Empire State was launched and put on the route, and in 1854 the mammoth Metropolis, the most superb boat of her period on eastern waters. Both of these boats were paid for out of the earnings of the line, which was indeed such a success as in 1850 to pay six per cent. monthly dividends for ten successive months.

In 1864, dissatisfied with his connection with Boston via the Old Colony Railroad, Col. Borden obtained an act of organization and set about a second through route to Boston, starting from the west side of Mount Hope Bay, opposite Fall River. During the war of 1812 he joined the army as a private, and received his first promotion while yet in his minority. He rose step by step until he attained the rank of Colonel. The beautiful monument at the entrance of Oak Grove Cemetery and the Richard Borden Post of the Grand Army of the Republic, named in honor of his benevolence to the soldiers and their families in the trying days of the rebellion, remain to perpetuate his memory.

Personally, Richard Borden represented the best type of that pure straightforward, stalwart Saxon virtue which has proven New England's best inheritance from the mother country. His sympathies were given to all good things; he was a man broad in his views, true and steadfast in his convictions and feelings. A sincere, outspoken Christian in early life, identifying himself with those observant of the Sabbath, the public services of the sanctuary and the requirements of the gospel, he became, in 1826, a member of the First Congregational Church of the city, and afterwards one of the leaders of the Central Congregational Church, which, to his energy, liberality, piety and judicious counsel, is largely indebted for the success that has marked its subsequent history. In the mission Sabbath-chool work he engaged with his characteristic energy, for a long time going even miles out of the village for this purpose. His interest in this deperatment of work continued so long as he lived. The benevolence of his nature flowed out as a deep and silent stream. He gave as to him had been given. None sought aid from him in vain, when they presented a worthy cause. He was always willing to listen to the appeal of the needy, and sent none such empty away. "Home and foreign charities alike found him ready, yea, often waiting to attend on their calls, and among our institutions of learning not a few are ready to rise up and call him blessed for the timely aid rendered in the hour of their greatest need. Thus he came to be looked upon as the foremost citizen of the place, and his death left a void in the community which no one man will probably ever fill again. Generous, noble-hearted, sagacious, enterprising, of untiring energy and spotless integrity, far-seeing, judicious, ever throwing his influence and his means on the right side, he presents a character for admiration and example, which is fragrant with all the best qualities of our New England life."

The cursory sketch of his business career which space has permitted will suggest the conspicuous qualities of Col. Borden's mind and temperament, as the world saw them and events caused them to develop. It is doubtful, however, if any qualities of his can be termed more conspicuous than others, among those who really knew him, so well rounded was his nature. His achievements were many and great, a few of them extraordinary in view of his resources and experience, yet he did not possess one spark of the so-called genius, to which exceptional successes are generally ascribed. His brain was like his body, robust and full of forces; his mental process direct and simple.

Col. Borden, as he was always called, was born on the 12th of April, 1795. What is now Fall River was then a portion of the town of Freetown, and he was in his eighth year when Fall River was incorporated, in 1803. Atfer the period of boyhood, his early years were spent as a farmer, and to the end of life he continued his interest in that honorable pursuit. But, step by step, he became identified with all the different leading business interests of the rapidly-growing town, village and city. He was early identified with the maritime interests of the place, and gave fresh impulse to the local shipping pursuit, when as yet it was but a rural village. While still a young man he ran a grist mill (1812-20), which stood just west of the present Annawan Mill, where the corn of the whole region was ground. In company with his brother Jefferson, it was his custom to go down to Prudence and Conanicut Islands in the sloop Irene and Betsey, which carried about 250 bushels of corn, and, having secured a load, to return to Fall River and tie up at a little wharf within the creek, and discharge directly into the mill. The Irene and Betsey was also a sort of packet between Fall River and the neighboring places, and the surplus meal was sold at Warren, Bristol or Providence, and a return freight secured, of provisions, groceries, cotton, etc. Another mill was placed on the north bank of the creek at the next fall above, where the Annawan Mill is now, and a tramway had been constructed from this mill (known as the Davenport Mill, but owned by Richard Borden, the uncle of Col. Richard), to the shore, and a car run up and down this incline, drawn by a rope. This rope was wound on a drum, which connected by gearing with the water wheel, and thus the water power was made to do double service. The great strength of the colonel was always a marvel to the small boys, sent on horseback with a grist to grind, it being his ordinary feat, afotr putting two or three two-bushel bags of meal on the horse with the greatest ease, to take the boy and lift him to his place on top of all. It was about this period he joined Maj. Durfee in the construction of several small vessels, the lumber for which was prepared in a saw mill adjoining the grist mill. Here, too, the strength of the colonel found development, as single-handed, he would roll into position great white

JEFFERSON BORDEN.
731.

oak or mahogany butts two feet through and twenty feet long. He had, moreover, a thorough self-reliance and self-assertion, yet was not over-sanguine. The possession of such a mental structure always assures excellence of judgment and consequent success, if combined with a suitable temperament, and such was the fact in the present instance. Col. Borden's nerve was strong and undisturbed by sudden or severe trials. Exceedingly honest of purpose, he was wonderfully peristent when his judgment supported his efforts, never giving up when legitimate means and thorough industry could compass an end he had started for. His industry was his conspicuous quality—if he had one. He was an indefatigable worker while the day lasted."

This account of Richard Borden is taken from Mr. Earl's book "Fall River and Its Industries."

729. MARY, born April 7, 1797; married David Anthony Dec. 21, 1822.

730. JOHN, born February 5, 1799; married Almira Manchester June, 1826; died October, 1856.

731. JEFFERSON, born February 3, 1801; died August 22, 1887; married Susan Elizabeth Easton April 15, 1825. He was one of the most enterprising men of the Borden name, and built more mills, railroads and steambots than any man Fall River ever had as a citizen.

732. MARIETTA, born September 5, 1803; married Ralph Crocker June, 1824.

393. RICHARD, Fall River,

733. ABRAHAM B., born July 8, 1798; married (1) Phoebe Davenport; (2) Phoebe Wilmarth, November 23, 1854.

734. AMY, born February 11, 1802; married (1) William Grinnell; (2) Jeremiah Wilcox.

735. HANNAH, born December 5, 1803; married William Cook November 24, 1824; died September 28, 1891. In 1819 Mrs. Cook started the first power loom put up in Fall River. The loom was the invention of Dexter Wheeler, and weavers were hired at $2.50 per week. It was Mrs. Cook also who discovered the skeleton which was celebrated by Longfellow in his poem "The Skeleton in Armor." The skeleton was discovered in a sandbank, and was afterward destroyed in the Athenaeum fire in 1843.

736. RICHARD, born December 22, 1805; married Lucy Cook October 1829.

737. ROXANNA, born February 8, 1808; died March 14, 1835.

738. COOK, born January 18, 1810; died September 20, 1880; married Mary Ann Bessey, born August 19, 1810; died October 6, 1891.

739. LADOWICK, born March 14, 1812; married (1) September 8, 1833,

Maria Briggs; (2) January 28, 1843, Eliza Darling; (3) February 29, 1866, Eliza T. Chace.

740. ZEPHANIAH, born July 18, 1814; married at Little Compton (1) Mary Perry, after her death he married Lydia Shearmann; no children by this marriage.

741. ANDREW, born December 28, 1816; died young.

398. JONATHAN, Tiverton.

742. HANNAH, born September 1, 1790; married David Babbitt.

743. ABRAHAM, born July 20, 1812; died October 28, 1864, in Westport, Rhode Island.   He married Phoebe Barker January 17, 1815.

744. PHOEBE, born September 1794; married Benjamin Durfee.

745. THOMAS, born September 18, 1796.

746. RHODA, born March 21, 1798; married Pardon Cook.

747. ISAAC, born January 8, 1800; married January 11, 1821, Abby Borden, daughter of William.

748. ELIZABETH, born November 3, 1803; married Thomas Tasker March 2, 1833.

400. SIMEON, Fall River.

749. ANN, born September 14, 1790; married Clark Chase of Portsmouth, R. I.

750. JUDETH, married Hon. Job Durfee of Fall River.

"He was the son of Thomas Durfee, Esq., of Tiverton, who for many years was Chief Justice of the Court of Common Pleas for the county of Newport. He was born September 30, 1790, at the house of his father on Tiverton Heights, near the Stone Bridge which connects Rhode Island with the main land. His death occurred July 26, 1847, at his own residence on Nannaquaket Neck. He pursued his studies preparatory to entering upon a collegiate course at Mt. Hope Academy in Bristol, R. I., under the tuition of Abner Alden, A.M., one of the most popular teachers of that day, and the author of several school books which were extensively used until the manufacture of such works fell into the hands of a class of empyrics, who were quick to discover the failings of all that had preceded them, but could never discern their own. At this school Mr. Durfee was considered a young man of fair promise. He was modest in his deportment, temperate, studious and persevering in his application to acquire knowledge. He at that early day had proposed to himself some point of distinction among his fellow-citizens, which he aimed at, and which he hoped to gain by his personal exertions. He carried this feeling through his collegiate course. While a student at college he composed and delivered a poetical address at the celebration of American Independence at Tiverton.

He graduated in 1813 and immediately after commenced the study of

SEVENTH GENERATION. 171

law in the office of his father. In 1814 he was chosen a representative to the Legislature; and thus he began to realize something of that distinction which he had so long looked forward to to stimulate his exertions. This was an introduction to the great political arena where battles were fought and victories won, and though it might seem a trifling elevation, yet by proper management it might prove to be the vestibule to the great White House at Washington. This thought must be very inspiring to an aspiring politician. Mr. Durfee continued his law studies and attending the Legislature until he was admitted to the bar, increasing his knowledge and extending his acquaintance among his political associates. He felt the necessity of doing something which would attract public attention to himself. Hitherto the course of business of the Legislature had been so tame that it had afforded no opportunity for any oratorical display. And to gain the point he aimed at it was desirable that the subject introduced should be of general interest and not a mere party issue, in order to insure its success, as defeat would be disastrous. After some time spent in consultation with some known friends in Tiverton who advised him that the most popular thing he could do was to get the bank process for the collection of debts abolished. By the law no person could attach the personal property of another while he was in the county, nor his real estate while he was in the State. But by the "bank process" the banks could attach the property of their debtors before their faces or behind their backs, which was most commonly done. All doubtful debtors were sure to be looked after and their property kept under secret attachment till their debts were paid. If any balance was left it belonged to the debtor; his creditors could not touch it. This bank process was highly obnoxious to all those who were not concerned in banks.

Having fully prepared himself for his task, Mr. Durfee first introduced a resolution into the House to tax the banks for the exclusive privilege granted them by the bank process for the collection of debts, and supported his resolution by a well-digested speech in which he successfully exposed the inconsistency of such a law with republican principles on which our government was founded, its baneful influence on private individuals who had money to lend, but could not loan it with safety while this law existed, and many of whom had determined never to lend a dollar to any man engaged in trade, or who was known to loan his funds to those who were employed. This state of feeling prevaded the whole State, and would call for the repeal of this law. But while the law is continued in operation Mr. Durfee thought the banks should be taxed for such a privilege, which, in fact, made them perferred creditors in all cases where they had any demands. This speech was triumphant. The bank advocates confessed they had been taken by surprise and were unprepared to attempt an answer to it, and asked for a continuance to the next meeting of the Legislature, which was readily granted. At the next meeting the subject was discussed fully and the opposition to the law being very stron, the subject was referred to a committee of which Mr. Durfee was chairman, to report to the next meeting. This committee, after due consideration, recommended the repeal of the law granting to the banks the privilege of attachment, etc., which being embodied in a resolution was passed without opposition.

This success was gratifying to the majority of the people, and peculiarly so to Mr. Durfee. He had now made the acquaintance of the political leaders of the State, and this success had made him known to all the people. This gave him a prominence among the literary men of the State, which was, recognized by the people during his life, and first mainfested by electing him to Congress in 1820. During his four years' service in Congress he was an attentive listner to the debates by older and more skillful politicians; but when it was proposed to cut down the representation from Rhode Island one-half, Mr. Durfee came forward to protect and defend its claims, and by a well-timed appeal to the House, he succeded in getting the apportionment so far changed as to allow the usual number of representatives from Rhode Island. His effort on this occasion was highly commended by both parties in the State, and would have secured his reëlection to Congress if he had remained silent the remainder of the term; but when the tariff bill came up in the House, he made another speech in opposition to the protective system of duties, displeasing to many of his political friends, though his speech was in strict accordance with the professed views of his party throughout the country, who following the lead of southern politicians, actually aided them in their attempts to destroy the industrial pursuits of the North. But some of the Democrats in the State had vested their fortunes in manufactories and therefore were as much interested in protective duties as their political opponents, and these withdrew when the contest came on, and suffered Mr. Durfee to be defeated. This affair seems to have wrought a change in the direction of his thoughts. In 1826 he represented his town in the Legislature, and was Speaker of the House. He then declined a reëlection, and devoted himself to agricultural pursuits, occasional reading and study. His plan for the poem "What Cheer" was formed and executed during this period, and was published in 1832. The object of this poem was to present in an attractive form an account of the trials and sufferings endured by Roger Williams when banished from Massachusetts.

Extract from the letter of Rev. John Foster, dated May 17, 1839:
" 'What Cheer' is truly, as I have described already (in his review) a remarkable performance in every sense; the wild and gloomy scene, the strange adventures, the singular character of Williams, the eminent fact of his being the first protestor for absolute liberty, and the great poetic power in which the whole is displayed—a power in always giving a distinct and visible reality to every sense, object and transaction; of investing striking incidents, of discriminating character, of carrying a

pervading intellectual tone of interesting sentiment and reflection through what is, in substance, so wild, stern and rugged; and of clear narration, really rapid, while apparently progressing with a quiet, unmitigated movement. Some time ago the Electric Review had rather a long article on 'What Cheer,' much in as analysis of the story, adapted to exhibit the work as of extraordinary interest. It would, I think, rather please the poet than otherwise, if he happened to see it. Can this be the only production of a man of such conspicuous talents? It should not be so. It is well, at all events, that such distinguished ability is in exercise for the public service when not in a literary one."

In his review he said of this poem: "The narration is perspicuous and consecutive, maintaining a close and natural connection in the train of events. It is also in fact rapid—many particulars being told in the fewest words. There is also a very singular cast of sobriety in the language that bears us through the changes, even when it relates matters of strongest excitement, so much in contrast with the hurried, precipitate and sonorous diction often assumed by poetic narration. In the power of description the poet excels eminently. The wild aspect of nature, in both its permanent and its changing phases; the gloom of a solemn desolation, with the beauties which here and there sparkle into life; the ominous incidents, the situations of alarm and relief; the external signs of the passions, the appearance, manners and imposing spectacles of the savage tribes; are all presented with a graphic reality, by combinations of expression, discriminatively chosen from an ample command of language."

Such commendation of this work coming from such a literary and practical critic as the Rev. John Foster of Stapleton, England, is of great value in estimating the literary talents of our poet, the Hon. Job Durfee of Tiverton. And we may safely anticipate that, at some future day, it will reappear in all its native beauties and hold the rank among the poetical works of America, to which it is entitled by its own merits.

Mr. Durfee was not engaged in public service for four years after his return from Congress. In 1833 he was elected in May to the Legislature, and at a meeting of the two houses that year he was chosen an assistant justice of the Supreme Judicial Court, the Hon. Samuel Eddy being Chief Justice. In June, 1835, on the retirement of Mr. Eddy, Mr. Durfee was chosen to the position of Chief Justice, which place he held by annual elections under the old charter, until May, 1843. When the government was organized under the new Constitution, he was again elected to the office of Chief Justice, which he held until his death.

With the exception of four years, Judge Durfee was in the service of the State in various departments from 1814 to the time of his death. I need not dwell on the ability and integrity with which he discharged the various duties which devolved upon him, for the unwavering confidence which was reposed in him by the public is the best of all proofs of his honest integrity and moral worth. The attempt to establish a new Constitution by force and fraud, greatly tested his moral courage and noble daring. Though receiving letters threatening him with assassination for his opposition to the conspiracy, he nevertheless made every effort in his power to arouse the people and prepare them to resist the threatened danger. In closing this article I will quote Mr. Hazard's remark: "But all his services in the legislative halls of the State or Nation, though highly creditable to him, appear unimportant when compared with those which for fourteen successive years, he rendered on the bench." Judge Durfee died July 26, 1847." S.

751. SIMEON, was born in Fall River January 29, 1798. His family removed to Tiverton, R. I., in 1806, and settled down upon the Nannaquaket property, an estate which had descended to his mother from her brother and sister (Nathaniel Briggs). Of course Simeon grew up as farmer's boys generally do, without any particular object to attract his attention or excite his ambition. He received a very limited education, such only as country schools afforded, and his father dying in 1811, and his mother having the estate of her husband to settle, which involved her in a maze of perplexities, Simeon found himself at the age of 13 years loaded with the cares and anxieties of an extensive farmer. His mind matured rapidly, and he became a co-worker with his mother in the mangement of the farm. After her death in 1817 he took the principal care of the estate, and with the aid of his brother-in-law, Mr. Clark Chase, completed the adjustment of the estates of both his father and mother. At this time Mr. Borden prepared himself for the business of land surveyor, and labored in it as occasion demanded. But finding the graduations upon his compass were not strictly accurate, he determined to go to Fall River and make one that should be strictly correct. He did so, and in a short time he brought home an instrument of which he was justly proud. His ideas appeared to rise above the drudgery of a farmer's life. He possessed naturally a mechanical and mathematical genius, which seemed to be developing itself with every succeeding year of his life. In 1821 the maternal estate was divided, each of the heirs agreeing that as Simeon and his sister Judith had so long and so faithfully officiated as father and mother in the family they should have the choice of lots. Simeon took the south part of the farm, and Judith the north. At this time his mind was not made up as to his future course in life. The reason he assigned for choosing the farm was he must have a home, and if he decided to leave it for some other employment, he could rent it, and as a tenant could not spend it all, the risk he would run was the loss of the rent. But he did not remain long undecided. He soon ceased to be a farmer, and entered the Pocasset machine shop, and in 1828 he took the full charge as superintendent. While engaged in this shop he constructed a measuring rod for the State of Massachusetts to measure the base line of the contemplated trignometrical survey of that State. The apparatus was fifty feet in length—was in-

SIMEON BORDEN.
751.

closed in a tube, and was compendated as to remain of an invariable length in all temperatures. Four compound microscopes were mounted upon trestles, having motion in every direction. At that time it was considered the most accurate and convenient instrument of the kind extant, and it is now only surpassed by one, that of the United States Coast Survey.

It is well known to practical mechanics that some metals will expand and other contract by the influence of heat, and to determine the quantity of each to be used in the construction of the measuring rod, so that the points at each end should be stationary in all temperatures, was the most difficult part of the work. Having nothing to guide him, no similar work having been performed in this country, he was forced to rely upon his own resources, and by repeated experiments and many computations, he succeeded to the admiration of all. He said to the writer of this notice: "An error of one-tenth of an inch in the length of this rod would so affect the base line as to render the whole triangulation worthless." The Governor and Council were so much pleased with his success that they appointed him assistant engineer. He then proceeded with Mr. Stephens, the chief of the corps, to the measurement of the base line, which was soon accomplished, and the triangulation of the State commenced. In 1834, Mr. Stephens having retired from the work, Mr. Borden took charge of the survey and carried it forward as rapidly as could be done and assure the accuracy of the work, which was an indispensable point with him. In the course of his labors he frequently had to pitch his tent upon some high point of land a week at a time, to get the bearings of other points lying in various directions from his own position and distant as far as the horizon would permit his vision to extend. But when he had accomplished his object, he had extended his triangulation over several hundred square miles of territory; and when he had accomplished this he felt that he had made "a great success," and was well paid for his labor and exposure. When some supercilious people thought he was wasting the money of the State by his delays, which to them no doubt, seemed very unaccountable, but the truth is that the refraction of the rays of light on account of the density of the atmosphere at certain portions of the day is so great that no accurate observations can be taken by the best adjusted and most perfect instruments, and all an observer can do is to remain at his encampment and watch for a favorable time to make his observations. Every mariner knows this to be true, and consequently never attempts to take an observation of the sun to get the true time at his ship to ascertain his longitude by his chronometer except at about 8 o'clock in the morning and 4 o'clock in the afternoon, when the sun is rising or falling the fastest; and if at these hours the sky is obscured by even a thin haze, he will wait till a more favorable opportunity presents, simply because he can make no dependence upon an observation taken under such circumstances.

One who for a time accompanied Mr. Borden in this laborious work has said in print: "Mr. Borden's genius and resources (and he might have added patience, too) were tried to the uttermost in the progress of this work. With limited means, and imperfect instruments, he proved his ability by doing good work with poor tools." The correctness of his survey along the sea coast has been tested at many points by the coast survey, working with the best instruments, and found to be correct. This speaks volumes in his praise, and proves him to have been a very accurate observer. The triangulation and the map was completed in 1841, and generally distributed in 1842.

His next engagement was on the railroad from Fall River to Myrick's station on the Taunton and New Bedford Railroad, where his judgment and skill in an entirely new business to him, was proved by making an excellent road. He was next employed to run the line between the States of Rhode Island and Massachusetts, and construct a map of the same, to be used in the case of Rhode Island vs. Massachusetts, argued before the Supreme Court of the United States in 1844; and after the decision he was employed to trace and mark out the line according to the decision. His next work was the Cape Cod Railroad. He commenced the survey November 27, 1844, the contracts were awarded March 25, 1845, and the road opened for travel the same year—length of road from Middleborough to Sandwich Village—twenty-seven miles. After this Mr. Borden was engaged on several railroads in New Hampshire and Maine, unknown to the writer. In 1851 he accomplished a feat of engineering by suspending a telegraph wire across the Hudson River, a distance of over a mile, from Ft. Washington to the Palisades on the west side of the river, and elevated upon masts 220 feet in height, allowing the largest ships to pass under it.

His reputation as a scientific man rests chiefly upon his eminent success in conducting the Ffirst trignometrical survey ever completed in this country. An account of this survey was prepared by Mr. Borden and published among the Transactions of the American Philosophical Society, of which he was a member, Vol. 9, p. 34. He was also a member of the American Academy of Arts and Sciences, and some other learned societies. In 1851 he published his formula for running curves upon railroads. During the last fifteen years of his life, Mr. Borden was several times called to testify in courts, as an expert in mechanism. These were cases touching the originality of inventions and the validity of patent rights. For this business his natural taste and acquired knowledge of the principles of mechanics, together with his plain common sense, eminently qualified him. On examination of a piece of mechanism he could easily distinguish between the true and the false—what was original and what was borrowed from other men's labors; and his honesty and truthfulness would not allow him to withhold an opinion on such occasions, which he had formed, whoever the parties might have been; nor, on the contrary, could he be induced by any consideration to give an opinion that he did not fully believe he could substantiate by a direct appeal to his authorities or the mechanism itself.

NATHANIEL BRIGGS BORDEN
752.

In the years 1832-33-44-45 and '49 Mr. Borden respresented the town of Fall River in the Legislature of Massachusetts. Mr. Borden was a bachelor, and died at the house of his brother, Hon. Nathaniel B. Borden, Esq., in Fall River, October 28, 1856, aged 58 years.   S.

752. NATHANIEL was the youngest son of Simeon Borden Esq., of Tiverton, R. I., and his wife, Amey, the daughter of Capt. Nathaniel Briggs of Tiverton, and brother of Simeon, Jr., the civil engineer. He was born in Fall River at the old family mansion of Joseph Borden, his great grandfather, who was accidentally killed in in his fulling mill December, 1736, at the age of 34 years.   His father removed to Tiverton in 1806, soon after the death of Maj. Nathaniel Briggs, the only surviving brother of his mother.   This removal of Mr. Borden brought his family so near to that of the writer that at times we attended the same school, which led to the formation of an attachment on both sides which neither time nor different views in politics or religion or any other subject has ever been permitted to disturb.   And I may add further, that amid all our boyish sports and recreation, which so often test the temper and disposition and moral principles of boyhood and youth, I never saw him angry with his playmates, nor heard him use profane language or indulge in low, vulgar conversation so common among boys.   He seemed to possess great command over his temper and will, which qualified him to be an arbitrator and peacemaker among his associates.   In his case I have often reflected upon the striking coincidences of his character when a boy, as compared with his course as he appeared in his business transactions and social intercourse with his fellow-men, and I must say that the temperament of the boy was clearly indicative of the character of the man.

To a common school education he added a few months' attendance at the Plainfield Academy, Connecticut, but soon returned home, having abandoned the idea of acquiring a liberal education, to which he had been urged by his brother and sisters.   He could not see the propriety of spending so many years in hard study, to acquire knowledge which might be of little practical use to him.   But young men know not what will be the course of their future lives.   "Man proposes, but God disposes."   A good education prepares a man to act well his part in any situation, and, if used aright, will greatly add to his influence and give him an introduction to the best society; and could Mr. Borden have anticipated the course Divine Providence had marked out for him, he would have persevered in obtaining a liberal education.   Still, I never heard him xpress any rgret on that account.

He was married to Miss Sarah Gray, daughter of Pardin Gray, Esq., of Tiverton, March 16, 1820, before he was twenty years of age.   In the fall of the year he removed to Fall River and occupied the old family mansion in which he was born.   The following year the family estate was divided, when most of the Fall River property belonging to it fell to him and his sister Ann, wife of Clark Chase.   Being now in the full possession of his property, he united with others in organizing a new manufacturing company, to which they gave the name of Pocasset, taken from the Indians, who formerly occupied this section. On this occasion he was very active and very successful in securing some property for the company, which has since become very valuable to them. By his diligent attention to business, his amiable deportment and promptness on all occasions, he soon secured the confidence and esteem of all who knew him. As the largest portion of the inhabitants had but lately settled in the town, and were ignorant of its former statistics and history, they often resorted to him for information relating to various disputed points respecting boundaries, streets, roads, and divers other matters which were occurring almost daily; and finding him so well posted in what related to town affairs, they urged him to take part in the management of town affairs. This he positively declined. But determined to force him into it, the next year finding him absent from the annual meeting, they chose him town clerk. This he deeply regretted at the time, but told his friends that he would serve that year, as he considered every citizen bound to do that much; but no longer; and when all the citizens had taken their turn, he would serve them again. This appointment was of little moment in a man's life, but the mischief was that it was the entering wedge which led to a succession of other calls, which continued during the whole of his after life. He was four times elected to represent Fall River in the Legislature of Massachusetts; and five times he prevailed upon his brother Simeon to accept the nomination in his place, as the party would be equally well satisfied with either of them, but would not consent to leave out both. The following years show the time of service of both, 1831-32-33-34-44-45-49-51-64. No doubt other men might have been as trustworthy and would have done as well for the town, but there seems to have existed, among the majority of the voters, great confidence in their inteligence and honest integrity, which they wished to honor, and improved every opportunity to do so. In truth, no party will long adhere to a defeated candidate however intelligent and honest he may be; availability is the standard of politicians, and it covers a multitude of deficiencies. But there was a tide of good feeling which accompanied these two brothers through life, alike honorable to those who gave it and those who were the objects of it.

Mr. Nathaniel B. Borden was chosen a representative to the Congress of the United States three times, Twenty-fourth, Twenty-fifth and Twenty-seventh Congress. The service of the two first elections extended from 1835 to 1839 inclusive; the last commencing in 1842, and closing in 1844; he resigned after the close of the first session on account of the position of his private affairs. In regard to his course in Congress I have very little to say. He was elected as an anti-slavery man, and his repeated election to the same position for so long a period shows that he possessed the confidence of his political friends and associates; and that he could have longer retained the position if his circumstances had permitted. He was not fitted by his early training for legislative debates, nor for oratorical display anywhere; but he possessed good common sense and was quick to apprehend; diligent in investigation, and persevering in effort; and when his mind was made up, on any principle or fact, he had the moral courage to avow his convictions and support them both by his speech and his vote under all circumstances when duty required it.

Mr. Borden was called to part with his wife, who died May 22, 1840. He subsequently married her sister, Louisa G., who died June 4, 1842. He afterwards married Sarah G. Buffum, daughter of Arnold Buffum, Esq., July 12, 1843; who also died September 10, 1854, by the Asiatic cholera, his daughter Sarah having died of the same day previous, September 9. These family afflictions, following in such rapid succession, made a deep impression on his mind; his spirits were much depressed, and for a long time any allusion to his afflictions affected him very sensibly. He had personally witnessed the death-bed scenes of each of these dear friends, and when any person alluded to them, it seemed to recall to his mind the same impressions which he had then felt, all combined in one. Those only who have passed through similar trials, can enter into his feelings. But these were not all the afflictions which fell upon him about this time. In 1843, July, occurred the great fire in Fall River, which destroyed the center of the town, burning over about twenty acres. Mr. Borden's dwelling and barn escaped destruction, but nearly every other building of his was consumed, including a long row of stores on South Main and Pleasant streets and the Pocasset Block, the last valued at $20,000. Some of these buildings were partially insured. The Pocasset House had been insured for $16,000, but the policy expired three days before the catastrophe occurred. I have never seen any estimate of his losses, but they must have been very heavy. Several years afterwards the Pocasset Block was again destroyed by fire. He rebuilt it again promptly, and though not destroyed, it has been injured again by the same cause. But the severest calamity and the most trying to his feelings was the failure of the house of N. Borden & Co., when he was forced to acknowledge himself the victim of misplaced confidence; and he might, with propriety, have taken up the lamentation of David when overwhelmed with affliction: "Yea, mine own familiar friend, in whom I trusted, which did eat of my bread, hath lifted up his heel against me." Psalms 41:9.

At different periods Mr. Borden has filled the following positions, viz: Treasurer of the Pocasset Company, treasurer of the savings bank, president of the Fall River Railroad, director of the Union Bank, Mayor of Fall River; and at his death he was President of the Savings Bank and the Fall River Union Bank, Alderman of the Second Ward, elected first in 1859, and continued by annual elections. From this exhibit and the foregoing enumeration of the different positions which Mr. Borden has held satisfactorily to the majority of his fellow-citizens, I feel warranted in saying that he has been eminently a public man; that he has enjoyed a full measure of their confidence and esteem, and that he deserved it; otherwise it would not have been continued to him until his spirit had taken its flight. This generous confidence was highly honorable to him and gratifying to all his friends and commendable in those who have for so long a period bestowed it upon him; but it must now be seen and felt by all, that their unbounded confidence and esteem has proved to have been a heavy tax upon his estate. If we now add to his public services the innumerable consultations with other people concerning their own affairs, in which often occurred questions of law, of farm boundaries, town history, religious society affairs, anti-slavery politics, etc., one is ready to ask what time there was left him to attend to his own private business? The answer must be, surely not enough to guard and protect his own interests. No man is obligated by our civil compact to do so much, and it should not be required of any one, however popular and useful he may be. If now we add to the foregoing the cares and perplexities growing out of his own private affairs, and his heart-rending afflictions which had fallen upon his family during the most active portion of his life, we must be constrained to admit that Mr. Borden possessed a very capacious mind, an unusual versatility of talent, and great power of endurance. Those who have seen him riding quietly through the streets on his daily routine of duties, may have thought him in the pursuit of recreation, when his mind was actually overburdened with thought and oppressed by cares, which were fast bringing upon him, prematurely, the tokens of old age. He possessed, naturally, a happy cheerful disposition which made him a pleasant companion, and when among his associates he often joined in their jocose merriment to relieve his mind from depressing cares; but his mirth was confined within due bounds and never indulged to assail the character or wound the feelings of others.

Mr. Borden was strictly temperate through life. His house was often the home for lecturers on the various subjects connected with the moral improvement of society. He was also emphatically a friend of the poor, whom he assisted by his counsel, his influence, his money, and by various kindly acts which caused the remark at the first announcement of his death: "The city has lost a faithful servant, and the poor their best friend." S.

753. SARAH, born December 27, 1803. Married August 27, 1820, Rev. Pardon G. Seabury, to whom the entire Borden family is deeply indebted for the service he rendered in collecting and arranging so much geneological matter concerning the early members of the family, and without whose manuscript this Borden record would have been far from complete, for it would have been impossible to have otherwise gathered the vast fund of information which he had saved by his untiring efforts. Most intelligent people are interested enough to wish to know of their ancestors, yet few are willing to work year after year

SEVENTH GENERATION. 181

as the one must who succeeds in gathering family history and genealogical data. Mr. Seabury worked for years on this line entirely unrewarded, except as all Bordens may now with their gratitude of heart, raise an invisible cenotaph to his memory.

401. PERRY. Fall River, Mass.

754. ABRAHAM, born January 29, 1787, died unmarried.

755. PHILIP, born November 5, 1788.

756. ELIZABETH, born July 8, 1791. Married George Collins of Uniondale, Dutchess County, New York.

407. BENJAMIN. Fall River, Mass.

757. RICHARD, born January 31, 1796, died April 13, 1861. Married Sarah T. Chace, daughter of Benjamin Chace of New Bedford, Mass.

758. ISAAC, born —, died February 19, 1866. Married, first, Lucy Willcox, September 2, 1824; she died June 18, 1852. Second, Delia Edwards, June 12, 1855; no children by last marriage.

759. JOSEPH, born March 13, 1802, died 1865. Married Hannah K. Westgate of Rochester, N. Y., May 26, 1822.

760. JOHN, born —, miller at Globe Village, R. I. Married Ella T. Willcox.

761. ELIZA, married Joseph Gardiner.

762. RHODA, married Daniel Willcox.

411. SAMUEL, Tiverton, Mass., and Nova Scotia.

763. SAMUEL, never married.

764. ELIZABETH, married a Mr. Hance.

412. JOSEPH. Nova Scotia.

765. ABIGAIL, born 1799, died young in Nova Scotia.

766. AMEY, born 1801, married Daniel Rodic.

767. CHARLOTTE, born 1804, died 1856. Married Ezra Harrington.

768. ELIZABETH, born 1806, died 1819.

769. ANN, born 1808, died 1833.

770. SAMUEL, born 1813.

771. HEZEKIAH, born 1817. Married Margaret Matheson, 1844.

413. LEMUEL. Horton, Nova Scotia.

772. REBECCA, married James Lockhart, 1815.

773. PEARL, married John Gould, 1827.

774. MARY ANN, married Stephen Gould, February 23, 1824.

775. LEMUEL PERRY, born Avonport, Nova Scotia, September 30,

1805; died September 15, 1861. Married Margery Cummins, born January 10, 1815, in Cornwallis, died May 16, 1880.

776. RUBY, married Gordon Cox, Horton, Nova Scotia.

777. DAVID HENRY, married Annie M. Fuller, 1837.

414. DAVID. Horton, Nova Scotia.

778. SARAH, born October, 1795. Married Isaac Phinney, March 18, 1819.

779. MARY, born May, 1798. Married Charles Johnson, November, 1834.

780. ADOLPHUS K., born January 7, 1801; he was a physician. Married Lucy Ann Brown, 1820.

781. ELIZABETH, born March, 1804. Married Leonard Newcomb, June 12, 1828.

782. EDWARD, born January, 1807. Married Olivia Martin, June, 1838.

783. HARRIET, born October, 1810. Married Robert Childs, 1844.

784. OLIVIA, born 1813, died 1827.

415. JONATHAN. Horton, Novia Scotia.

785. JONATHAN, born December 24, 1814.

786. MARY ELIZABETH, born September 6, 1816. Married Isaac Dickey.

787. JAMES, born February 24, 1820; settled in Avonport. Married Sarah Adelaide Dickey of Cornwallis, Nova Scotia, June 30, 1858.

788. JOHN, born December 14, 1822, died May 14, 1842.

789. ANNIE A., born January 1, 1823, died April 20, 1842.

416. PERRY. Horton, Nova Scotia.

790. JONATHAN, born June 14, 1809, in Horton Township; he was a physician of great reputation. Married Maria F. Brown, September 24, 1845.

791. WILLIAM, born May 25, 1811, died May 23, 1845.

792. ANDREW, born February 14, 1816, at Grand Prie, Nova Scotia. Married, first, Catherine Sophia Fuller, December, 1841; had two children. After her death he married Eunice Leard, October, 1850.

793. AMANDA, born March 20, 1820. Married John Caldwell, July 31, 1843.

794. ARDELIA ANN, born June 20, 1821. Married, first, Lewis Gilmore; second, John Fisher.

795. THOMAS, born November 25, 1822, died in the United States, leaving a family of three sons; he married Jane Cochrane.

SEVENTH GENERATION.

417. JOSHUA. Horton, Nova Scotia.

796. SOPHIA CHARLOTTE, born October 23, 1809, died September 13, 1811.

797. JOSHUA WELLINGTON, born October, 1813. Married in Horton, Nova Scotia, Lavinia Greenough, February 17, 1837. He died November 30, 1891.

798. GEORGE W., born December 20, 1816, in Horton, Kings county, Nova Scotia. Married Mariam Susannah Crane in Winsor, Nova Scotia, 1845.

799. SILAS HIRAM, born September 9, 1818. Married Mary Card, January, 1841.

800. CHARLOTTE ANN, born April 28, 1822, died February 28, 1828.

418. WILLIAM. Cornwallis, Nova Scotia.

801. WILLIAM, born February 13, 1805, in Cornwallis, Nova Scotia. married Rebecca Rand, March 21, 1826.

802. SAMUEL, born December 16, 1807.

803. CATHERINE ANN, born April 13, 1809. Married Charles Dawson, 1844.

804. HENRIETTA, born September 13, 1811. Married William Buckley, July 12, 1854.

805. MARGARET, born August 23, 1813, died young.

806. MARY JANE, born November 22, 1815. Married Nelson Patterson, December 3, 1856.

807. JOHN, born December 15, 1817, died in infancy.

808. JOHN ALEXANDER, born March 20, 1820. Married Maria Bentley, March 10, 1857.

809. ELIZA, born July 19, 1823. Married George Ells, May 1, 1850.

419. BENJAMIN. Cornwallis, Nova Scotia.

810. MARY, born October 21, 1802, died young.

811. PERRY, born August 20, 1804. Married Eliza Knowles, June, 1843.

812. JOHN WELLS, born November 30, 1806. Married Sarah Malvina Pineo, January 18, 1838.

813. MARY ANN, born October 2, 1808.

814. BENJAMIN, born September 7, 1810, died 1889, in Cornwallis, Nova Scotia. Married Mary Jane West, October 25, 1843.

815. JUDAH, born November 11, 1812. Married Nancy Pineo, May 8, 1837.

816. LEVI, born November 20, 1814.

THE BORDEN FAMILY.

817. MARTHA JANE, born October 11, 1816.

818. JAMES NEWTON, born December 25, 1818. Married Emma Fogg, May, 1857.

819. ELIZA, born September 24, 1820, died May, 1822.

420. EDWARD. Cornwallis, Nova Scotia.

820. EUNICE, born August 7, 1815.

821. TABITHA, born June 1, 1817, died June 28, 1860.

822. NATHAN, born August 29, 1819, died August 31, 1826.

823. ROXANNA, born June 14, 1821.

824. RUTH, born May 12, 1823.

825. NAOMI, born June 21, 1825.

826. LUCILLA, born October 16, 1828.

827. CATHERINE, born January 1, 1830.

421. ABRAHAM. Cornwallis, Nova Scotia.

828. JOHN D., born October 26, 1818. Married February 14, 1850, Althea Armstrong.

829. SAMUEL SMITH, born August 24, 1820. Married Laura Crane December 13, 1855.

424. RESCOME. Westport, R. I.

830. SUSAN, born September 10, 1837. Married Samuel T. Sanford, September 10, 1858.

426. EDWARD. Westport, R. I.

831. ELIZABETH, married Isaac Haskell.

832. RESCOME, married Jane Davol, daughter of Ruben Davol.

833. SQUIRE, married Matilda Lapham.

834. MARY S., married James Jones; was divorced from him.

428. SAMUEL. Dom. Quebec, Canada.

835. ASA, born about 1800.

836. SILAS.

837. SAMUEL.

838. JAMES, born in Lee, Mass., September 25, 1811, died in Canada, September 19, 1878.

839. GAIL.

840. LAURA, married a Kingsley at Oswego, N. Y.

841. LORENA.

842. NATHANIEL, died October, 1898; left one son, Henry, perhaps others.

SEVENTH GENERATION. 185

435. JOHN. Scituate, R. I.

843. AULDIS, born March 13, 1812. Married Mary M. Stone, February 3, 1833.

844. ANNA, born June 7, 1814.

845. JOHN H., born August 7, 1816. Manufacturer at Scituate, R. I. Married Ann E. Harrington, January 5, 1842.

846. MARY C., born September 10, 1818. Married Salah Bennett; had two daughters.

847. JOB W., born May 15, 1821. Married in Scituate, R. I., first, Mary Weaver; second, Mary E. Walker.

848. EARL D., born August 1, 1823, in Scituate. Married Lucinda Bowen, 1845.

849. KNIGHT H., born December 21, 1824; he was a mariner of Scituate. Married Mary Harrington, February 26, 1846.

850. RICHARD E., born April 29, 1827; he owned a sawmill at Scituate. Married Martha Hopkins June 12, 1857.

851. BETSEY P., born April 20, 1829, died April, 1849.

852. SAMUEL, born April 3, 1831.

437. ISAAC. Scituate, R. I.

853. HARLEY P., born April 20, 1819. Married Anchurlinda Luther October 7, 1847.

654. WILLIAM, born September 2, 1822; married Susan C. Peck, January 29, 1843; she was born May 19, 1827.

855. RHODA, born February 24, 1825; married, (1) Samuel Baker; (2) Alexander Cameron.

856. NANCY, born December 29, 1827; married James W. Slocum; had three children.

857. LOUISA, born August 31, 1831; died young.

858. ARDELIA, born December 25, 1833; married Franklin Searle, November 26, 1850.

859. ANNE E., born July 12, 1839; married Allen F. Cameron.

438. MARY, Providence, R. I.

860. MARCEY KING BRADFORD, married John Turpin; they had twins, John M., and Adelaide. John M. Turpin married Nellie ——; their son was Bradford Turpin.

861. RUTH KNOWLTON BRADFORD.

439. GAIL, Norwich, Chenango county, N. Y.

862. GAIL, born November 9, 1801; died January 11, 1874; he married February 28, 1828, Penelope Mercer in Amitte county, Miss. She died September 5, 1844. She was the mother of all his children. After her death he married Mrs. A. F. Stearns.

Mr. Borden was married a third time in 1860 to Emeline Eunice Church, nee Eno. By her first marriage Mrs. Church had two children—Alfred B., and Samuel Mills Church. Their father was Hiram Church. Samuel Mills Church was born August 7, 1842, in Vernon, Oneida county, New York. He has never married, and now resides in Brewsters, Putnam county, New York. Alfred B. Church was born in Vernon, Oneida county, New York, May 10, 1844. He married October 11, 1871, Miss Mary Elizabeth Peterson, born at Rockton, Winnebago county, Illinois, December 20, 1842. Their first child, Mary Borden Church, was born March 31, 1875, and died July 24, 1876, in Elgin, Ill. A second child, Alfred Whiting Church, was born in Elgin, Ill., May 18, 1877.

The following is taken from the Columbian Biographical Dictionary of Eminent Men of the United States:

"None of the famous men of America whose energy and genius have left an impression upon its rapidly developing civilization are more deserving the gratitude of mankind than Gail Borden, whose inventive genius brought to him a world-wide reputation and proved of material advantage to every civilized nation on the face of the globe. Expeditions have failed because of an inadequate supply of food with which to sustain the lives of those participating therein, and through all ages the question of preserving foods has been a momentous and much considerd one. After deep research and study Mr. Borden solved this problem and by close research and investigation has given to the world articles of nourishment which can be taken into every region and to every clime.

Every important discovery is attended with more or less contest by claimants appropriating to themselves sole credit for its origin or perfection. The period immediately following the discovery and practical working of the vacuum principle as a means of condensing milk for preservation was no exception to the general rule, but it soon became established as an indisputable fact that the late Gail Borden was entitled to all the credit attached to this invention. The principal authority, the Encyclopedia Britannica, so awarded the credit. The United States granted him patents on the following dates: August 18, 1856; May 13, 1862; February 10, 1863; November 14, 1865; and April 17, 1866; Complete foreign patents were not taken out; consequently parties abroad early attempted to appropriate Gail Borden's invention. Accurate statistics upon this branch of the dairy business have never been compiled; therefore estimated figures are the only available ones. From a small beginning, in 1856 and 1857, it has grown to be one of the most important branches of dairy industry."

GAIL BORDEN.
862.

SEVENTH GENERATION. 187

In 1815, when Gail Borden was 14 years old, his father moved to the vicinity of Cincinnati, Ohio. At that time Covington, Ky., was a farm with only one farmhouse and a barn. The son assisted the surveyors in laying out that place, and in that year cultivated corn where the City Hall of Covington now stands.

In 1816 the family moved to Jefferson county, Indiana, before the territory had become a state. The land was just beginning to be settled. Gail, Jr., attended such schools as the back woods afforded, not being able to go more than six or ten weeks at a time, the whole period being less than a year and a half. He was uncommonly fond of hunting, and had few equals in the use of the rifle. Owing to his possessing a decidedly military turn of mind, he was elected captain of the Hoosier Company of 100 men (before he was 21). Soon after leaving school, as a pupil, he taught a school in the back woods for two years, at the expiration of which time, his health became very much impaired and he was advised by his physicians to go South.

At the age of 21 he went to Amite county, Mississippi. Here he had charge of a school for six years.

In 1829 he went to the Republic of Texas. His first employment here was farming and stockgrowing. He was elected a delegate from La Vaca district to the convention held in 1833, at San Felipe, to define the position of the colonies, and to petition the Mexican government for separation from the State of Coahuila.

Appointed by Gen. Austin to superintend the official surveys, he compiled the first topographical map of the colonies, and, up to the time of the Mexican invasion, had charge of the land office at San Felipe, under direction of Samuel M. Williams, then Colonial Secretary.

As the war came on which led to a separation from Mexico, Mr. Borden, with two others, procured a press and printing materials, and published the only newspaper issued in Texas during the war. Its chief mangement devolved upon him, and he directed earnest efforts toward resisting the establishment of the Central Government by Santa Ana.

The Republic of Texas being founded and revenue departments established, Mr. Borden was appointed first collector of the port of Galveston. His first dwelling there was a rough board structure, on the bay shore, erected by two carpenters in half a day, his office being in what had been the Mexican custom house.

In 1839 he was appointed agent of the Galveston City Company, a corporation holding several thousand acres on which the city is built. This position he held for over twelve years. Toward the close of this period his attention was drawn to the urgent need of more suitable food supplies for the emigrants and travelers across the plains, the want of which involved great suffering and even loss of life. His experimental labors with this end in view resulted in the production of a "meat biscuit," to which reference will again be made. Its merits soon became recognized, so that he felt warranted in embarking all his means in its manufacture, and he did so.

"This article gained for him 'the great council medal' at the London Fair in 1851, and he was elected an honorary member of the London Society of Arts. Meeting opposition from army contractors, he was unsuccessful in the manufacture of this food, and discontinued its production in 1853. In this undertaking he lost his entire property, and he then removed to the North, where he turned his attention to the study of a method for preserving milk. The result of his investigation and labors is known in the condensed milk so widely used today. He applied for a patent for "producing concentrated sweet milk by evaporation of same." He gave the question much study and at length took out 75 per cent. of the water, and with the milk added a sufficient quantity of pure granulated sugar to preserve it.

Recognition of the merits of his invention, however, came slowly, and it was not until 1856 that he could secure a patent, and subsequently the New York Condensed Milk Company was formed. During the war condensed milk was extensively used by the army and navy, and the sale of this product has continually increased until it is now used by every civilized country on the face of the globe. Pecuniary as well as manufacturing seccess in due time crowned his labors. Many, however, were the discouragements and disappointments which he had met. It was three years after he first applied for a patent that it was granted him. He claimed that the method of evaporation through means of a certain vacuum was the important point of the discovery, but this was not conceded by the patent officers until it was firmly established by scientific men. Even when a patent was secured it was some time before the business was established on a firm financial basis.

Mr. Borden also experimented with condensed meat juices and produced an extract of beef of superior quality, which was first manufactured in Elgin, but later an establishment was erected especially for the purpose in Borden, Texas, where the industry was continued after his death. Subsequently he produced an excellent preparation of condensed tea, coffee and cocoa. In 1862 he patented the process by means of which the juice of fruits, such as apples, currants and grapes could be reduced to one-seventh of its original bulk. His labors were conducted with the utmost care and perseverance, and his success was obtained only through long, tedious and expensive experiments; but his intense energy, unyielding tenacity and great ingenuity enabled him to perfect his inventions which have so largely contributed to the good of humanity. While Justus von Liebig, surrounded with the elaborate apparatus of his well-appointed laboratory at Giessen, was experimenting and prosecuting those researches into the nature of flesh and animal juices, which culminated many years later in the production of "Extractum Carnis," Gail Borden, in the wilds of Texas, was independently investigating the same problem without scientific

THOMAS H. BORDEN.
863.

apparatus, and his labors resulted in bringing him the great council medal before mentioned."

The following newspaper clipping will explain Gail Borden's sentiments during the Civil War:

"To the Editor of the Amenia (New York) Times:

"Dear Sir—As it is right that every man should show his flag in this time of peril to our good government, I call your attention to an article in the Winstead Herald of the 19th inst., to show you that I have 'run up my colors.' Although not a Texan Ranger, my father and brothers did good service in the Revolution, while I was engaged in conducting the only newspaper published during the conflict from October, 1835, until after we had expelled the enemy and into 1837. The press and all the materials were destroyed by Santa Ana four or five days before the battle of San Jacinto (just twenty-five years ago yesterday), and reëstablished in August, four months after.

"Yes, my best possessions are in Texas, that misguided State, where I had hoped to spend my last days; yet I love my whole country and government more, and wish to do what I can to sustain them.

"Yours truly, "GAIL BORDEN, Jr."

"Wassaic, April 22d, 1861."

The Winstead (Ct.) Herald of April 19, 1861, contained the following paragraph:

"THE RIGHT SPIRIT—The heart of our good friend, Gail Borden, Jr., is in the right place. A crowd was standing around last Monday morning's telegraphic bulletin containing President Lincoln's proclamation, and a certain political lawyer was improving the opportunity to add to an already unenviable reputation by contemptuous abuse of the President and sneers of his supporters. Just then friend Borden (himself an old Texas Ranger), stepped up. He said: "My father was in Texas, my brothers are in Texas, my children are in Texas and the bulk of my earthly possessions are in Texas. But we must have a government. I did not vote for Mr. Lincoln, but I would vote for him today—aye, and fight for him, too."

"His powers of observation were keen, critical and appreciative; his facility for devising and adapting means to ends was remarkable. As a citizen, he was devoted to the good of his whole country, no sectional limits ever curbed his efforts. An ardent Christian, he was eminently catholic in his sympathies, loving all true disciples because followers of his Lord and Master, and no more nor less because of any peculiarities of creed.

His religion was eminently the life, manifesting itself less in professions than some, but abundantly in kindness and courtesy to all, and in active philanthropy and hearty co-operation with hand and purse in every good work.

As the sun went down on the 11th of January, 1874, he departed this life, from his Texan home on the banks of Harvey's Creek—entering into a much-desired rest from the wearisome burdens of earth, and upon the endless activities of the higher life."

863. THOMAS HENRY, born January 28, 1804; died March 17, 1877, in Galveston, Texas. He married (1) Demis Woodward, June 4, 1829. She was born in Vermont in 1808; died in Houston, Texas, August 4, 1836; (2) Lovisa R. Graves, 1838. In 1824 Thomas H. Borden went to the Republic of Texas, being followed several years later by his father and brothers. All of them settled Austin's Colony, and for several years were engaged in such industrial and business pursuits as the condition of the country permitted. During the troublous times that preceded the Revolution, from Mexico, he, with his brothers, warmly espoused the cause of the settlers. Thomas H., and Gail, who had established the first newspaper in the colony, advocated with vigor a separation from Mexico. Thomas H. Borden was in the "Grass Fight," and also took part in the storming of San Antonio.

The Galveston News of March 17, 1877, said of him:

"He passed through all the vicissitudes attendant upon the settlement of the country by Anglo-Americans, and died at his residence in Galveston on Friday morning, March 16, at 6 o'clock, aged 72. Like many others, once well-known in the country, he had spent so much of the later years of his life in retirement as to be little known to the majority of the present generation. He was a brother of Gail Borden, deceased, whose invention of meats, milk and vegetables, have conferred so many benefits on mankind, and also a brother of John P. Borden, first commissioner of the General Land Office of the Republic of Texas, who still survives. Gail and Thomas H. Borden were the founders of the Telegraph (newspaper), established in Texas in 1835, and the office of which was destroyed at Harrisburg by Santa Ana, previous to the battle of San Jacinto, in 1836. The paper was afterward revived and sold to Cruger & Moore, who continued to publish it at Houston, where it has just been discontinued. Thomas H. Borden, like his brother Gail, was a man of ingenious and inventive turn, and was the patentee of a steam gauge that was long in use on boats throughout the United States, and from which he realized such profits as for a time placed him in independent circumstances.

He was the builder of the first great mill for grinding grain in Galveston, and as early as 1840, had a mill on Postoffice street, opposite the present postoffice."

The above is not entirely correct. He was not the patentee of a steam gauge, as he never took out a patent. He had a mistaken notion that the granting of patents was wrong in principle, and hence never applied for one. We claim for him that which the Patent Office would prove, but for this peculiar idea of his, that he was the inventor not of a steam gauge, but of THE STEAM GAUGE; that he was the first to measure the amount of steam in a boiler by mechanical device. It did not require a great length of time for the world to find out its value, nor that its inventor and manufacturer had failed to protect

PASCHAL P. BORDEN.
864.

SEVENTH GENERATION.

the result of his genius. Hence his pecuniary benefit, together with the recognition as the inventor of so useful a device, both soon left him. It is a satisfaction for the editor (his granddaughter) to thus let those of his own household know that the world owes him the praise for this invention.

864. PASCHAL PAVOLO, born Dec. 1806; died April 28, 1864; married July 19, 1842, Martha Ann Stafford, born July 16, 1827; died July 13, 1852. Paschal went to Texas from Indiana in 1829 with his father. He took part in the war for Texas independence, belonging to Sam Houston's army, Mosley Baker's company, Burleson's regiment, which did great service in the battle of San Jacinto, April 21, 1836.

865. JOHN PETTIT, born 1812; died November 12, 1890; he married Hary S. Hatch in 1843, in Brazoria, Texas; she died in 1893, and was buried in Weimar, Texas, beside her husband. John P. Borden was the last survivor of the commissioned officers, who led the triumphant Texans in the splendid victory achieved over the Mexicans at San Jacinto April 21, 1836. He was a lieutenant in the company of Moseley Baker in Burleson's regiment. Liuet. Borden's company, with several others, were detached from the main army by Gen. Houston and were placed on the east side of the Brazos River, at San Felipe, to prevent the passage of Santa Ana at that point of the river. So gallantly did they perform their duty that the Mexican army was compelled to deflect down the river to Richmond, where they effected a crossing. At the organization of the civil government of Texas, following the successful termination of the revolution, Mr. Borden was appointed the first commissioner of the General Land Office, and kept the records with great care and accuracy, proving that he was admirably fitted for the position. Mr. Borden was an ardent Christian, and member of the Methodist Episcopal Church.

866. ESTHER, born February 3, 1816; died September 27, 1826.

440. LUTHER, Newport, R. I.

867. SQUIRE, born August 27, 1823, on an incoming sea vessel; died September 7, 1882, in Minneapolis, Minn. He married (1) Rebecca Randall after her death he married February 10, 1864, Gertrude Fedge, born in Bergen, Norway, May 12, 1846.

441. MARTIN E., Warren, R. I.

868. LEMIRA, born March 4, 1828; married Charles Jones.

869. JANE, born March 4, 1832; married Franklin Howard.

870. FANNY, born December 24, 1835; married March 25, 1859, George Howard.

442. DANIEL, Fall River.

871. RUTH, married Thomas Borden.

872. BETSEY, married Joseph Durfee of Tiverton, Feb. 9, 1804.

THE BORDEN FAMILY.

873. SUSAN, married John Brown.

874. HENRY, married Mary Borden, daughter of Stephen; no children.

875. THOMAS, born ——; died September 11, 1810; married Elizabeth Church, November 23, 1797; she died December 26, 1824.

876. HANNAH, married Earl Borden, son of George.

877. GARDNER, married Elizabeth Bennett, June 29, 1802.

443. BENJAMIN, Fall River, R. I.

878. ELIZABETH, married John Simmons of Cape Cod, Mass.

879. MARY, married Abraham Simmons of Cape Cod, Mass., March 23, 1794.

880. THOMAS, married Lucy Durfee, daughter of Benjamin Durfee.

444. JOHN, Tiverton, R. I.

881. HOPE, born September 7, 1774; married Dr. John Turner of Fall River.

882. REBECCA, born August 3, 1776; married Nathan Simmons.

883. STEPHEN, born December 14, 1778; married Martha Earl, daughter of Robert Earl of Westport, March 21, 1802.

884. SAMUEL, born January 16, 1885; married Mary ——

885. JOSEPH, born May 11, 1787; died young.

446. NATHAN, Tiverton, R. I.

886. PRUDENCE, born January 16, 1784.

887. STEPHEN, born July 20, 1787; married Martha Evans, November 16, 1806.

888. SARAH, born June 15, 1789; married Benjamin Sweet of Somerset.

889. BENJAMIN, born December 24, 1793; married Hannah Coggeshall.

890. POLLY, born 1795; married Abial Davis.

891. JOHN, born 1799; married Mary Davis of Somerset.

453. ISAIAH, Fall River, Mass.

892. THOMAS, born August 3, 1783; died unmarried, October 13, 1864.

893. ISAAC, born January 30, 1785; married Sarah Winslow, daughter of Luther Winslow, December 16, 1810.

894. AMOS, born July 11, 1786; married in Fall River (1) Clarissa Borden, October 8, 1809; no children; (2) Mary Hathaway, May 21, 1812.

895. HANNAH, born January 12, 1788.

896. ABRAHAM, born April 12, 1790.

JOHN P. BORDEN.
865.

SEVENTH GENERATION. 193

897. BRADFORD, born October 21, 1791; married Hannah Jones of Dartmouth, January 17, 1813.

898. STEPHEN, born January 7, 1795; married Prudence Brightman, December 12, 1816.

447. GEORGE, Fall River, Mass.

899. HOLDER, born June 29, 1799; died September 12, 1837; never married.

900. DELANA, born May 5, 1801; married Dr. Nathan Durfee, April 24, 1827.

901. SYLVIA, born October 30, 1803; married Juseph Durfee, December 28, 1826.

902. FIDELIA, born April 17, 1803; married Matthew C. Durfee, December 22, 1826.

461. PELEG, Fall River, Mass

903. JEREMIAH, married Alice Handy of Westport, Aug. 11, 1825.

463. ADAMS, Fall River, Mass.

904. GEORGE, born February 5, 1807; married Rebecca Reed July 21, 1831.

905. REBECCA, born February 13, 1809.

906. ADAMS, born May 13, 1812; married (1) Justinia B. Savery, November 30, 1831; (2) Aseneith Alice Forbes.

907. HENRY, born October 24, 1814.

908. MARY S., born May 16, 1815; married D. D. Petty, December 27, 1834.

909. LYDIA, born October 25, 1816; married Spencer Macomber, October 15, 8135.

910. GENERAL, married Sarah E. Borden, daughter of Stephen, March 15, 1845.

464. AARON, Fall River, Mass.

911. DAVID, born June 1, 1796; married Miss Atwood of Taunton.

912. PHILADELPHIA, born October 16, 1799; married John Church, November 18, 1819.

913. SUSAN D., born January 31, 1803; married William Brownell, December 22, 1823.

914. AARON D., born March 30, 1806; died unmarried.

915. CHARLES E., born March 30, 1809.

THE BORDEN FAMILY.

467. JOSEPH, Fall River, Mass.

916. ELIZABETH, born May, 1793; married Samuel Borden, son of Edward, September 29, 1816.

917. LYDIA, born July 13, 1795; married Israel Coggeshall.

918. SUSAN, born September 15, 1797; drowned in Fall River.

919. JAMES, born May 31, 1799; died December 24, 1866; married Louisa Sherman, September 29, 1825.

920. RUTH, born July 21, 1801; died young.

921. STEPHEN W., born July 4, 1803; lost at sea.

922. JOSEPH, born March 12, 1805; married Abby Waldron, September 19, 1831.

923. PEACE, born October 13, 1807; married Philip Gardner, January 25, 1831.

924. ALEXANDER, born December 27, 1809; married at Fall River, Deborah Crapo, March 1, 1843.

925. CHARLES L., born November 25, 1811; was a deacon in the Congregational Church in Fall River. He married (1) Phoebe Hathaway, October 16, 1836; (2) Peace Bassett in 1860.

926. SUSANNAH, born May 17, 1814; married Charles Coolidge, May, 1834.

927. EVELINE, born May 23, 1816; married George W. Read, December 21, 1841.

468. PARKER, Fall River, Mass,

928. PARKER, born April 7, 1836.

929. CHARLES HENRY, born April 7, 1836.

930. MARY ELIZA, born January 23, 1838.

931. DOLLY CHURCH, born January 12, 1840.

469. ABEL, Fall River, Mass.

932. ABNER, born April 16, 1793; died young.

933. HANNAH, born October 8, 1795; married David Elsbree, January 16, 1820.

934. ABRAHAM, born October 22, 1797; married Rhoda Weaver, January 21, 1820.

935. MAJOR, born May 10, 1803; married Elizabeth P. French, December 4, 1828.

936. RUTH, born July 6, 1800.

937. SARAH ANN, born November 26, 1806.

SEVENTH GENERATION. 195

938. LEFAVOUR H., born August 12, 1809; he was a physician; married Priscilla Dwelley, October 2, 1831, in Patterson, N. J.

939. ISRAEL, born March 1, 1811; married Hannah S. Gardner, September 24, 1843.

940. ABEL, born July 22, 1814; died January 24, 1852; married Juliann B. Nye, November 25, 1838.

470. ABNER, Fall River, Mass.

941. ABEL, born 1792; married Betsey Read, no issue.

942. NATHAN, born January 7, 1796; died at sea.

943. PARKER, born April 11, 1799; died at sea.

944. CAROLINE M., born January 23, 1801; married (1) John Oswell; (2) James Brown.

945. WILLIAM, born January 20, 1806.

946. EMELINE, born January 3, 1809; married Benjamin LeValley, November 20, 1826.

473. WILLIAM, Fall River, Mass,

947. EARL, born November 14, 1798; died 1813.

948. JOHN, born July 23, 1799; died 1819.

949. ABBEY, born April 3, 1801; married Isaac Borden, son of Jonathan, January 4, 1821.

950. PRUDENCE, born February 3, 1803; married Laban Borden, son of Thomas, May 23, 1824.

951. ANN, born April 9, 1805; married Charles Crary.

486. LEMUEL, Tiverton, R. I.

952. ELIZA, married (1) Robert Brightman; (2) Robert Westgate.

953. THOMAS, "while yet a young man, went to the western country, and after a few years worked his way to Paris, Bourbon county, Ky., where he married, and located himself under favorable circumstances. He owned a good farm and soon became possessed of a comfortable living. He is now dead, but has left six sons to enjoy the benefit of his early wanderings, enterprise and labors. Their names, as given me by Bailey E. Borden, Esq., who once visited his brother in his distant home." S.

954. BENJAMIN, maried in Pawtucket, Mrs. Roby Sweet, nee Gardner.

955. PATIENCE, married Job Munroe of Providence.

956. SARAH, married A. Moffit of Smithfield.

957. SUSAN,

THE BORDEN FAMILY.

958. JOANNA,

959. LUTHER E., married (1) Ruhama Bishop, June 1, 1816; (2) Alice Hart.

960. ALMIRA,

961. ABBY, married Leonard Carlton of Slatersville.

962. WILLIAM, died unmarried.

963. LYDIA, married Ira King; had two sons and three daughters.

964. BAILEY E., lawyer in Providence; married Ardelia Carpenter.

965. RUTH, married Thomas H. Stranger of Providence.

490. SETH, Fall River, Mass.

966. HARRIET, born August 3, 1795; married Coggeshall Peckham.

967. ARNOLD, "born September 3, 1795; died March 7, 1845, in Goldsboro, N. C. He settled there in 1820, commenced business as a keeper of a hotel, and by a strict attention to his affairs he soon was able to purchase a plantation in that vicinity. Later in life, in connection with another, he purchased an estate in the northern part of Alabama. After a few years of prosperity he and his wife visited his relatives in Fall River, and on his return his brother Abraham accompanied him; he also married there, but has since died, leaving no issue. He was by trade a sailor." S. He married Ann Brownrigg in 1824; she was born in 1804; died March 7, 1872.

968. SUSAN, born September 22, 1797; married Isaac Whitman.

969. SARAH, born March 4, 1801; married (1) John Sanford; (2) William Young; her son is Arnold B. Sanford.

970. ABRAHAM, born November 22, 1805; died at Goldsboro.

491. ARNOLD, Fall River, Mass.

971. STEPHEN,

972. CHARLES,

973. JOHN,

974. MARY,

975. PRUDENCE, married Arnold Philips, Valley Mills.

976. ANGELINE,

497. THOMAS COX, Nova Scotia.

977. GEORGE WELLINGTON, born May 17, 1821; died April 5, 1876.

978. SUSANNA, born February 1, 1823; married —— Thomas.

979. JOSIAH, born January 4, 1825.

980. LUCY ELIZABETH, born February 13, 1827.

SEVENTH GENERATION. 197

981. THOMAS HENRY, born June 25, 1829; married at Port Williams, Nova Scotia, Mary Ann Best, January 25, 1859; he then removed to Santa Ana, California.

498. HENRY, Kings county, Nova Scotia.

982. JANE R., born October 10, 1819.
983. MARY, born January 5, 1822; died July, 1842.
984. LAVINIA, born February 23, 1824; died 1880, in California; she married William Loomer.
985. CHARLES HENRY, born October 14, 1825; married in Town Plot, Nova Scotia, Charlotte D. Woodworth.
986. BYARD W., born March 26, 1827.
987. ELIZA A., born December 12, 1829; married Mr. Welton of Port Williams.
988. SAMUEL B., born December 22, 1831.
989. SIDNEY D., born January 18, 1833; married Margery Margeson, September 25, 1867.
990. JOHN ALEXANDER, born March 20, 1836; Windsor, Nova Scotia.
991. FREEDOM, born January 3, 1839; married Johnson Bishop of Wolfville, Nova Scotia.
992. BENJAMIN, born May 3, 1840.
993. GEORGE W., born May 1, 1844.

507. EARL, Fall River, Mass.

994. EARL, Jr., married Hannah Read, June 13, 1824.

508. ABRAHAM, Fall River, Mas.

995. LAZARUS, born June 22, 1808; died April 23, 1869. Married Julia Ann Elsbree, born May 13, 1822; died April 26, 1873.
996. SOPHRONIA, born October 31, 1809.
997. ABRAHAM, born July 5, 1811; died at sea.
998. LYDIA, born September 4, 1812; died December 4, 1822.
999. EARL, born March 15, 1814; died at sea.
1000. HARRIET, born December 15, 1816.
1001. WILLIAM, born March 4, 1818; a bachelor.

509. THOMAS, Fall River, Mass.

1002. LABAN, born December 4, 1802; married Prudence Borden, daughter of William, May 23, 1824.
1003. MELVIN, born March 2, 1805; married (1) Phoebe Potter; (2) Eliza B. Lawton.

THE BORDEN FAMILY.

1004. PHILANDER, born March 11, 1807; married Nancy G. Shearman.

1005. LEANDER, born March 11, 1807; died July 24, 1894; married (1) Joanna Edson, born November 23, 1811; died September 16, 1849; (2) Adrienne Durfee, January 27, 1853; she was born June 19, 1828.

1006. SUSAN, born July 4, 1809; married Isaac Brightman, Oct. 1827.

1007. ALONSO, born April 16, 1812.

1008. ERASTUS, born December 11, 1815.

1009. AVERY, born 1820.

575. GEORGE G., Fall River, Mass.

1010. JEROME BONAPARTE, born May 30, 1817; married Rebecca Ricketson, January 11, 1843.

1011. LUCETTA, born January 24, 1822; died young.

1012. TRUMAN, born November 14, 1823; died Dec. 18, 1862.

1013. LUGENIA, died in infancy.

1014. MARY ANN, born March 14, 1825.

1015. TIMON, born March 18, 1827; married Sarah Myrick, June 10, 1854.

1016. LUGENIA, born December 28, 1828.

1017. ALPHONSO, born July 17, 1832; married Eliza Davis.

1018. LYSANDER, born June 11, 1830; married Elizabeth Young, November 12, 1859.

1019. PRINCE SEARS, born April, 1835; was an officer in the navy during the Civil War.

513. ISAAC, Fall River, Mass.

1020. ISAAC, born April 7, 1810; married Lydia Waldron, December 31, 1829.

1021. NANCY, born December 11, 1811; died June 10, 1828.

1022. LAURA, born September 1, 1814; married Frederick Chandler, April 3, 1835.

1023. CLARISSA, born September 15, 1816; married Alfred Atwood, March 15, 1838.

1024. MERRIL H., born May 6, 1818.

1025. WILLIAM N., born May 26, 1821; married Ann Maria Bailey, 1840.

1026. RUTH, born January 13, 1824; died in infancy.

SIXTH GENERATION.

521. STEPHEN. Fall River, Mass.

1027. PEACE, born August 12, 1820; married (1) Sylvester Bassett; (2) Deacon Charles Borden.

1028. LUTHER, was lost at sea.

1029. BENJAMIN D., married Amey Ann Sekel; died June 12, 1859.

1030. JOB, married Charlotte Harris July 9, 1849.

1031. STEPHEN, died young.

1032. WILLIAM A., married Harriet Bassett.

1033. ELIZABETH, married (1) Sylvester Davis; (2) George Lawton.

1034. SUSAN JANE, married Joseph Tallman of Portsmouth, R. I.

1035. ASA, died in infancy.

1036. MARY FRANCIS, married Gilbert Brownell of Portsmouth.

1037. GEORGE.

522. ELIHU, Fall River, Mass.

1038. PELEG, married Cynthia King, daughter of Capt. Godfrey King.

1039. PARDON S., married Abby Lawton, November 10, 1855.

1040. ISAAC,

1041. ANDREW,

1042. ANN,

1043. SARAH,

531. JOHN, Fall River, Mass.

1044. AMASA,

1045. ABDELLA,

1046. NANCY,

532. ABRAHAM, Tiverton, R. I.

1047. ISAAC, born April 17, 1813.

1048. PATIENCE, born April 15, 1815; married Ashley Ross of Connecticut.

1049. MARY ANN, born July 26, 1817; married Albert Booth.

1050. RUTH, born November 11, 1820; married William Alford of New Bedford.

1051. JOHN C., born March 12, 1823; died March 21, 1848; from injuries received on board a whaleship, and was buried at the Seychell Islands in the Indian Ocean.

THE BORDEN FAMILY.

1052. LYSANDER, born September 21, 1826; served in the army of the Potomac under Gen. Grant.

1053. ABRAHAM F., born June 27, 1829; married Anna B. Jenney, October 11, 1852. He was killed while engaged in telegraphing on board of an United States armed vessel on the Sabine River. He enlisted in the Forty-first Regiment of Massachusetts Volunteers, but was detailed to serve on the telegraphic corps. He was killed while faithfully serving his country, and left a widow and two small children, besides other friends to mourn his loss. He was a mason by trade, and his family were entirely dependent on his labors for support.

1054. PRISCILLA, born April 6, 1832; married Stephen Grinnel of Tiverton.

NATHAN, Pompey, N. Y.

1055. NATHAN,
1056. WILLIAM,
1057. JAMES.

540. JOSHUA, Pompey, N. Y.

1058. PARKER, born November 6. 1807; married Jemima Hanchett, September 15, 1833.

1059. BETSY, born June 4, 1809; married Watson C. Hanchett, March 14, 1832.

1060. SARAH, born January 6, 18913; married Alexander P. McCormick, November 23, 1849.

544. TIMOTHY, Madison county, N. Y.

1061. OLNEY Lorn 1792.
1062. CALEB,
1063. HIRAM,
1064. JOSEPH.

546. JOSEPH, Cazenovia, N. Y.

1065. THOMAS FREEBORN, born 1799; died January 6, 1874; married Susan Esther Prentice, born March 14, 1808; died 1886 at Elgin, Ill.

1066. BATEMAN, died in Cazenovia, N. Y.

1067. JOSEPH,

1068. CLARISSA, married Mr. Knowlton; died in 1894.

1069. ALMIRA, married Caleb Skiff; died in Cazenovia, N. Y.

SEVENTH GENERATION.

547. JOSEPH H., Madison county, N. Y.

1070. ELIZABETH, born September 22, 1808.
1071. PHILURA ANN, born November 7, 1814.
1072. LUCINDA, born November 17, 18916.
1073. L. WELLINGTON, born November 13, 1817, in Kalamazoo, Mich.
1074. JEANNETTE, born February 22, 1822.
1075. WILLIAM DEAN, born April 13, 1824; name of his first wife unknown; (2) Elizabeth Horn Chappell.
1076. JOHN B., born July 8, 1826.
1077. SAREPTA, born September 2, 1830; married John W. Hoxie, February 23, 1856.
1078. JOSEPH H., born May 18, 1833; lives in Milford, Neb.

555. JOHN, Fall River, Mass.

1079. THOMAS H., born May 26, 1817; married Lucy Durfee, daughter of Hon. Job Durfee of Tiverton.
1080. WILLIAM W., born August 18, 1823; he is president of a college in Borden, Ind. He married Emma Dunbar of New Albany, Ind., November 13, 1884.
1081. JOHN, born April 23, 1825; he is a lawyer in Chicago. He was educated at Brown University, Providence, R. I., and married (1) Alice Jane Wood, September 2, 1849; she died July 15, 1852. John Borden married a second time, but the name of his wife is not known to the editor.

556. ISAAC, Fall River, Mass.

1082. FINIS G., born 1826; married Joseph E. Macomber.

559. ASA, Fall River, Mass.

1083. MARY,
1084. WILLIAM, born August 12, 1826; married Susan Chace.
1085. JOHN, born February 21, 1829; married a Mrs. Bowen.
1086. BENJAH, born January 22, 1833; married Ruth Sherman.
1087. LYDIA, born June 5, 1834; married Charles Slocum.

561. WILLIAM, Fall River, Mass.

1088. CATHERINE T., born March, 1833; married John Mott.
1089. JOHN LEVI, born November 8, 1851; married Ruth A. Barker, 1886.

569. JOHN C., Fall River, Mass.

1090. WILLIAM H.

THE BORDEN FAMILY.

1091. FREDERICK,

1092. FERDINAND,

1093. JOHN C.,

572. THOMAS, Lowell, Mass.

1094. THOMAS H., born August 13, 1852; living at 17 Park Place, New York.

597. HENRY, Portsmouth.

1095. HENRY,

1096. THOMAS,

601. WILLIAM, New Bern, Ala.

1097. ELINOR HULL, born May, 1830; died in 1868, unmarried.

1098. MARIA WEBB, born December, 1835; died in 1870; married John S. Telfair, March, 1854.

1099. CATHERINE HALE, born April 4, 1839; married William P. Brown, February, 1874; no children.

1100. ANNIE HAWKS, born August 5, 1842; married Noborn Jackson, 1866; died in 1867; no children.

602. BENJAMIN, Green county, Ala.

1101. JOSEPH, born January 27, 1828; married Fannie Scott Gray, 1850.

1102. THOMAS JAMES, born 1830; died in Mobile, Ala., 1876; he married Elizabeth S. Byrn.

1103. MIRANDA, born March 8, 1839; in New Bern, Green county, Ala.; married Maj. Thomas Crawford Clark, April 16, 1862.

1104. WILLIAM ALFRED, born January 27, 1847; died January, 1892; married Alice G. Moore, November 1, 1871.

1105. MARY ESTHER, born January 23, 1849; married (1) William Thomas Cheney, November 25, 1870; died ——; (2) Edward Fenwick Campbell, November 13, 1880; died ——

1106. BENJAMIN CLAYBORN, born November 8, 1850; died March 4, 1887; married Roberta Moore, sister of the wife of William Alfred, his brother.

1107. JAMES PENNINGTON, born November 19, 1852, in Hale county, Ala., married Melissa Caroline Parham, November 12, 1879. He is a physician.

603. DAVID WALLACE, Alabama.

1108. ELIZABETH GRAHAM, born December 8, 1828; died September 16, 1856; married George Lovic Pearce.

JUDGE RHODES BORDEN.
1114.

SEVENTH GENERATION.

1109. MARY JAMES, born March 28, 1831; died March 16, 1869; married David Grace; left one son David Borden Grace of Burmingham, Ala.

1110. HANNAH WARD, born June 16, 1835; married (1) George Lovic Pearce, New Bern, Ala., December 3, 1857; (2) William Kirke Wallace at Montgomery, Ala., January 8, 1879.

1111. JOSEPH A., born January 3, 1826; died in 1838.

604. JOSEPH, Borden, Fresno county, Cal.

1112. THOMAS PENNINGTON,

1113. MARY ESTHER,

1114. RHODES, born November 16, 1850; died December 2, 1898; Judge Superior Court, San Francisco, Cal.

His father was Dr. Joseph Borden, who emigrated from North Carolina to Alabama when a very young man, and acquired a competence, which was afterwards almost wholly dissipated by the Civil War. In 1868, broken in health and fortune, he removed to California with his family, consisting of his wife, one daughter and four sons, of which Rhodes was the eldest. They settled in the San Joaquin Valley, then a vast uncultivated plain, and began the usual frontiersman's struggle with hardship and privation. His father's health failing, the burden of providing for the family fell upon the subject of this sketch, then a youth of seventeen, and he bore it with courage and unselfish devotion to duty, for many years. He had attended the University of Kentucky at Lexington, Ky., up to the time of the family's removal to California, and the education he there acquired, especially in mathematics and civil engineering, stood him in good stead, during the ten years of frontier life which followed.

Determined to enter one of the learned professions, he went to San Francisco in 1881, and commenced the study of law, reading in the office of the eminent attorneys, Garber, Thornton & Bishop, and attending the Hastings College of Law, at that time presided over by Prof. J. N. Pomeroy.

From the time of his graduation, in 1884, he continued to practice law in San Francisco and adjoining counties, his patience and resolution in no wise daunted by his late start and the difficulties with which he contended. His brothers had become self-supporting, but the care of his invalid mother and unmarried sister still devolved upon him. Gradually he climbed up the toilsome road to professional success. He became known as a man of sound legal judgment, a tireless worker, and possessed of the deliberative temperament which enabled him to look calmly upon all sides of a question before making up his mind. Above all, he was regarded as a man to be depended on, of perfect frankness and unswerving integrity. His first employment in the public service was as chief assistant to the City and County Attorney of San Francisco, to which he was appointed in 1892. In that capacity he was actively engaged in important litigation, affecting large interests, and added much to his reputation as a lawyer, as well as reflecting credit upon his principal. In 1896 he was a candidate for the office of Superior Judge of San Francisco; he received both the Democratic and nonpartisan nominations; and, although he was defeated, owing to party dissensions, the endorsements he received from the bench and bar were of the most flattering character. He continued to occupy the office of assistant to the City and County Attorney, until April, 1898, at which time he was appointed by Gov. Budd to the office of Superior Judge of San Francisco, to fill a vacancy occasioned by the resignation of Judge Slack. Thus elevated to the bench he discharged the judicial functions with the same patient and painstaking care which marked his entire career, and during the nine months he lived after his appointment, he earned the respect and good-will of all with whom he came in official contact.

In November, 1898, he was reëlected for a term of six years, and success at last seemed to have come to him, after long years of labor and waiting. But it came too late, and the end was near, how near no one, not even himself, suspected. In the evening of December 2, 1898, after presiding in court during the day, he spoke of feeling indisposed, and ten minutes afterward was found dead upon the floor of the bath room. Death was caused by the rupture of a blood vessel in the brain.

This brief history of his life can be most fittingly closed by the subjoined quotation from the press of San Francisco. The San Francisco Examiner of December 5, 1898, said editorially:

"A pure and able man expired when Rhodes Borden passed away; a man whose life made for dignity and virtue distinctly, though unobtrusively. He loved the quiet way, and moved along most unassumingly. No community can have enough of that calm, pure influence men of his class exert. San Francisco learned to know him best since his elevation to the Superior bench, where his fairness and knowledge came into general view, and the brief experience we had of him as a judge was of a character to cause deep and general regret that his career is at an end."

From a communication to the San Francisco Call of December 7, 1898, the following excerpt is taken:

"And so, amidst all the political evils which infest the State, Rhodes Borden died, a recognized scholar, an accomplished lawyer, a pure and competent judge, a refined and well-mannered gentleman, a faithful citizen, beloved and respected by his family and by his friends, and lamented wherever he touched humanity. Finis coronat opus. His life is worth the study and the imitation of young men."

SEVENTH GENERATION.   205

The San Francisco Chronicle of December 6, 1898, referred to the deceased in the following terms:

"Prior to the adjournment of court yesterday morning, Judge W. T. Wallace referred to the death of Judge Borden, alluding to his ability, uprightness and commendable habits. He deplored his untimely death, and ordered the court adjourned until Wednesday morning as a mark of respect."

Following is Judge Wallace's tribute to the dead jurist:

"The court has learned with regret of the death of the Hon. Rhodes Borden, lately a member of this court. Judge Borden has been a member of this court for a considerable length of time. He was a man of ability, uprightness and studious habits, and was an ornament to the bench where he sat. He had lately been reëlected by his fellow-citizens to continue his services here. The court desires to bear witness to his eminent fitness and to sympathize in the loss which the community must feel at his untimely death, and in token of respect to his memory, it is now ordered that the court do adjourn until Wednesday morning next, and these proceedings be entered upon the minutes of the court."

1115. NATHAN LANE, born March 21, 1855; married Rinnie Lee Borden, daughter of Joseph, in 1882.

1116. SHELDON, born March 21, 1855; married in 1885, Frances Margery Burnett, daughter of Alexander Burnett, and granddaughter of Gen. Alexander Burnett of Kentucky; she was born November 5, 1868, in San Francisco.

1117. IVEY LEWIS, born February 5, 1864; married October 4, 1888, Hetty Bell Thompson, daughter of Capt. R. Thompson, of San Francisco.

1118. ANNA HELEN, born May 6, 1868.

605. THOMAS R., New Bern, Ala.

1119. CAROLINE SNEED, born June 4, 1832.

1120. LYDIA JONES, born July 16, 1834; married Charles Frederick Sheppard.

1121. JAMES WALLACE, born September 16, 1835; married Maggie Neld.

1122. SARAH COART, born May 3, 1837; married William Hughson Burr, born in Camden, S. C.; his father, Aaron Burr, named for Col. Aaron Burr, Vice-President of the United States, being of the same family of Burrs.

1123. FREDERICK ASA, born December 30, 1838.

1124. BENJAMIN, born November 16, 1843; married Lulu Knox.

THE BORDEN FAMILY.

1125. THOMAS SYDENHAM, born December 19, 1845; married Fannie McGee; his name can be found enrolled in the gallant body of J. E. B. Stuart's cavalry. He was also a member of Jeff Davis Legion. At the close of the Civil War Thomas Sydenham was presented with a beautiful horse by Gen. Morton.

1126. JOSEPH LANE, born December 19, 1845; married Olive Sheppard.

1127. HARRIET CAROLINE, born October 5, 1853.

606. JAMES WALLACE, Ft. Wayne, Ind.

1128. ESTHER ANNA, born at Fairfield, Herkimer county, N. Y., February 29, 1832; she received an excellent education, attending first St. Mary of the Woods, a young ladies' seminary near Terre Haute, Ind., at that time one of the best schools in the West; she then went to Fairfield Academy. In 1852 she was married to George H. Aylesworth of Troy, Ohio; she died March 8, 1854.

1129. MARY EMELINE, was born at Fairfield, Herkimer county, N. Y., April 8, 1834; and died at Richmond, Ind., February 12, 1837.

1130. REBECCA KENYON, was born at Richmond, Wayne county, Ind., September 4, 1836. On November 25, 1856, she married Charles E. Grover. She died January 16, 1866, at Cairo, Ill.

1131. JOSEPH JOHN, born at Richmond, Wayne county, Ind., November 29, 1839; he died at Fort Wayne, Ind., August 2, 1840.

1132. WILLIAM JAMES, was born at Richmond, Wayne county, Ind., November 29, 1839; he was educated at Fairfield Academy; served in the Civil War, and on September 11, ——, he married Lavinia Freising of New York city; he died inJersey City, November 14, 1898.

1133. GEORGE PENNINGTON, born at Fort Wayne, Ind., April 24, 1844; he is a captain in the Fifth United States Infantry; he served under Nelson A. Miles (now General and Commander in Chief of the United States Army), in the Indian wars in Montana. He was married to Elizabeth Reynolds of Fort Riley, Kas. He is now (1899) stationed at Moro Castle, Santiago de Cuba.

1134. EMELINE GRISWOLD, was born at Fort Wayne, Ind., February 15, 1847; she was married in 1876 to Charles E. Hargous, who was at his death in 1890, a captain in the Fifth United States Infantry.

1135. DAVID HENRY, born in New York May, 1863, and died at Borden, Cal., September 3, 1893; he married Mary Edgerton Nelson of Piqua, Ohio, July 25, 1890.

607. MARY WALLACE, Orange, N. J.

1136. MARY FRANCES SHELDON, born April 9, 1832; married (1)

WILLIAM J. BORDEN.
1132.

SEVENTH GENERATION.

Dr. William Watson Woolsey, son of William Woolsey and Catherine Bailey; (2) to Col. Woolsey Rogers Hopkins.

1137. CATHERINE JOSEPHINE SHELDON, born 1840; married William Franklin of New York.

608. ISAAC P., Green county, Ala.

1138. WILLIAM, born 1845; married Mary Brittain of Green county, Ala.

1139. ANNA, married William Given of Green county, Ala.

1140. EMMA, married Samuel Brown.

614. ELI, Bledsoe county, Tenn.

1141. DANIEL, died in Kentucky, in 1843.

1142. JOHN, died in 1894 near Borden Springs, Ala.

1143. MASSEY,

1144. SALLIE,

1145. CATHERINE,

1146. MARTHA, married William Bryant.

615. HAWKINS, Walker county, Ala.

1147. JOHN,

1148. FRANKLIN, died in 1892, in Denton county, Texas.

1149. HAWKINS, lives in Alabama.

616. JOHN, Tennessee.

1150. REBECCA, born June 26, 1817; died July 28, 1851; she married Mr. Alexander, leaving two sons and three daughters.

1151. GEORGE H., born October 24, 1819; died in 1865, in Palestine, Texas, leaving two sons, John and William.

1152. ANNIE, born September 8, 1821; died December 22, 1888; she married a Mr. Alexander, leaving one son and seven daughters.

1153. ELIZABETH, born November 5, 1825; died September 5, 1851.

1154. EUPHEMEA, born January 4, 1828; died September 16, 1866.

1155. WILLIAM JOSEPH, born May 14, 1830, in Benton county, Ala.; married in New Orleans, Emma Gabriel Gosson, born of French parents; she died December 4, 1880.

1156. MARY CATHERINE, born May 2, 1832, in Benton county, Ala.; married a Mr. Bacon; had no children of her own, but adopted a child, Julia Ann Whatley.

THE BORDEN FAMILY.

1157. ANDREW CAMPBELL born November 15, 1835; married (1) Fannie Knighten, January 1, 1856 (2) Fannie Buford of Dallas, Ga., December 1, 1859.

1158. JOEL ELI, born August 12, 1838; died in Hope, Ark., in 1891, leaving one son, Patrick Donnelly, and three daughters.

617. JOEL, Calhoun county, Ala.

1159. ELI, lives in Texas,

1160. JOSEPH,

1161. GEORGE,

1162. WILLIAM,

1163. MITCHELL,

1164. JAMES, lives in Mississippi.

1165. NANCY,

1166. POLLIE,

1167. MARGARET,

626. JACOB VAN METER, Botetourte county, Va.

1168. WILLIAM STEELE VAN METER, born April 29, 1817, in Charleston, W. Va.; was educated at Oxford College, Ohio; married Mary Elizabeth Shrewsbury, December 21, 1844; died January 10, 1884; his widow died June 14, 1893.

1169. MARY JANE VAN METER, born September 29, 1819; married February 23, 1837, to William Cooke.

1170. JULIA ANN VAN METER, born October 18, 1820; married November 23, 1836, to Atwood G. Hobson, who died in Bowling Green, Ky., January 4, 1898.

1171. CAROLINE EVE VAN METER, born July 11, 1822; was married June 23, 1841, to George Bradley Adams of Bowling Green, Ky.; he was born September 29, 1819; he died June 30, 1854.

1172. SAMUEL KIRK VAN METER, born March 26, 1824; married March 5, 1860, Cessna Jane Sharp of South Carrolton, Ky. He was a prominent physician in Bowling Green, Ky.; died in 1893.

1173. CHARLES JOSEPH VAN METER, born May 22, 1826, in Bowling Green, Ky. After spending some years in the management of one of his father's plantations, he, in partnership with his elder brother, William, engaged in steamboating on the Green and Barren rivers in 1856, and continued it until the breaking out of the Civil War, in 1861. He then entered the quartermaster's service of the Confederate States Army, and continued in that duty until 1865. After the close of the

SEVENTH GENERATION.

Civil War he and his brother William resumed steamboating, and, in connection with it, engaged in the lumber business. In 1868 they joined a syndicate, known as the Green and Barren River Navigation Company, and leased from the State of Kentucky the Green and Barren rivers for thirty years. This purchase was sold to the Federal government some ten years before the expiration of the lease. In the same year, 1868, he and his brother purchased Grayson Springs, in Grayson county, Ky., and managed it themselves until 1884. On the 1st of October, 1878, he married Mrs. Kate Moss Overall of Paducah, Ky., who was born in Greenburg, Ky., March 2, 1838. They are now residing on a farm near Bowling Green, Ky. They have no children.

1174. SARAH FRANCES VAN METER, born October 25, 1828; died January, 1883; she married Manoah P. Clarkson, May 14, 1856.

1175. CLINTON CLAY, born July 20, 1834; educated at Center College, Danville, Ky.; was a civil engineer by profession; died January 30, 1875.

627. JOSEPH VAN METER, Botetourte county, Va.

1176. ROBERT LOGAN, born 1818; died in Arkansas, December 15, 1862. He was twice married, but left no children.

1177. MARTHA HESTER, born October, 1820; died May, 1825.

1178. ELLEN MARY, born March, 1822; died April, 1895.

1179. MARGARET JENNINGS, born February 26, 1824; died July, 1832.

1180. WILLIAM ALFRED, born October, 1825; died in Knoxville, Tenn., September, 1854; never married.

1181. SARAH ELIZABETH, born August, 1829; married January 6, 1880, to John E. Helms of Morristown, Tenn.

1182. IDA VIRGINIA, born June, 1831; died March, 1833.

628. SALLIE HAWKINS VAN METER, Wytheville, Va.

1183. ELIZABETH ANN SWEETLAND, born September 12, 1815; she died in Wytheville, Va., July 1, 1892.

Mrs. Elizabeth Ann (Sweetland) Obenchain, great great granddaughter of Benjamin Borden, the elder, of Virginia, inherited the excellent qualities and the courage of her pioneer ancestors, as the following incident will show: In the summer of 1864, Gen. Hunter marched with a large army up the Valley of Virginia, to attack Lynchburg from the rear. Her two elder sons were in the Confederate army Her husband, who was a member of the Home Guard, was on duty in the fortifications at Lynchburg Her youngest son, then but 15 years of age, had ridden the family horse to the country, to save it from capture. Her eldest daughter was away at school. She was left alone, then, with only her five younger daughters, ranging in age from four to thirteen years. Her eldest son had left at home some eight or ten pounds of sporting powder. When Hunter's advance guard appeared on the opposite side of the James River from Buchanan, Mrs. Obenchain, fearing that her house would be searched by Federal soldiers as soon as they entered the town, and wishing to save her son's powder, caried it over to the church yard of St. John's, which adjoined her premises, and concealed it under some rank, matted grass near an old tombstone in rear of the church, where, from the solemness of the place, she supposed no soldier would go, and that the powder would be safe. Great, then, was her surprise and amazement when, going out on the back porch at about 10 o'clock at night she saw several fires burning in that part of the churchyard, and soldiers lying around them on the ground. She realized at once the situation. "Should fire get to that powder," she thought, "and cause it to explode and do any injury, the soldiers, supposing it was intentional, would become infuriated and burn the town. The mere thought of being the cause of such a calamity, though innocent, was more than she could bear. Immediately she called her housemaid, and said, pointing to the churchyard: "Hannah, look at those fires over there, and the soldiers lying around them; you must go over there at once and get that powder away."

"La, Miss Lizzie," said Hannah, with a look of terror on her face, "I wouldn't go dar among dem Yankees for de worl'."

"Then I'll go myself," said her mistress, and she set out at once.

"And I'll go with you," said faithful Hannah, trembling in every limb.

Followed by the servant, she went out through the garden and crept cautiously up to the dividing fence. Soldiers were stretched out on the ground, here and there, on the other side, fast asleep. Some of the fires were spreading slowly into the grass.

Thinking not of self—not of her own danger, but only of what might happen to others, she whispered to the servant to remain where she was, climbed the fence noiselessly, crept steadily among the sleeping soldiers, got the powder, and returned safely with it to the house.

When told afterward that she was in great peril at the time; that if she had been detected when coming out with the powder in her possession, the soldiers would have believed she was attempting to do the very thing she had gone there to prevent, she replied: "I did not once think of that." She was married June 14, 1840, to Thomas Jefferson Obenchain of Botetourt county, Va. His ancestors, Palatinates, from the Upper Rhine, were among those who were driven from their country on account of their religious belief, and who came to this country in the first half of the eighteenth century. The name was originally Abendschön. Thomas J. Obenschain was born in Botetourt county, Va., June 20, 1814. He was a merchant in Buchanan, Va. For nearly half his life he was a prominent Justice of the Peace in

SEVENTH GENERATION.

Virginia, and he held that office, again and again given to him without his asking it. At the time of his death, which occurred in Wytheville, Va., May 31, 1895.

1184. MARY HESTER SWEETLAND, born September 14, 1816; she was married in Pattonsburg, Va., March 1, 1837, to George Walter Strickland; she died in Botetourt county, Va., June 30, 1846.

1185. CHARLES GOULD SWEETLAND, born April 10, 1818; died in Sweetland, Cal., November 24, 1858; was never married.

1186. SAMUEL MC FERRAN SWEETLAND, born February 3, 1820, in Pattonsburg, Va.; he went to Memphis, Tenn., in 1846, and engaged in the mercantile business. He was twice married (1) Martha Virginia Abernethy; (2) Mary Jane Abernethy. He died April 17, 1856.

1187. MARTHA H. SWEETLAND, born September 27, 1823; married Elijah Walker in Pattonsburg, Va., May 4, 1848; died in Kentucky March 22, 1885.

1188. ISAAC VAN METER SWEETLAND, born April 21, 1821, in Pattonsburg, Va., June 2, 1844, he married Martha Russell. He removed to what is now West Virginia, and engaged in the mercantile business; died in Hamblin, Lincoln county, W. Va.

1189. HENRY PETTIT SWEETLAND, born July 29, 1827; went to California in 1849, was a member of the Lower House of the California Legislature in 1853, and married Augusta Ladd in 1855; died in California in 1877; no children.

1190. WILLIAM ALBERT SWEETLAND, born April 27, 1829; was a captain of cavalry in the Confederate service, and was killed near Gettysburg in 1862.

1191. SALLIE E. SWEETLAND, born in Pattonsburg, Va., June 16, 1831; she married Luke Powell in Greenup county, Ky., April 25, 1861.

1192. JAMES OTIS SWEETLAND, born June 14, 1833; he married in Nevada county, California, January 3, 1856, Martha Virginia Scott, daughter of John W., and Ruth Scott. James O. Sweetland was elected to the Lower House of the California Legislature September, 1879, and again in 1882.

1193. CAROLINE SWEETLAND, born February 12, 1835; married in Greenup county, Ky., January 26, 1860. Sylvanus Howe Wolcott; she died October 20, 1890.

1194. MARGARET SWEETLAND, born in Pattonsburg, Va., August 17, 1837; she was married in Greenup, Ky., March 25, 1862, to J. N. Powell.

624. ELIZABETH VAN METER, Virginia.

1195. JAMES CARPER.

THE BORDEN FAMILY.

1196. GEORGE CARPER.

1197. JOSEPH CARPER, married Ann West.

629. PETTETIAH.

1198. DANIEL, born October 25, 1786; died August 31, 1848; married October 26, 1806, (1) Mary Avery, born October 22, 1786; died August 23, 1824; (2) Catherine Bellinger, in 1825, December 18. She died October 19, 1878; is buried at Cave Hill, Louisville, Ky.

1199. MARY, born August 13, 1788.

1200. JOHN, born September 16, 1790.

1201. LEAH, born September 7, 1792.

1202. REBECCA, born September 26, 1795.

1203. WILLIAM, born November 26, 1797; died March 6, 1817.

1204. ELEANOR, born May 23, 1800.

1205. ANDREW, born November 10, 1804.

1206. JAMES, born July 12, 1802.

1207. HARRIET, born May 9, 1807.

1208. PETTETIAH, born October 14, 1809.

1209. NELSON, born May 28, 1812.

645. HON. EZEKIEL TAYLOR COX, Zanesville, O.

1210. COL. THOMAS JEFFERSON COX, born March 19, 1823; died September 17, 1866; married April 29, 1846, Lucy Ann, daughter of James M. and Elizabeth Van Zant.

1211. HON. SAMUEL SULLIVAN COX, born September 30, 1824; died September 10, 1889; married Julia A. Buckingham Mr. Cox was one of the many picturesque figures that have adorned our national life during the last fifty years. President McKinley has said of him: "He was my personal friend, a man of rare attainments, with whom I was associated many years in Congress. He was loved by his associates, and all Americans, for his noble qualities, and for thirty years of faithful service to his country."

A life of Samuel Sullivan Cox has been written by his nephew, Mr. William VanZant Cox, and his friend, Milton Harlow Northrup, which is said to be a "valuable additon to American political history," and a "book which will have a charm for readers of all kinds in every part of the country."

1212. LAVINIA COX, born October 22, 1826; married Dr. R. H. Sledgewick.

1213. ALEXANDER S. COX, born January 28, 1830; died December 16, 1867; married Elizabeth Gardiner.

SEVENTH GENERATION.

1214. EZEKIEL TAYLOR COX, born July 24, 1834; married Sarah Ewing.

1215. MARIA MATILDA COX, born February 10, 1832; married Edward Van Ranselear.

1216. ANGELINE S. COX, born October 11, 1837; married Thomas S. Sites.

1217. AUGUSTUS C. COX, born October 12, 1842; married Ida Monserrat.

1218. ELIZABETH COX, born Oct. 19, 1844; married John B. Taylor.

1219. MARY S. COX, born May 29, 1847; married Col. T. F. Spangler of Zanesville, O.

EIGHTH GENERATION

EIGHTH GENERATION.

648. JOHN ALLEN, Philadelphia,

1220. ELVIRA, born in Philadelphia, October 1, 1826 ied February 4, 1870; married Joseph T. Linnard, born April 15, 181., died April 13, 1874.

1221. THEODORE, born September 6, 1828; married Emma Margaret Page, daughter of William T. and Anna Page, June 27, 1867; she was born September 3, 1845, at Mt. Carmel, Ill.

1222. JOHN, born in Burlington county, New Jersey, December 25, 1830; married Elizabeth Noston, born December 11, 1837, daughter of William Noston.

1223. SARAH, born in Burlington county, New Jersey, May 19, 1833; taught a finishing school for young ladies for many years, until failing health forced her to give up her life work and retire to the invalid's room. She has rendered great assistance in securing data for this work on Borden Genealogy.

1224. HAMILTON, born in Burlington county, New Jersey, December 25, 1830; married June 26, 1869, Charlotte Maria Page, born at Mt. Carmel, Ill., May 18, 1847.

1225. SIDNEY DAVISON, born in Burlington county, New Jersey, February 12, 1838; died August 3, 1842.

1226. SELINA, born in Burlington county, New Jersey, May 8, 1840; died July 29, 1842.

1227. WILLIAM DAVISON, born in Burlington county, New Jersey, September 15, 1843; died March 21, 1866.

649. THOMAS J., Burlington, N. J.

1228. WILLIAM H., died 1868.

1229. MARY, married Joshua Fennimore.

1230. LUCIA, married John Wesley Adams,

1231. THOMAS JEFFERSON, killed by the railroad cars at 14 years of age.

1232. JOB, born July 18, 1835, in Burlington county, New Jersey; married Mary E. Asay, daughter of James Asay.

1233. STEPHEN COMMODORE DECATUR, born August 14, 1840; married Sarah M. Bonham, March 24, 1874; she was born in 1857, was the daughter of Rev. Moses Bonham,

1234. MARTHA, married Charles Lorce.

1235. CHARLOTTE, married (1) Samuel Carr; (2) Harry J. Fillman of Bordentown, N. J.

1236. CHARLES T,. born April 25, 1845; died June 23, 1868.

650. PETER, Burlington, N. J.

1237. AMELIA, married John Canes; left one daughter, Alice.

1238. LAVINIA, married John Price.

1239. ANNIE, born 1862; died 1880.

657. CHARLES, Burlington, N. J.

1240 CHARLES, lawyer in Burlington, N. J.

658. WILLIAM, Burlington, N. J.

1241. SARAH, married William Barton.

1242. ANN,

1243. ELIZABETH,

1244. MARY, married Sylvester Ellis.

1245. CHARLES,

1246. JOHN.

661. AARON. Monmouth county, N. J.

1247. DANIEL SCHENCK, born 1841; lives in Monmouth county, N. J.

1248. JAMES ALEXANDER, born 1841; lives in Monmouth county, N. J.

1249. JOHN W., born May 16, 1843; married in Manasquan, N. J.

662. DANIEL SCHENCK, Red Bank, N. J.

1250. CHARLES H., born August 22, 1846.

1251. ELEANOR, born September 23, 1851.

1252. AMOS S., Iorn March 16, 1856.

663. AARON, Emleytown, N. J.

1253. PHOEBE ANN, born July 2, 1829; died July 13, 1894; married Henry Emley, March 7, 1855.

1254. JOSIAH, born October 16,1830; married Louise Ridgeway, October 16, 1861.

1255. MARY R., born August 8, 1832; died July 15, 1894; married Edward Emley, December 1, 1852.

1256. BEULAH, born February 27, 1834; married Charles S. Bullock, February 6, 1859.

EIGHTH GENERATION.

1257. JOHN E., born March 9, 1837; married Emily Curtis, March 22, 1860.

1258. RACHAEL, born December 1, 1840; married Dr. James M. Bean, January 3, 1861.

1259. ELIZABETH A., born May 4, 1844; married G. W. Warren, March 26, 1874.

1260. WALTER E., born March 28, 1851; married Joanna R. Wainwright, daughter of John Wainwright, November 25, 1873.

664. MARY R., Emleytown, N. J.

1261. CLARK TILTON,

1262. EMILY TILTON,

1263. JOSIAH BORDEN TILTON, born May 15, 1837; married Elizabeth Bullock, January 1, 1863.

1264. ELIZABETH TILTON, born February 2, 1842; married Joel Wainwright, January 1, 1867.

1265. JOHN, died unmarried.

665. APOLLO W., Emleytown, N. J.

1266. ANN POPE, born November 30, 1831; died April 21, 1871; married Charles S. Collier. He was born in 1820.

1267. GEORGE WASHINGTON, born June 22, 1832; married Caroline Pomset, January 14, 1863; she was born September 8, 1838.

1268. AMANDA, born October 1, 1838; died February 28, 1839.

1269. CHARLES WESLEY, born November 16, 1840; married Josephine Augusta Conover, January 1, 1863; she was born October 22, 1843.

666. EDWARD, New Jersey.

1270. MARY, born March, 1857; died November 9, 1863.

1271. GEORGEANNA, born 1855; married Joseph T. Parr, December 6, 1878; he was born January 12, 1854.

667. CHARLES S.

1272. MARY E., born January 21, 1837; married Pearson Scott, October 10, 1858.

1273. MARTHA, born June 28, 1839; married Joseph Painter, August 4, 1872.

1274. WILLIAM, born May 1, 1841; married Mary Bussom, October 3, 1866; she was born March 17, 1842.

1275. SARAH, born September 17, 1843; married Benjamin Bussom, November 20, 1867; he was born October 11, 1840.

1276. EDWARD, born March 29, 1846; married Susan Lewis, February 22, 1868. She was born September 23, 1851.

1277. CHARLES WESLEY, born November 17, 1860.

1278. ELLA, born January 1, 1874.

669. SAMUEL W., New Jersey.

1279. HELEN, born August 1, 1847; married October 21, 1869; married Maj. William Henry Lloyd, son of William Lloyd, and his wife, Elizabeth Spackman.

1280. FRANCES STRAWBRIDGE BORDEN, born March 3, 1844; married Eugenia Reeve, daughter of Samuel Reeve and Constantia Preuss, his wife, April 17, 1869.

1281. HENRY, born June 9, 1850; married Ada C. Duhring, November 8, 1876.

670. ELIZABETH A., Emleyton, N. J.

1282. WALTER S. EMLEY, born September 26, 1848; died May 24, 1850

1283. BUELAH W. EMLEY, born July 23, 1850; married Scott Logan, January, 1873.

1284. MARY ELLA EMLEY, born October 28, 1851.

671. JOHN H., New Jersey.

1285. LAVINIA B., born March 10, 1845; married Richard Y. Cook, March 6, 1869, of Philadelphia, Pa.

1286. WINFIELD E., born December 16, 1848.

1287. MARTHA E., born February 12, 1852; died ——

1288. JOSEPH E., born May 9, 1854; married Henrietta Sebastian Evans, February 7, 1891.

1289. JOHN, born November 2, 1857; married Mrs. Edith Twenlon of Liverpool, Eng., December 26, 1889.

1290. FLORENCE, born July 23, 1861.

672. WILLIAM D., Burlington, N. J.

1291. TENBROECK,

1292. ISABELLA, married Frank Ommic.

674. EDWARD, Burlington, N. J.

1293. RHODA, born July 19, 1827; married Nathan Gaskill, January 1849.

1294. JOHN S., born March 2, 1829.

1295. JULIA S., born January 15, 1830; married Harper S. Gillet.

1296. EDWARD T., born November 15, 1832; married Sarah White.

EIGHTH GENERATION. 221

1297. RICHARD P., born October 21, 1834; died August 17, 1864.

1298. DANIEL S., born February 12, 1837; died February 28, 1852.

1299. HARRIET, born October 7, 1842.

1300. EMILY S., born September 20, 1844; married S. Phillips, February, 1868.

1301. MARIA F., born February 23, 1846.

1302. DANIEL S., born September 17, 1852.

680. SIDNEY PARKER, Allentown, N. J.

1303. ASHER, born November 20, 1820; died April 5, 1894.

1304. JOHN HANCE, born at Allentown, N. J.; married Caroline Yeager, daughter of J. P. Yeager; she died February 15, 1897.

1305. PARKER, born in Allentown, N. J.

1306. ALEXANDER, married at Mt. Holly, N. J.; name of wife unknown.

1307. MARGARET, born in Allentown, N. J.; married Mr. Atkinson.

1308. ANN ELIZA, born in Allentown, N. J.; married Daniel Zelley

1309. MARY, married —— Taylor.

1310. JENNIE, married —— Conover.

1311. ANDREW JACKSON, enlisted on a whaling vessel, and never returned to his home in New Jersey; the last time heard of by his family he was in California.

1312. JULIA, married —— Walker.

1313. REBECCA, married —— Dolen.

681. GEORGE, New Jersey.

1314. RANDALL, name of his wife not known.

1315. FRANK.

1316. JOSIAH, name of his wife not known.

1317. GARRETT,

1318. CHARLOTTE,

1319. JOHN.

686. EDWARD P., Philadelphia.

1320. LETITIA ERWIN, born 1844; died at the age of 5 years.

1321. FRANCES HOPKINS, born July 21, 1847; married Robert Mercer Parker, March 19, 1872.

1322. ANNA ELIZA, born November, 1848; died in infancy.

1323. EUGENIE DAVENPORT, born July 24, 1853; unmarried, and living in Washington, D. C.

1324. EDWARD PARKER, born January 9, 1857; lives in Washington, D. C.

691. JOHN, Monmouth county, N. J.

1325. JACOB PARKER, born 1831.

1326. WILLIAM HANCE, born 1833.

1327. RICHARD, born 1835.

1328. JOHN WHITE, born 1838.

1329. CAROLINE, born 1841; married ——Pontin.

1330. GEORGE EDWARD, born 1843.

1331. CHARLES CARROLL, born 1846.

1332. SARAH FRANCES, born 1848.

692. RICHARD, Red Bank, N. J.

RICHARD (Doctor)

668. JAMES, Sharptown, Salem county, N. J.

1333. JOHN, born 1824(about), in Sharptown, N. J.; married Mary Apple Smith, born in Philadelphia, June 13, 1825. John Borden died in Woodstown, N. J., December 2, 1892.

1334. CLEMENT ACTON.

690. THOMAS.

1335. JOHN FRANCIS, born 1849; died March 17, 1889; married Margaret Corlally in New York, June 1, 1869. He was a designer for the Meriden Britania Company for twenty years.

1336. JENNIE, married Edward Henry Caffrey of New York city; he was born in 1846; died in New York, January 1, 1899.

1337. KATE,

1338. THOMAS.

695. EDWARD W., Battle Creek, Mich.

1339. ANNA AMELIA, born October 8, 1843; died in infancy.

1340. HANNAH CHAMBERS, born October 8, 1843; married Dr. Elmore Palmer, 1864; she served as a hospital nurse in Tennessee during the Civil War; her husband being hospital surgeon. Mrs. Palmer has been prominently connected with the W. C. T. U. in Buffalo and elsewhere. Dr. Palmer is now connected with the R. V. Pierce Sanitarium, Buffalo, N. Y.

1341. ALMIRA CLAYTON, born June 18, 1845, at Battle Creek, Mich.; married Charles C. Rice, October 6, 1868; he served as a private in

EIGHTH GENERATION. 223

New York Vlounteers during the Civil War, and died a pensioner in 1894.

1342. EMMA LOUISA, born July 14, 1848; died May 28, 1848, while at college in Olivet, Mich.

1343. EDMUND JAYNES, born March 10, 1852.

1344. ORCELIA A., born February 4, 1855; died June, 1856.

1345. GEORGE WEBSTER, born November 9, 1857; graduate of the University of Michigan, 1880; Colporteur of the American Bible Society, and teacher, 1881-2; graduate of Union Theological Seminary, New York, 1884; pastor of Presbyterian Church, Salem, Neb, 1884-6; Marion, N. Y., 1886-91; Gladwon, Mich., 1892-4; Auburn, Neb., 1894-9.

1346. FLORA ELLEN, born June 6, 1864; married Cornelius B. Exelby of Britton, Mich.

696. THOMAS H., New York City.

1347. HENRIETTA, born September 2, 1847; died August 4, 1849.

1348. MALVINA AUGUSTA, born November 21, 1850; married Sylvanus Seeley of Shelton, Ct., December 23, 1868; died October 2, 1898.

1349. MARCUS HENRY, born August 14, 1857.

1350. THOMAS ADOLPHUS, born October 26, 1859; died February 21, 1862.

1351. GEORGE EMERY, born August 10, 1861; died July 10, 1862.

1352. PERCIVAL ERNEST, born April 28, 1868; married Florinda Jane Murry, March 5, 1891.

1353. EVELYN, born April 26, 1870.

1354. HORATIO SEYMOUR, born January 16, 1873; died July 27, 1873.

699. RICHARD ALBERT, New Jersey.

1355. ALBERT,

1356. ADA

700. DANIEL WILLIAMS, Trenton, N. J.

1357. CHARLES, born November 9, 1873.

1358. IDA, born July 16, 1872.

701. CHARLES E., New York City.

1359. JOHN HARVEY, born April 22, 1878; graduated from Yale with honors, June, 1899.

1360. GLENTWORTH DE GRAUW, born March 29, 1880; is in the banking business with the firm of Strong & Co., New York.

703. GEORGE F.

THE BORDEN FAMILY.

1361. SUSAN, married Lewis Brown; has one child.
1362. WILLIAM, married Carrie Curtes.

705. ELIZABETH C., Eatontown, N. J.

1363. HARRY WARDELL,
1364. SUSAN BORDEN WARDELL,
1365. ELIZABETH WARDELL,

707. HANNAH, Eatontown, N. J.

1366. THOMAS HARTSHORN, died in infancy.
1367. MARY HARTSHORN, died in infancy.
1368. WALTER HARTSHORN, died in infancy.
1369. WILLIAM HARTSHORN, lives in Jersey City; not married.
1370. MARGARET HARTSHORN, married Lawrence Hartshorn, a cousin.

716. BENJAMIN HANCE, Pullman, Ill.

1371. ELIZABETH HANCE.
1372. SUSAN HANCE, married Harry Haviland.
1373. HOWARD HANCE.

720. JOSEPH, Fall River, Mass.

1374. SETH, born January 26, 1802; married Edith Tompkins.
1375. BAILEY H., born August 12, 1804; married (1) Mary D. Gurney, December 2, 1827; (2) Rachael D. Hathaway, August 12, 1873.
1376. ISAAC, born October 5, 1806; married Martha Hathaway.
1377. ARDELIA, born August 17, 1808; married Joseph Brow, October 28, 1827.
1378. MARY R., born June 17, 1810; married James Brow, May 20, 1835.
1379. JOSEPH C., born September 26, 1812, at Fall River; died July 12, 1895; married Amey Hathaway, born April 30, 1814; died April 4, 1893.

722. WILLIAM, Fall River, Mass.

1380. MALINDA, born January 15, 1805; married (1) Rev. Augustus B. Read; (2) Nathaniel Eddy.
1381. MARY ANN, born October 14, 1809; married John Read of Fall River.
1382. SARAH.

723. ISAAC, Fall River, Mass.

1383. ADRIENNE, born June 8, 1813, married Deacon Leander P. Lovel, November 28, 1828.
1384. WILLIAM, born July 19, 1819.
1385. LUCY, born October 27, 1821, died in infancy.

THOMAS JAMES BORDEN.
1398.

EIGHTH GENERATION.

724 CAPTAIN THOMAS, Fall River, Mass.

1386. STEPHEN, born July 8, 1812, died young.

1387. ANDREW, born February 22, 1814, in Fall River, married Ann Eliza Dean October 22, 1835, died December 20, 1844.

1388. PHILIP D., born May 11, 1816, married (1) Sarah F. Bennett, daughter of Samuel Bennett of Fall River; (2) Caroline Seabury.

1389. SARAH D., born June 12, 1818, married Elijah Williams September 25, 1843.

1390. LYDIA, born February 12, 1827, married John N. Swan, January 11, 1845.

1391. THOMAS, born June 19, 1834, died in infancy.

1392. THOMAS R., born December 17, 1836, died February 27, 1841.

1393. ISAAC, born November 11, 1833, married Abby C. Allen, October 21, 1862.

725. SARAH, New York.

1394. LORENZO Luther, born August 2, 1811, died September 10, 1811.

1395. IRENE LUTHER, born May 14, 1813; died June 7, 1876; married (1) Ashael N. Bliss, Dec. 1, 1831; (2) September 5, 1836, to Edward S. Keep.

1396. SARAH BORDEN LUTHER, born April 3, 1815; died July 1, 1888; married Henry Abbott Newhall, born May 3, 1811, died March 28, 1872.

728 RICHARD, Fall River, Mass.

1397. CAROLINE, born September 20, 1829.

1898. THOMAS J., born March 1, 1832, in Fall River, married Mary E. Hill February 20, 1855.

"Thomas J. Borden was thoroughly educated in private schools, and when 16 years old entered upon what has proved to be an active business career by taking a position in the office of the Fall River Iron Works, where he remained for one year. Then, in order that he might better fit himself for the chosen labors of his life, he entered Lawrence Scientific School at Cambridge, where he completed a two years' technical course, studying engineering under Prof. Eustis, and chemistry under Prof. Horsford. The knowledge thus gained proved invaluable in his business career thereafter.

"In 1851 Mr. Borden resumed employment in the office of the Iron Works company, where he remained until 1853. In midsummer of that year Col. Richard Borden, Jefferson Borden, Oliver Chace and others bought the Globe Print Works, and changed the name to the Bay State Print Works. Thomas J., who had just attained his majority, was appointed agent and treasurer of the new corporation. In the crash of 1857, three of the partners failed, and financial considerations induced Col. Richard and Jefferson Borden to consolidate the Bay State Works with the American Print Works, with Thomas J. as manager of the mills. In February, 1860, he was appointed agent and treasurer of the Troy Cotton and Woolen Manufactory. When he assumed control the company had 9408 spindles and 252 looms, and manufactured about 2,400,000 yards of cloth annually. In less than a month he prepared and submitted plans for the enlargement of the factory to four-fold its then existing capacity. The plans were immediately adopted, buildings erected, and in less than ten months the works entire contained 38,736 spindles and 932 looms, all in operation, and manufacturing more than ten million yards of cloth annually.

"In 1858 the Mechanics' Mill Company was formed, and Mr. Borden was elected president and agent. At the end of eighteen months he accepted the post of treasurer, and with it the entire management of the business. This was a serious addition to his cares, some idea of which may be formed when it is stated that the new concern contained 53,712 spindles and 1248 looms, and was operated with a capital of $750,000. He organized, in 1871, the Richard Borden Manufacturing Company, with a capital of $800,000, principally subscribed by members of his own immediate family. While the edifice was in process of erection, the company sold sites for the Chase & Tecumseh Mills, constructed ninety-six tenements, and in other ways greatly improved the locality. The energy and judgment of Mr. Borden in less than two years transformed what had been waste lands, belonging to his father, into the basis of a thrifty and growing settlement. He was elected treasurer of the corporation, and held the position until 1876. He had been a director from its formation, and its president from 1874. All of the three great institutions which were under his control prospered financially. Every detail, however minute, passed under his watchful supervision; and the Napoleonic capacity for the smallest detail, as well as the conception of gigantic plans, has seldom been more thoroughly exemplified by any man. In February, 1876, Jefferson Borden, then in his seventy-fifth year, and who for thirty-nine years was agent of the American Print Works, desired to retire from the post he had so long held. Yielding to the numerous inducements brought to bear upon him, Thomas J. Borden finally consented to relinquish the management of the three great corporations with which he was so closely identified, and where he had exhibited the possession of such financial genius, and to devote his whole time and attention to the American Print Works in the relation of treasurer, agent and director. His connection with that great concern continued up to December, 1887, when he sold his entire interest therein. While thus engaged in enterprises that might well command the entire care of one man, Mr. Borden has found time to give attention to other matters and to make himself especially felt in the railroad world of Massachusetts and New England. He has been treasurer of the Watuppa Reservoir Company since 1864; and when the Metacomet Bank, now a

EIGHTH GENERATION.   227

National bank, was organized in 1854, he was elected a director, which position he has held until the present time. In January, 1874, he was elected a director of the Old Colony Railroad Company; in 1877, a director of the Fall River, Warren and Providence Company; and in June, 1874, of the Old Colony Steamboat Company. In his railroad labors he has ever shown the same good sense, business enterprise and financial skill that have been the marked features of his whole business life. Mr. Borden has also been a director of the Boston Manufacturers' Mutual Fire Insurance Company since 1876; of the Worcester Manufacturers' Mutual Insurance Company since January, 1879; of the State Mutual Fire Insurance Company since 1878; of the What Cheer Mutual Fire Insurance Company of Providence, R. I., since 1873; of the Fall River Manufacturers' Mutual Insurance Company since 1870, and is now president and treasurer of the latter. All of these corporations insure mill property only.

"During the War of the Rebellion Mr. Borden was commissioned first lieutenant of the Fall River Light Infantry, Massachusetts Volunteer Militia, his appointment taking effect on September 3, 1863. On May 4, 1864, he was made first lieutenant of the Fifth Unattached Company of Massachusetts Volunteer Militia, United States service, which served in the neighborhood of Boston for the term of ninety days. He was made captain of Co. K, Third Regiment Massachusetts Volunteer Militia, September 16, 1864; lieutenant-colonel of the same regiment on September 3, 1866, and its colonel on June 25, 1868. He resigned the latter position in 1871. Since the close of the war he has given much time and labor to the elevation of the standard of state militia.

"Like his father, Col. Borden has taken little part in the practical conduct of public affairs, although taking a deep interest in questions affecting the interests of the people at large. In 1874 he was made a member of the Fall River Common Council, and in 1875, elected president of the same. His great interest in the security of life and property was evidenced by eight years of efficient service in the fire department of Fall River, from 1865 to 1872, inclusive. For the first five years he acted as assistant engineer, and the last three as chief. He applied to this branch of the public service the methods of discipline and good management so conspicuously displayed in his business life, and when he retired had the satisfaction of knowing that he had greatly improved its morale, and that he was leaving it in excellent condition. The various connections of Col. Borden with the business and benevolent institutions of Fall River can hardly be enumerated here. On April 4, 1866, he was made trustee of the Fall River Savings Bank, and in conjunction with Guilford H. Hathaway, supervised the construction of the fine building now occupied by that institution. In 1879 he became a director of the Borden Mining Company, of which his father was one of the originators. Among the characteristic traits inherited from his father is that of a deep religious devotion, and on the father's decease, he took his place in church affairs, and proved himself the generous son of a generous sire. As chairman of the

Building Committee, he superintended the erection of the new edifice built by the Central Congregatonal Church on Rock street. He has been, since 1877, a corporate member of the American Board of Commissioners for Foreign Missions; and through many private and public sources has made his wealth a source of good to others, giving freely to any and all worthy objects brought to his attention. A genial, courteous gentleman, unostentatious and approachable, quick to perceive and prompt to act, with great busines ability and an honesty that is carried into every relation of life, he is one of the progressive men of New England, and has well earned the universal regard in which he is held."

1399. RICHARD B., born February 21, 1834; married Ellen M. Plumber October 15, 1863.

1400. EDWARD P., born February 12, 1836, married Margaret L. Durfee September 29, 1863.

1401. WILLIAM H. H., born September 13, 1840, married Mrs. F. J. Bosworth September 25, 1867. Captain William H. H. Borden died January 3, 1872, at Mentone, France, whither he had gone for his health. From early boyhood, he manifested a great interest in everything relating to navigation, especially steam navigation. He made several voyages to Europe for the purpose of perfecting himself in this branch. During the rebellion, he was in command of the steamers Canonicus and State of Maine, transporting troops on the Potomac and James rivers, as headquarters boat at Port Royal, and conveying wounded soldiers from City Point to Point Lookout and Washington. After the close of the war, he commanded the State of Maine, on the Stonington line, and the Canonicus, running between Fall River and Providence, and as an excursion boat on Narragansett Bay. Over exertion during the war, and an injury to the heart occasioned thereby, brought on rapidly failing health which a milder foreign climate could not alleviate, and he died in Southern France, in his thirty-second year.

1402. MATTHEW C. D., born July 18, 1842, married Harriet M. Durfee, daughter of Dr. Nathan Durfee, September 5, 1865.

A sketch of M. C. D. Borden was sought in vain. It seems that he systematically refuses all information about himself. Ascribing this to modesty and believing he will not be offended, we take the liberty of reprinting below a portion of a clipping from the New York Press:

"Mr. Borden has had a remarkable career during the past nine years. Manufacturing was bred into him. In the forties, when he was a little shaver, his father was one of the most prominent men in the town of Fall River. Matt Borden was the fourth son, and, while his father was operating the print works, building railroads, organizing

EIGHTH GENERATION.   229

steamboat lines and serving in the State Legislature, between times Matt went to Yale and graduated, the most popular man of the class of 1862. Later he entered the employ of Low, Harrison & Co., and after his father had turned over the print works to the eldest son, Col. Thomas J. Borden, the youngest brother was the American company's representative in the house of Bliss, Fabyan & Co.

The business connection followed the failure and reorganization of the American Printing Company in 1879. Still later Thomas J. and Matthew Borden operated the American, together as their father and brother had done in years gone by, and one day Col. Thomas Borden was surprised to receive an offer from his brother for his share in the concern. This was in 1888, and M. C. D. Borden entered upon the ownership of the printing company. Since then he has built a plant of four fine cotton mills, the output of which is handled by his company. Fall River mill men gave Matt Borden one or two tight squeezes, and this led him to bend his energies to erecting the factories which stand as monuments to his ability. In December of 1893 he came into the Fall River market and bought everything in sight, and, not having enough goods to suit his desire, operated in the Providence market and secured, all told, 700,000 pieces. This bold stroke simply astounded the conservative manufacturers of Fall River, and they saw Mr. Borden a factor whose importance to the trade they had not given full credit. The silver agitation had a most depressing effect on the print cloth market, and 4,000,000 spindles were stopped from eight to thirteen weeks, but the market didn't mend and the surplus stock didn't disappear, and the only ray of hope that struck Fall River mills was when in November Mr. Borden made a purchase of 300,000 pieces at 2 5-8 cents. The market gradually went off at 2 1-2 cents, the lowest point in the history of the trade. The stock of 1,500,000 pieces was still in evidence, and this prevented printers from showing very much interest in the market. Finally on January 7, this year, it was decided that another curtailment of print cloths was necessary. Mr. Borden is given credit of making the now famous agreement which is in operation, and which so materially assisted him in clearing out the market. A peculiar provision was that all print cloth mills in Fall River should shut down before the agreement to run but forty-two of the fifty-eight hours a week became operative. Mr. Borden headed the list, and on February 1 the 3,500,000 spindles necessary were pledged. The market was at 2 1-2 cents, but went up a point, to 2 9-16 cents. The production of the Fall River print mills was bound under the signature of treasurers to cut off 60,-000 pieces a week for thirteen weeks. However, the agreement was only five days old when Mr. Borden made one of his personal visits to Fall River. His presence was kept a secret. The market would jump if the sellers knew Mat Borden was in the city. He met three mill treasurers and told them he would buy all the spot regulars they could find in the market. These men made haste to discover the 720,-

THE BORDEN FAMILY.

000 pieces of "extras" that were labelled in the reports as being in the warehouses. They could find but 500.000 pieces, but agreed to sell these and make contracts for 250,000 more at 2 9-16 cents. The deal was completed last Monday morning, and caused the biggest trade sensation in years. By clearing the market of spot goods and contracting ahead Mr. Borden put his competing printers in the class of men who wanted goods quickly and couldn't get them, and the market rose in two days to 2 11-16 cents."

1403. SARAH W., born May 13, 1844, married Alphonso S. Covel May 19, 1869.

730. JOHN, Fall River, Mass.

1404. ALEXANDER.

1405. JOHN FRANCIS, born 1827.

1406. MARY Ann, born 1837.

731. JEFFERSON, Fall River, Mass.

1407. ELLEN, born December 23, 1830, married Walter Paine August 13,1850. He was born August 7, 1827; died October 13, 1897.

1408. ELIZA OWEN, born May 20, 1835, married George B. Durfee December 6, 1855, died July 23, 1892.

1409. EUDORA SEXTON, born March 20, 1839, married George W. Dean, (born November 4, 1825, Fall River,) July 2, 1873. George W. Dean died January 23, 1897.

1410. E. CORINNA, born November 16, 1841; married Dr. W. W. Keen December 11, 1867; died in Philadelphia, July 2, 1886.

1411. JEFFERSON, born June 2, 1844, married Ellen Westall in Fall River.

1412. SIDNEY, born July 29, 1846, died young.

1413. SPENCER, born June 10, 1849, married Effie A. Brooks at Montpelier, Vermont, August 9, 1871.

1414. NORMAN EASTON, born December 31, 1850, died January 22, 1884.

1415. CHARLES S., born October 21, 1852.

733. ABRAHAM, Fall River, Mass.

1416. ANDREW J., born September 13, 1822, married December 26, 1845, (1) Sarah J. Morse; (2) Abby D. Gray, daughter of Oliver, June 16, 1865.

1417. CHARLOTTE, born November 3, 1824.

1418. LAURA ANN, born August 10, 1826, married Hiram Harrington 1854.

1419. PHOEBE ANN, born February 19, 1829.

EIGHTH GENERATION.

736. RICHARD, Fall River, Mass.

1420. HIRAM C., born July 23, 1822, married Betsey J. Borden, daughter of Isaiah Borden, May 15, 1862.

1421. CHARLES WILLIAM, born December 26, 1835, married Jane B. Durfee April 18, 1854.

1422. NELSON C., born April 15, 1835, married Ellen Durfee.

738. COOK, Fall River, Mass.

1423. MARY JANE, born May 10, 1833, died October 10, 1833.

1424. MARY J., born September 1, 1834, married Dr. James Hartley.

1425. THEODORE W., born August 25, 1836; married Mary Louisa Davol.

1426. AVIS ANNE, born September 14, 1838, died September 19, 1839.

1427. PHILIP H., born June 8, 1841, married Ruth Dennis April 8, 1861.

1428. JEROME, born October 5, 1843, died May 1, 1844.

1429. JEROME COOK, born September 30, 1845; married Emma Eliza Tetlow June 28, 1870.

739. LADOWICK.

1430. MARIA, born October 22, 1844, married Samuel B. Hinckley October 2, 1866.

740. ZEPHANIAH, Fall River, Mass.

1431. MARY C., married Thomas Brayton February 25, 1862.

1432. ANGENETT.

1433. CHARLES RICHARD, born February 21, 1850.

742. ABRAHAM, Fall River, Mass.

1434. CHRISTOPHER, born October 20, 1815, married Lucy H. Davis February 11, 1840.

1435. RHODA, born October 12, 1820; married Abial Davis December 25, 1820.

1436. MARIA B., born December 4, 1826, married Weston Jenney November 28, 1861 in New Bedford.

747. ISAAC, Fall River, Mass.

1437. LOUISA, born November 20, 1821, died in infancy.

1438. ALANSON, born January 7, 1823, married Mary C. Topham, daughter of Capt. William Topham June 22, 1852.

1439. FREDERICK A., born August 24, 1824, married (1) Sarah Welsh; (2) Huldah Beach.

1440. AMASSA G., born June 22, 1826, married Mary Flery, Venice,

1441. FERNANDO, born November 19, 1827. Cayuga County, New York.

1442. AMANTHA, born December 13, 1829, married Volney Tupper January 6, 1856.

1443. LYSANDER, born August 3, 1831, married Sarah Dodel of Rockford, Illinois.

1444. ISAAC N., born in Cayugo County, New York, May 16, 1834, married Abigail Buttler, March 5, 1860, in Rockford, Illinois.

1445. DANIEL W., born April 10, 1841, killed in the battle of Fredericksburg, Virginia, December 13, 1862.

749. ANN, Portsmouth, R. I.

1446. SIMEON B. CHACE, born October 13, 1812; died November 7, 1832.

1447. AMEY ANN CHACE, born July 9, 1814; married Humphrey Almy January 15, 1838.

1448. BORDEN CHACE, born April 5, 1816; married Elizabeth Thomas, daughter of Joseph Thomas of Portsmouth.

1449. PHILIP CHACE, born February 3, 1818, married Sarah Cook, daughter of William of Portsmouth.

1450. SARAH CHACE, born February 17, 1820, married Stephen Davol, son of Abner.

1451. ELIZA CHACE, born May 23, 1822, married Charles Fowler of Brooklyn, New York.

1452. CHARLES CHACE, born February 2, 1824, married Fannie Pearce, daughter of George Pearce of Bristol.

1453. NATHANIEL B. CHACE, born November 1, 1825, married Louisa Pierson.

1454. ALFRED CLARK CHACE, born March 21, 1832, married Ruth Anthony, daughter of William Anthony.

750. JUDITH, Fall River, Mass.

1455. LUCY DURFEE, born July 1, 1821, married Thomas Borden November 19, 1846.

1456. AMEY DURFEE, born January 18, 1824.

1457. THOMAS DURFEE, born February 6, 1826

1458. MARY DURFEE, born October 23, 1827.

1459. SIMEON BORDEN DURFEE, born September 2, 1829, died February 23, 1858.

1460. SARAH ANN DURFEE, born July 16, 1831.

SIMEON BORDEN.
1465.

EIGHTH GENERATION.   233

1461. JULIA MARIA DURFEE, born May 24, 1834; died September 24, 1845.

*723. NATHANIEL B., Fall River, Mas.

1462. AMEY, born January 3, 1821, died August 16, 1871.

1463. SIMEON, born December 26, 1824, died September 15, 1825.

1464. SARAH, born August 27, 1826, died September 9, 1854.

1465. SIMEON, born March 29, 1829, died March 9, 1896. A Fall River, Mass., newspaper printed the following the morning after his death:

"Mr. Borden was born in Fall River on the 29th day of March, 1829, and has always lived here. He was the elder son of the late Hon. Nathaniel B. Borden, who was one of the most promising citizens in Southern Massachusetts. The subject of this sketch came of a long line of distinguished ancestry. He was a descendant in the eighth generation of Richard Borden, who came from England in 1635, and settled in 1638 in Portsmouth, R. I. Mr. Simeon Borden's great-great-grandfather was Capt. Nathaniel Briggs, and his great-grandfather on his mother's side was Col. Pardon Gray, an officer of the revolution. His uncle, Simeon Borden, whose honored name he bore, was the foremost civil engineer and mathematician in this country. Job Durfee and his son, Thomas Durfee, both of whom were Chief Justices of the Supreme Court of Rhode Island, were Mr. Borden's kinsmen.

"Mr. Borden's early education was obtained in the public schools. He was fitted for college by that eminent instructor, Mr. Belden of Fruit Hill, near Providence. He entered Harvard University in 1846, and graduated with honor in the class of 1850, being the first native of Fall River to graduate therefrom. Among his classmates were John Noble, Clerk of the Supreme Judicial Court of Suffolk Conuty; Charles Hale, editor of the Boston Advertiser and Consul-General to Egypt, also Assistant Secretary of State; Mr. Everett C. Banfield, Assistant Secretary of the Treasury under Mr. Boutwell; Gen. William A. Burt, postmaster of Boston; the Rev. Dr. Joseph H. Thayer, Bussey professor of sacred literature in Harvard College, and also professor at Andover Theological Seminary; the Rev. Dr. Howard Osgood, a member of the American committee to revise the Old Testament; Mayor George W. Richardson of Lowell; the Hon. T. Jefferson Coolidge, United States Minister Flenipotentiary to France; the Hon. James C. Carter, the acknowledged leader of the New York bar; Horatio Hathaway of New Bedford and others who have become eminent in various walks of life. Mr. Borden was the first president of the Harvard Club of this city.

"On graduating Mr. Borden entered the Law School at Cambridge, where he remained two years, and received the degree of Bachelor of Laws. Among his classmates in the Law School were the Hon. Edward L. Pierce, the biographer of Charles Sumner, and Hon. John Winslow of Brooklyn, N. Y. Mr. Borden then entered the office of William Brigham, Esq., an eminent lawyer of Boston; after studying there a year he was admitted to the bar, and began the practice of his profession in this city. He was naturally of a judicial temperament and his mind was enriched by long and faithful study. He at once commanded the respect and esteem of his fellow-citizens and of his associates at the bar.

"Mr. Borden was associate counsel and prepared with great thoroughness and ability the case before the committee of the Legislature defending the conventional line, which subsequently became the boundary line, as it now exists, between the States of Massachusetts and Rhode Island, and its final determination settled forever a long controversy between the two States and from which our city has reaped great advantages. In this case were associated with him Judge Benjamin F. Thomas, Judge William G. Choate, now of New York, and William W. Blodgett of Pawtucket.

"He was of counsel in the important trial in which was involved the will of Mr. Allen Mason. Mr. Borden prepared this case with great skill and care and received the commendation of Gov. Clifford and Judge Reed, who were associated with him at the trial before Judge Merrick and a jury. The case was on trial a week, and the jury returned a verdict sustaining the will. Counsel for the contestants were Judge Thomas, Hon. T. D. Eliot and Mr. Eliab Williams.

"Mr Borden was a member of the Common Council two years, and its president one year. For seven consecutive years he was a member of the Board of Aldermen and two years city solicitor. He was a member of the fire department under the old organization, being foreman of the Niagara Engine Company. For two years he was a member of the Massachusetts House of Representatives from Fall River. For seventeen years he was trustee of the Public Library, and he was also one of the commissioners of the sinking fund, and a trustee of the Fall River Savings Bank.

"Upon the resignation of Mr. John S. Brayton as clerk of the courts, in January, 1864, Mr. Borden was appointed by the justices of the Supreme Judicial Court to hold that office until the annual election, when he was elected to fill out the unexpired term, and by repeated re-elections held the office till the day of his death.

"Mr. Borden carried into the performance of public duties the same conscientious spirit and high standard which he exhibited in regard to those of a private nature. He illustrated in civil life the very best New England examples. He had a sound legal training, and his ability, fidelity, and unfailing courtesy and affability, with which he discharged the arduous and responsible duties of the office of clerk of courts for nearly a generation, won the unstinted approval of judges and the members of the bar, and of all with whom he came in contact.

EIGHTH GENERATION.

"The Commonwealth has lost a most valuable and efficient public servant, and Fall River will mourn for one of its most highly respected citizens.

"Mr. Borden leaves a widow and two children, Simeon Borden, Jr., the assistant clerk of the courts, and Sarah, an unmarried daughter.

"A meeting of the Fall River Bar Association will be held at the Superior Court room, at 4 o'clock this afternoon, to take action in respect to the death of Simeon Borden, Esq.. In speaking of Mr. Borden today a prominent member of the Fall River Bar said:

"'He had a long and varied experience as an auditor. He was entirely free from prejudice, of a judicial temperament, eminently fair, and was often selected by parties as an arbiter or referee. Until his health began to fail no man in Bristol county sat as often as he did as auditor and master."

"The distinguished experience of Mr. Borden's life, however, was his remarkable aptitude for the important office of clerk of courts. Many times judges had stated to the speaker that Mr. Borden was a model clerk—naturally fond of detail and the nicest accuracy. Members of the bar used to say that it was hard to see how Mr. Borden could be improved upon as clerk. Patient, amiable, obliging always, ready to assist, and exceptionally well versed in practice and of mathematical accuracy, he was recognized as eminent in the duties of his office. His counsel was sought by other clerks and by judges.

"Everything in his office was arranged with the utmost care. He hardly ever was known to make a mistake, though often called upon to act very quickly and attend to a multiplicity of details.

"He was the friend and adviser of all young lawyers; a confidant in their troubles, a friend in their prosperity. Never jealous or envious, always free to award praise, always putting the best construction on one's conduct, thinking no evil of any one, he made the duties of courts, jurors and counsel as smooth as possible. He was without exception the most popular member of the Bristol County Bar and occupied a personal and fraternal relation with every member of it. No one can be more missed in death, the loss of no one can cause deeper regret.

"Mr. Borden was a very public-spirited man. He was always in favor of improvements which would embellish and beautify the city. In fact, the great business error of his life, the building of the fine Borden Block, was prompted by two leading motives—to build a lasting memorial to his honored father, and to give to Fall River one of the finest buildings which then existed in New England. This noble block was somewhat in advance of the times and caused Mr. Borden financial loss, but, after all, it was a source of much pride to him.

"In politics Mr. Borden was always a strong anti-slavery advocate and a Republican. He brought into his politics the same liberal sentiments and breadth of view which characterized all his other relations in life.

He was very hospitable and frequently entertained judges of the court and members of the bar at his house.

"He was a great lover of out-door life, especially yachting, and for many years made it a point to spend several weeks on the salt water.

"About twenty years ago Mr. Borden built the beautiful house on Highland avenue, now occupied by Miss Sarah S. Brayton, which he sold to her about four years ago.

"Mr. Borden was always proud of his connection with Harvard College, and was the first president of the Harvard Club of Fall River. He always attended the commencements when his duties would permit."

1466. NATHANIEL, born October 21, 1832, died November 3, 1833.

1467. LOUISA GRAY, born January 14, 1836, married Dr. James M. Aldrich October 23, 1862, died October 24, 1897.

1468. NATHANIEL BRIGGS BORDEN, son of Nathaniel Briggs, and Sarah Gould (Buffum) Borden, was born at Fall River, Mass., February 23, 1844. His maternal grandfather was Arnold Buffum, the first president of the New England Anti-Slavery Society. Mr. Borden was educated in the public schools of Fall River until 1862, when he went to the Phillips Exeter Academy at Exeter, N. H., staying there until 1864, when he entered Harvard College, in the class which graduated in 1868. His father died in 1865, and in the fall of that year went to Peacedale, R. I., and was employed in the woolen mills of R. G. Hazard & Co., for the purpose of learning the woolen manufacturing bus'ne s. After wrids he was at the Carolina Mills in Rhode Island. While at Peacedale and Carolina Mr. Borden familiarized himself with the practical working of the machinrey of the various departments. In 1869 he left Carolina and went to Valley Falls, R. I., and entered the employ of his uncle, Samuel B. Chace, in the counting room of the Valley Falls Company, which was engaged in the manufacture of cotton cloth. In the spring of 1870 he went back to Fall River to live and entered the employ of the Merchants' Manufacturing Company as book-keeper. On February 2, 1870, he was married to Miss Annie E. Brown, and their children are Nathaniel Briggs Borden, Jr., born March 4, 1871; Annie Brown Borden, born December 4, 1877; Arnold Buffum Borden, born March 19, 1882; Louise Gould Borden, born October 11, 1883. In 1871 Mr. Borden left Fall River again and went back to Valley Falls, R. I., and became the agent and superintendent of the Valley Falls Company's cotton mills of about 35,000 spindles. He continued to live in Valley Falls until October 1873, when he again returned to Fall River, and immediately began to solicit subscriptions for the formation of a corporation to engage in the manufacture of cotton cloth. As a consequence of his exertions the Barnard Manufacturing Company was organized, and Mr. Borden was elected treasurer. The

NATHANIEL B. BORDEN.
1468.

capital stock of the corporation was fixed at $400,000, but it was afterwards reduced to $330,000. Plans were matured during the following winter, and contracts for machinery and materials were placed, so that in April, 1874, ground was broken for the erection of a factory to contain 28,000 spindles and 768 looms, for the manufacture of cotton cloths of print cloth yarns. The mill was erected and equipped during the year 1874 under Mr. Borden's supervision. Weaving operations were begun in January, 1875. In December, 1895, Mr. Borden was authorized by his stockholders to increase the capacity of the Barnard Manufacturing Company by the addition of about 30,000 spindles, and the capital stock was raised to $490,000. Plans were immediately prepared and contracts made for building a weave shed capable of holding 1832 looms, contracts were also made for new machinery which was supplied during the year 1896, making the capacity of the mills 64,560 spindles and 1708 looms. Mr. Borden was president of the Cotton Manufacturers' Association of Fall River in the years 1889 and 1890. In 1890 he was elected a member of the Common Council of the city government, and was returned in 1891, and each year was chosen president of that assembly. He is vice-president and director of the Massasoit National Bank and a trustee of the Fall River Five Cent Savings Bank. He is president of the Children's Home of Fall River, to which office he was elected in 1889, having been reëlected each succeeding year.

753. SARAH, New Bedford, Mass.

1469. FREDERICK A. SEABURY, born October 3, 1821, died August 27, 1821.

1470. CAROLINE A. SEABURY, born April 16, 1823, married March 28, 1854, Miner S. Lincoln of Boston.

1471. SARAH L. SEABURY, born November 22, 1823.

1472. CHARLOTTE A. SEABURY, born March 30, 1832.

757. RICHARD, Fall River, Mass.

1473. WILLIAM G., born May 1, 1820, married Caroline K. Lake.

1474. JAMES H., born November 15, 1821, died May 17, 1864.

1475. HANNAH W., born September 3, 1823, married (1) Jeremiah Taber, April 10, 1842; (2) Benjamin Chace.

1476. BENJAMIN C., born May 14, 1825, married Mary D. Taber March 26, 1865.

1477. GILBERT B., born February 1, 1827, married (1) Amey Hambley; (2) Phebe A. Hobbs.

1478. FANNIE A., born May 3, 1830, married Bradford Durfee, son of Gideon Durfee of Tiverton.

1479. ABBY A., born December 21, 1831; married Peleg Stafford.

1480. JOSEPH V., born October 4, 1833, married Sarah J. Gray, daughter of John.

1481. RICHARD H., born June 5, 1835, died young.

1482. RESCOME, born June 5, 1835, married Elizabeth Waters.

1483. IRENE P., born December 31, 1836, married Edward Gray November 24, 1859.

1484. SAMUEL E., born December 12, 1838.

1485. HOLDER B., born 1840, married Mary Brocklehurst March 13, 1864.

1486. RICHARD B., born March 1, 1843, married Helen B. Chace, daughter of Silas Chace.

758. ISAAC, Fall River, Mass.

1487. BERNICE C., born January 24, 1826; married Tamsen J. Bassett.

1488. CYRUS H., born April 9, 1828, married Caroline M. Waters September 9, 1850.

1489. LUCY, born October 29, 1829, died young.

1490. ANN ELIZA, born November 5, 1830, married Daniel Shearman.

1491. ISAAC J., born August 9, 1831, died in infancy.

1492. ADELAIDE G., born April 5, 1833, died in infancy.

1493. LUCY JANE, born February 12, 1835, married Lyman M. Fisher of Vermont.

1494. SARAH P., born March 9, 1838.

1495. ISAAC A., born April 19, 1840.

1496. JOHN J., born April 24, 1842, married Caroline F. Setlers, December 31, 1863.

1497. MARY A., born April 24, 1845, married Nelson P. Cook April 24, 1862.

759. JOSEPH, Fall River, Mass.

1498. PRISCILLA W., born April 27, 1823 married Humphrey Wilcox September 7, 1845.

1499. BENJAMIN F., born August 23, 1825, married Hannah S. Jencks January 6, 1850.

1500. MARY C., born September 14, 1827; married Squire Crapo.

1501. LYDIA W., born July 18, 1829, married William Davis.

1502. WILLIAM S., born October 17, 1831, died at fourteen years of age.

1503. DELIA W., born May 22, 1834, married Jabez Manchester March 9, 1858.

HON. FREDERICK W. BORDEN.
1522.

EIGHTH GENERATION.

1504. HANNAH H., born June 15, 1836; married Perry Gifford, January 19, 1859.

1505. JOSEPH W., married Mary Manchester February 6, 1861.

760. JOHN, Globe Village, Mass.

1506. ADRIENNE J., born August 19, 1836.
1507. ANDREW, born January 29, 1839.
1508. MARY A., born May 8, 1845.

771. HEZEKIAH, Horton, Nova Scotia.

1509. AMEY, born 1850.
1510. REBECCA, born 1852.
1511. MARY JANE, born 1854.
1512. SAMUEL, born 1856.
1513. CATHERINE, born 1858.
1514. WILLIAM, born 1860.

780. ADOLPHUS K., Horton, Nova Scotia.

1515. GEORGE, born 1831.
1516. EDWARD, born 1833.
1517. MARY.
1518. HENRY, born 1845; is a physician in Boston, Mass.

787. JAMES, Avonport, Nova Scotia.

1519. BRENTON., born April 16, 1859.
1520. MARY ELIZABETH, born August 27, 1861.
1521. JOHN, born February 27, 1864.

790. JOHNATHAN, Horton, Nova Scotia.

1522. FREDERICK W., born May 14, 1847. He is a graduate of Harvard University and in 1896 represented Kings County at Ottawa. He is now Minister of Militia of the Dominion of Canada. He married October 1, 1873, Julia Maude Clarke, daughter of John H Clarke, Esq., of Canning, Nova Scotia; she died April 2, 1880. Dr. Borden married a second time June 12, 1884, to Bessie B. Clarke, a daughter of John H. Clarke, also.

Hon. F. W. Borden is a man of fine executive ability, managing the affairs of the office of Militia and Defense at Ottawa, Canada, with vigor and discretion.

1523. MARIA F., born May 15, 1864.

792. ANDREW, Grand Prie, Nova Scotia.

1524. THOMAS ANDREW, born October 29, 1842.

1525. SOPHIA AMELIA, born July 29, 1844, married Edward McLatch 1865.

1526. ROBERT LEARD, born July 26, 1854, member of House of Commons at Ottawa, Canada.

1527 JOHN WILLIAM, born October 10, 1856.

1528. JULIA REBECCA, born November 18, 1858.

1529. H. C. BORDEN, Barrister at Halifax in 1896. He was elected Chief Justice of the Province.

795. THOMAS

1530. JUNIETTA, born 1852, died 1854.

1531. PERRY AI ELBERT, born 1854.

1532. WILLIAM ALBERT, born 1854.

1533. AMANDA, born 1856.

1534. THOMAS JAMES, born 1859.

775. LEMUEL PERRY, Horton, N. S.

1535. ANNIE MARIA, born January 24, 1846, married October 28, 1863. Daniel S. Borden, son of Thomas Borden, who left Rhode Island during the revolutionary war and joined the British army and went to Canada.

1536. LEMUEL PERRY, born June 18, 1848, died May 14, 1895, in California. He married Lalia Bishop November 27, 1872.

1537. MARTHA DICKEY, born June 6, 1850, died June 3, 1856.

1538. ARTHUR CUMMINS, born May 9, 1858, married Annie Alcorn July 17, 1889, and went to Japan. His address is Toyo Elwa Gakko, Tokio.

777. DAVID HENRY, Horton, N. S.

1539. ELIZABETH E., born December 17, 1838; died April, 1842.

1540. ELIZABETH ANN, born April 7, 1841.

1541. CHARLES ALFRED, born May 30, 1843.

1542. AUGUSTA MARIA, born September 30, 1851.

797. JOSHUA W., Horton, Nova Scotia.

1543. CHARLOTTE ANN, born May 25, 1838; married Garland Cox November 17, 18758.

1544. WILLIAM JOSHUA, born April 30, 1840.

1545. MATILDA AMELIA, born April 29, 1842.

EIGHTH GENERATION.

1546. GEORGE FREDERICK, born August 2, 1844; married Mary Vaughn, Avonport, Nova Scotia.

1547. EDWARD PERRY, born July 17, 1846.

1548. JAMES MARTIN, born November 18, 1848.

1549. CASSIE BURBIDGE, born December 29, 1850, married Alfred Gilmore.

1550. HERBERT HUNTINGTON, born April 19, 1853.

1551. CAROLINE OLIVER, born July 28, 1856.

1552. ELLA ALBERTA, born January 19, 1858, married Abel Woodman.

1553. ARTHUR HENNIGER, born March 31, 1861.

798. GEORGE NEWTON, Avonport, Nova Scotia.

1554. ROBERT ALLISON, B. A., Barrister, born February ., 1846, married Annie Smith Oct. 11, 1870. He is City Solicitor at Moncton, New Brunswick, and received the degree of A. B. at Mount Allison University, Sackville, N. B.

1555. BYRON CRANE, B. A., principal Sackville College, Nova Scotia, married Alice Black in 1880. He is a graduate of Mount Alison University, Sackville, New Brunswick, where he received the degree of A. B.

1556. LILA LAVINIA, born March 8, 1859; married Ruben S. Smith, Soomerville, Mass.

799. SILAS HIRAM, Chelsea, Mass.

1557. WILLIAM PERRY, born February 19, 1843.

1558. AMANDA, born September 14, 1844, married George Lockhart January 14, 1863.

1559. BESSIE, born March 9, 1846, married Frederick Crocker of Providence, Rhode Island.

1560. ELFRIDA, born August, 1848, married William Darling.

1561. AUGUSTA, born May 10, 1852.

1562. MIRIAM ROSETHA, born February 29, 1854, married J. Kincaid, Chelsea, Mass.

1563. MORTIMER, born February 29, 1854, twin witn Miriam R.

801. WILLIAM, Horton, Nova Scotia.

1564. OLIVIA ANN, born March 17, 1827, married Leander Rand March 19, 1851.

1565. REBECCA, born May 9, 1828; married Arnold Burbidge, February 4, 1857.

1566. LYDIA, born August 16, 1829, married John Thomas, March 21, 1860.

THE BORDEN FAMILY.

1567. RUFUS, born April 16, 1832.

1568. SAMUEL, born December 1, 1833, died January 27, 1856.

1569. JONATHAN, born March 17, 1836.

1570. JOHN NEWTON, born March 13, 1838.

1571. HENRY POPE, born May 4, 1840.

1572. MARGARET ELIZABETH, born August 30, 1842, married Philip Rand.

1573. RICHARD HARDING, born December 16, 1844.

1574. CHARLES ALEXANDER, born March 23, 1847.

1575. LAVINIA JANE, born March 30, 1849.

808. JOHN ALEXANDER, Nova Scotia.

1576. MARGARET ELIZA, born October 10, 1859.

1577. JOHN NELSON, born September 23, 1861.

812. JOHN WELLS, Cornwallis, Nova Scotia.

1578. HENRY ALBERT, born November 12, 1838.

1579. CHARLES EDWIN, born December 15, 1839.

1580. MARTHA ELIZABETH, born April 2,6 1841, married Silas P. Tupper November 25, 1858.

1581. PRUDENCE LAVINIA, born May 24, 1847.

814. BENJAMIN, Cornwallis, Nova Scotia.

1582. MARTHA A., born August 1, 1844, died in infancy.

1583. ELIJAH C., born August 15, 17845, married Ruby Ann Cox March 1869.

1584. BENJAMIN H., born February 21, 1847.

1585. JAMES R., born December 25, 1851, died young.

1586. ANNIE E., born December 17, 1855.

815. JUDAH, Cornwallis, Nova Scotia.

1587. FREDERICK PERRY, born May 8, 1837.

1588. PAULINE LAVINIA, born November 22, 18378, died November 28, 1842.

1589. MARY ELIZABETH, born July 6, 1841.

1590. ADELAIDE, born February 11, 1844, died young.

1591. JAMES JUDAH, born January 29, 1846.

1592. SIDNEY ENNIO, born April 26, 1847.

1593. ANN, born April 3, 1851.

EIGHTH GENERATION.

1594. FRANCIS, born February 15, 1853.
1595. AUGUSTA MAUDE, born September 14, 1855.
1596. CLIFFORD, born May 7, 1858.

828. JOHN D., Cornwallis, Nova Scotia.

1597. CLEMENT A., born March 18, 1851.
1598. FRANCIS M., born October 14, 1855.
1599. WILFORD, born February 13, 1857.
1600. HENRY HAVELOCK, born February 1, 1859.
1601. ALTHEA MAY, born February 3, 1861.
1602. BARRY, born November 24, 1862.

829. SAMUEL SMITH, Cornwallis, N. S.

1603. ALFRED CRANE, born March 7, 1857.
1604. CLARENCE HILBERT, born July 12, 1860.
1605. CLIFFORD AUBREY, born January 21, 1864.

835. ASA, Phillipsburg, Canada.

1606. WILLIAM C., born August 18, 1821, wife's name not known.
1607. JOHN, born October 24, 1822.
1608. AUGUSTUS V., born October 2, 1826, left no children.
1609. ROMEO V., born September 30, 1828; lives in Reno, Nev.
1610. EDITHA M., born August 8, 1831, married Gibson Lind in Wisconsin.
1611. SOCRATES, born February 27, 1833, left no children.
1612. LUTHER, born July 30, 1835; married Amanda.

836. SILAS, Bedford, Quebec.

1613. WILLIAM, married and left one daughter at Bedford Province, Quebec.
1614. ANSON left no children.
1615. SARAH, married —— Gustin, lives at East Pepperill, Mass.
1616. JOHN left one daughter in Iowa.
1617. SILAS.
1618. ALBERT, killed in battle in Shenandoah Valley.
1619. ELLEN, married —— Beek, lives at Bedford, Province of Quebec.
1620. EDGAR, left one son, lives in Michigan.
1621. ELIAS, lives in Hartford, Connecticut; has three children.

THE BORDEN FAMILY.

837. SAMUEL, Philipsburg, Quebec.

1622. HORATIO, has one son, Sanford, lives at Philipsburg, Province
1623. ELIZA.
1624. WILLIAM,
1625. DELVINA, left some children in Massachusetts.
1626. AMOS, lives in Massachusetts.
1627. HULDAH,
1628. MARTHA PAULINE.

838. JAMES, Province Quebec, Canada.

1629. KEZIAH, lives in Chelsea, Mass.
1630. WILLIAM H., lives in Massachusetts; was born April 27, 1841.
1631. GEORGE, born November 9, 1855; died April 26, 1890.
1632. CORNELIA, died April 4, 1872, aged 19 years.
1633. NELSON, lives in Boston; born April 13, 1858.
1634. OLIVE,
1635. CHESTER, born March 2, 1843; died September 2, 1848.

839. GAIL, Oakland, Cal.

1636. GAIL,
1637. AMANDA,

842. NATHANIEL, Montreal, Can.

1638. HENRY, lives in Montreal,
1639. LUCY,
1640. HATTIE, married; lives in Massachusetts.
1641. ADELBERT,
1642. MARY,

943. AULDIS, Fall River, Mass.

1643. HENRY C., born September 27, 1835; married Tabitha M. Angel, May 20, 1855.
1644. JAMES A., born June 26, 1838; married Dora P. Barber, November 7, 1859.
1645. GEORGE R., born April 4, 1840; married Mary A. Barber, October 17, 1859.
1646. ANN M., born June 4, 1843; married Henry N. Capwell, April 13, 1863.
1647. MARY E., born November 24, 1846.

EIGHTH GENERATION.

1648. AULDIS, born May 3, 1849.
1649. LUTHER C., born January 16, 1851.
1650. NANCY M., born October 5, 1854.

845. JOHN H., Scituate, R. I.

1651. JULIET, born January 30, 1845; died August, 1849.
1652. JOHN H., born August 5, 1849; died September 12, 1850.
1653. RUTH A., born May 7, 1854; died September 3, 1854.
1654. ALLEN M., born August 4, 1859; died September 22, 1859.

847. JOB, Scituate, R. I.

1655. ARTHUR,
1656. JOHN, had a son, Abraham.
1657. WILSON W.,
1658. SARAH D.,
1659. JULIA,
1660. CLARA D.,

848. EARL D., Scituate, R. I.

1661. EUPCRA, born July 6, 1849.

849. KNIGHT H.,

1662. ISADORA F., born August 22, 1844.
1663. ELVIRA, born August 17, 1850.

850. RICHARD E., Scituate, R. I.

1664. CLARA J., born May 2, 1856.
1665. BETSEY P., born November 13, 1858.
1666. ANNA M., born November 16, 1861.
1667. WAITY MARIA, born May 22, 1865.

853. HARLEY P., Scituate, R. I.

1668. LOUISA M.,
1669. WAITY ANN,

854. WILLIAM,

1670. WILLIAM HENRY, born April 30, 1844.
1671. NANCY M., born June 16, 1848.
1672. ISAAC C., born February 9, 1850.
1673. SARAH A. L. E., born January 21, 1852.
1674. SAMOS D. W., born July 2, 1854.

THE BORDEN FAMILY.

1675. FRANKLIN J. A., born February 22, 1856.

862. GAIL, Elgin, Ill.

1676. MARY, born in Galveston, Texas, December 24, 1829; died August 28, 1833.

1677. HENRY LEE, eldest son of Gail and Penelope Borden, was born January 18, 1832, in Egypt, Wharton county, Texas. When he was yet a young man he identified himself with the condensed milk business, and is now president of the Borden Condensed Milk Company, a position held by his father and brother for so many years. His home is at T nti, Ill., where he has a large stock farm, which he calls his "plaything," and on which he spends his leisure hours. Mr. Borden lost his first wife and only child many years ago, and in 1886 married Mrs. Retta Buckley, a widow with two sons, Gerald M., and Bert, now adopted.

The monument Mr. Borden has erected to himself is in the hearts of those who know him. His lovable, whole-souled nature endears him to all with whom he comes in contact. Generosity and unselfishness characterize his life. He distributes his ample fortune so quietly and unobservedly as to be unknown even to those near him. If a record of all his kindly deeds was written out, verily a larger book than this would be needed. His generosity made this book possible.

In connection with this brief sketch of the president of a great company, it may not be inappropriate to print the following, taken from the American Grocer of May 10, 1899, under the head of "Genuine Progress:"

"In these days of trusts and combinations it is pleasing to learn of a genuinely progressive step by a representative concern in an important line of business. The first organization of Gail Borden's valuable invention took the title of Borden's Condensed Milk Company, and some years later a new organization was formed to take its place, which has been long and favorably known as the New York Condensed Milk Company. The Civil War precipitated upon the company very heavy demands for its product, and the northern armies were supplied as extensively as manufacturing facilities would permit. It was a happy day when the Southern soldiers were lucky enough to get access to a Union supply train, and thereby obtain what was eagerly sought for by all the soldiers, namely Borden's Eagle Brand Condensed Milk.

"During the late war with Spain, among the first food supplies to follow the American troops, whether in Cuba, Porto Rico or the Philippines, was Borden's Eagle Brand Condensed Milk, and the Red Cross outfit was not considered complete without a liberal stock of the Eagle Brand Condensed Milk and Peerless Brand Evaporated Cream. These popular brands of milk have been the standby of the day and the most valuable food products which the soldier in the field or the

HENRY LEE BORDEN.
1677.

sailor at sea can obtain. Since the days of Dr. Kane all the exploring parties and expeditions have deemed their most important item of food to be taken with them the Eagle Brand Condensed Milk. Lieut. Peary only recently testified very highly of its value, and his present expedition was well supplied with this indispensable article of food.

During the forty years of constantly increasing business the methods of the company have commanded not only the highest respect of the jobers, retailers and consumers, but of its competitors as well. The fame of the Gail Borden Eagle Brand has become world-wide, and represents the highest standard in this line of goods. We are pleased to observe the aggressive reorganization of the company under the title of Borden's Condensed Milk Company, capital $20,000,000. This entire amount of stock was taken by the stockholders of the old company, except a portion which was set aside to be sold to some of the prominent and most valued employés of the company. The increased capital insures the company being able to take care of its business upon a scale of confident expectation that has always placed this concern so far in advance of its competitors.

"Constant additions to its manufacturing facilities have been made year by year to keep pace with the increasing demand. New factories have been added, until now the company operates fifteen, among which are some that outrank in size, capacity and actual product manufactured therein any other plants of similar nature in the world.

"We are sincerely pleased to note that the reorganization is formed on broad-minded lines, and that the management will remain as heretofore. We congratulate it upon its past success and well-earned reputation for energy, integrity and discretion. A concern which transacts its business on the lines which have been followed by the New York Condensed Milk Company during its history deserves unqualified success, and its policy will surely win, and profitably to all concerned—jobbers, retailers and consumers, as well as the army of men which it employs."

1678. MORTON QUINN, born September 10, 1834; died October 31, 1846.

1679. PHILADELPHIA WHEELER, born May 24, 1837; died August 17, 1880, in Houston, Texas. She married Judge Jehu Warner Johnson, born August 10, 1830, in Clarksburg, W. Va.

1680. STEPHEN F. AUSTIN, born June 5, 1839; died March 24, 1844.

1681. MARY JANE, born September 28, 1841; married in 1859, Mills S. Munsill at Winchester, Conn. He died in Hartford, Conn., 1887.

Mrs. Munsill's generous, noble nature is well known at Hartford, Conn., and by a large circle of friends. Many there are who will rise up and call her blessed. She has given efficient aid in this record, prompted by her love of family.

1682. JOHN GAIL, born January 4, 1844; died October 20, 1891, at Ormond, Fla.; married Ellen L. Graves, daughter of Dr. Lewis Graves and Adaline Janes, December 14, 1865; Ellen L. Graves was born at Seneca Falls, N. Y., September 7, 1845.

John Gail Borden succeeded to the presidency of the New York Condensed Milk Company on the death of his father, Gail Borden, the inventor, and occupied the position for ten years, until 1884. A genius for invention very early showed itself in him, and found ample opportunity for utilization in the requirements of the large business under his management. Mr. Borden was a man of indomitable energy, with a large capacity for details. Courtesy and gentleness were his marked characteristics. His hat was always removed as quickly for a little girl as for a lady. One of his unwritten mottoes was "what is worth doing at all it worth doing well." Mr. Borden believed in rest and vacations for others, but not for himself. He worked incessantly, and was old before his time, dying at 47. He is buried on Home Farm, at Wallkill, N. Y.

863. THOMAS HENRY, Galveston.

1683. JOHN BOLDEN, born December 7, 1832, in San Felipe, Austin county, Texas; died 1864; he married Mary Jane McKee in 1855; entered the Confederate army in 1861, enlisting in Taylor's Battalion, with which he served for a year on the Rio Grande, when he returned home, raised a company and again entered the service, and died in it in 1864.

1684. JAMES COCHRAN, born January 18, 1835, at Richmond, Fort Bend county, Texas. He was educated at the Western Military Institute at Drennon Springs, Ky., being a student at that institute at the time James G. Blaine was a tutor there. Returning to Texas after completing his education, he engaged in railroading as civil engineer until 1858, when he went into the live stock business, and was so engaged until the beginning of the Civil War. He raised a company in the fall of 1861, of which he was elected captain, and which was enlisted into the Confederate service as Co. D, First Texas Cavalry. After a year's service on the Rio Grande his command was transferred to the coast country, and later to Louisiana, where it took part in the series of engagements, following Banks in the Red River campaign. Capt. Borden was dangerously wounded in the battle of Mansfield, April 8, 1864, which disabled him from further service. After the war he resumed the live stock business in Jackson county, until 1873, when he removed to his old home in Galveston, Texas. He was married November 9, 1858, to Palmyra Atkinson at Victoria. She was the daughter of Dorsey and Mary (Patton) Atkinson; born in Williamson county, Tenn., November 9, 1836; she died February 8, 1870; he married a second time Mrs. Jennie McClanahan nee Harrison; she died in Galveston, September, 1881; no children by this marriage; on May 3 1883, he married Clara V. Arnold of Galveston, Texas; she was the daughter of Lloyd and Mary (Smith) Arnold.

JOHN GAIL BORDEN.
1682.
At 30 years of age.

EIGHTH GENERATION.

864. PASACHAL P., Egypt, Wharton county Texas.

1685. MILAM, born August 24, 1844; died July 23, 1875; he married Ella Underwood, October, 1873; she was born in Columbia, October, 1852.

1686. GUY born November 6, 1846, at Richmond, Texas; married Fannie Quarles, daughter of Dr. James W. and Eliza (Cleveland) Quarles in Shreveport, La., September 18, 1883.

1687. WILLIAM JOSEPH, born January 1, 1849, in Richmond, Texas; he died in New York city July, 1898; married Emma Eliza Graves at Galveston, Texas, May 20, 1880. William Joseph Borden was blind from his birth; he was educated at Rochester, N. Y., and was a man of unusually brilliant qualities of mind.

865. JOHN PETTIT, Borden, Colorado county, Texas.

1688. THADDEUS H., born 1844; died 1863, from wounds received in the Confederate service.

1689. SIDNEY G., born March, 1846; married at San Patricio, Texas, Mary Sullivan, December 28, 1876; she was the daughter of John and Eliza Sullivan; they reside at Sharpsburg, Texas.

1690. JOHN T., born January 18, 1848; died November 7, 1873; was accidentally killed by a fall from a horse.

1691. GEORGE G., born 1850; died 1852.

1692. FIDELIA, born December, 1853; married William J. Duffel, 1870.

1693. MARY ELIZABETH, born June, 1855.

1694. ALICE G., born June, 1858; died 1872.

1695. KATE, born 1860; died 1866.

1696. LEE DE WITT, born 1865; April 5; married Mary A. Green, daughter of James W. and Susan (Cunningham) Green; she was born in Louisville, Ky., December 24, 1871.

867. SQUIRE, Minneapolis, Minn.

1697. LUTHER GRANT, born February 25, 1865; married Lillian V. Holt, August 8, 1888.

1698. IDA MINNETTA, born December 29, 1867; married Henry Leslie Blethen, February 21, 1884.

1699. ADMIRAL FARRAGUT, born August 31, 1869; married Marie Royer, in Santa Ana, Cal., May 9, 1898.

1700. EVA GERTRUDE, born December 12, 1873; married E. E. Marsall of Pomona, Cal.

1701. FREDERICK, born June 10, 1880; died February 18, '881.

875. THOMAS, Fall River, Mass.

1702. DANIEL, born October 1, 1798; was drowned young.

THE BORDEN FAMILY.

1703. HANNAH, born June 5, 1801; married George Crocker, January 1, 1822.

1704. ELIZA, born June, 1803; died in infancy.

1705. ANN, born September 21, 1806; married (1) Isaac Fowler; (2) Ezra Marvel.

877. GARDNER, Fall River, Mass.

1706. ROBERT, born September 11, 1803; never married.

1707. DANIEL, born October 8, 1805; married Mary B. Jenny April 24, 1834.

1708. HANNAH, born January 7, 1805; died May 30, 1827.

1709. MARCY, born October 20, 1809; married Thadeus Perry, New Bedford, September 27, 1835.

1710. BENJAMIN, born September 11, 1812; married Nancy G. Spooner, March 9, 1842.

1711. SARAH, born October 10, 1814; married Capt. Weston Jenny, May 27, 1838.

880. THOMAS, Fall River, Mass.

1712. PATIENCE, was drowned at the age of 12 years.

1713. BRADFORD D., married Charlotte Evans.

1714. HANNAH, died in infancy.

1715. SARAH ARLINDA,

1716. LUCY,

1717. CASSANDRA,

1718. LOUISA,

1719. MARY, died in infancy.

1720. BENJAMIN, died in infancy.

1721. SARAH, married Jonathan Harris.

1722. STEPHEN, married Nancy Burns of Chilocothe.

1723. SAMUEL, married (1) Sarah Church; (2) Sarah Workman; (3) Ruby Bishop; (4) Almira Keyser.

1724. BENJAMIN, married Caroline Thatcher.

1725. PATIENCE, married Allen Butler.

1726. MAJOR, married Elizabeth Kinsey.

1727. THOMAS, married Mary Jane Evans.

1728. HENRY,

1729. OLIVER,

EIGHTH GENERATION. 251

883. STEPHEN, Fall River, Mass.

1730. JOSEPH, married Eliza Walker, 1834.
1731. STEPHEN.

884. SAMUEL, Tiverton, R. I.
1732. BENJAMIN,
1733. CHARLES,
1734. MARY,

891. JOHN, Fall River, Mass.

1735. PATIENCE, married David P. Davies of Somerset, Mass., July 1848.

898. STEPHEN, Fall River, Mass.

1736. PRUDENCE, born 1807; died 1834.

1737. ROBERT E., born 1809; died 1859; married Rebecca Wade, 1832.

1738. ELIZA D., born 1811; married Harrison G. C. Hearsy of Grafton, November 29, 1832.

1739. PHILIP II., born May 20, 1813; married (1) Lucinda Morse; (2) Ann Durfee.

1740. THOMAS E., born December 14, 1815; married in New Bedford, Mass., Harriet Davis, November 15, 1840.

1741. MARY E., born March 26, 1818; married Charles Edmons, November 5, 1840.

1742. SALLIE C., born October 26, 1820; married Gen. Borden, March 15, 1845.

1743. FANNIE D., born February 9, 1825; died 1846.

1744. MARTHA H., born April 27, 1826; married Charles R. Cobb, January 25, 1847.

1745. BENJAMIN, born May 26, 1830; died at sea.

893. ISAAC, Fall River, Mass.

1746. HANNAH, born December 26, 1811; married Richard Smith of Bristol.

1747. HARRIET, born October 17, 1815; married Josiah Talbot.
1748. ISAAC, born October 14, 1818; died July 29, 1825.
1749. JULIANN, born October 5, 1821; died December 24, 1838.
1750. SARAH, born June 19, 1824; died September 10, 1861.
1751. ELIZABETH, born November 7, 1827; died April 27, 1842.

894. AMOS, Fall River, Mass.

1752. CLARISSA, born May 17, 1813; married Frederick Winslow, July 15, 1841.

1753. JOHN H., born December 13, 1814; married (1) Susan Albro; (2) Sarah Anthony.

1754. ISAIAH C., born April 11, 1816; died July 27, 1847.

1755. JAMES, born March 11, 1818; died August 13, 1842.

1756. THOMAS, born September 11, 1819; married (1) Jane Thomas; (2) Lucy Harrington.

1757. MARY, born September 29, 1821; married Thomas Winslow.

1758. ISAAC NEWTON, born April 6, 1826; died May 10, 1846.

1759. PELEG, born January 31, 1828.

897. BRADFORD, Fall River, Mass.

1760. FREDERICK P.,

898. STEPHEN, Fall River, Mass.

1761. SYLVIA W., born December 21, 1817; married William F. Almy, July 28, 1840.

1762. ISAIAH, born September 24, 1819; married in Fall River, Sarah Westgate.

1763. RUTH, born June 25, 1821; died in infancy.

1764. JOHN L., born October 29, 1822; married (1) Sarah Haskell; (2) Agnes Curran, March 3, 1860.

1765 STEPHEN, born March 21, 1826; married in Fall River, Sarah P. Brayton.

903. JEREMIAH, Fall River, Mass.

1766. JEREMIAH.

904. GEORGE, Fall River, Mass.

1767. LUCY J., born July 21, 1832; married Charles W. Bennett of Providence.

1768. MARY ANN, born August 13, 1834; married Capt. John B. Borden, October 1, 1857.

1769. SARAH R., born November 1, 1836; married J. P. Norvel, November 23, 1853.

1770. GEORGE H., born November 25, 1840.

1771. CATHERINE E, born March 15, 1846; married John G. Hambley, December 24, 1867.

1772. AMEY MARIA, born April 18, 1853.

906. ADAMS, Fall River, Mass.

1773. JAMES, born May 30, 1833; died young.

1774. SPENCER, born December 9, 1834.

1775. ALBERT T., born May 8, 1838; died in infancy.

1776. JOHN A,. born July 19, 1841; married Susan B. Watts.

EIGHTH GENERATION.

1777. JOSEPH, born February 13, 1846.
1778. JUSTINIA, born December 14, 1853.
1779. SILAS HENRY, born December 14, 1855.
1780. MARY EMMA, born October 4, 1858.

910. GENERAL, Fall River, Mass.

1781. NATHAN D., born November 22, 1846.
1782. PHILIP H., born April 9, 1850.

919. JAMES, Fall River, Mass.

1783. LEANDER A., born August 2, 1826; married Persis S. Hambley.
1784. ANDREW M., born June 18, 1828; died at New Orleans, December 26, 1848.
1785. ABBY JANE, born June 2, 1831; died December 25, 1851.
1786. SAMUEL, born November 29, 1834; died June 16, 1854.
1787. GEORGE H., born October 2, 1837; married Abby D. K. Borden, daughter of Philip, November 29, 1865.
1788. ELIZABETH A., born September 17, 1841.

922. JOSEPH.

1789. CHARLES LOTT, married Abby Bullock, November 30, 1860.

924. ALEXANDER, Fall River, Mass.

1790. ABNER D., born February 4, 1839.
1791. JAMES, born May 5, 1846.

925. CHARLES L., Fall River, Mass.

1792. AMANDA M., born December 20, 1841; married Nicholas Taylor, January 4, 1860.
1793. CHARLES A., born November 27, 1837; married at Fairhaven, Mass., Rebecca H. Lawton.
1794. HANNAH H., born September 24, 1844; died July 24, 1866.
1795. ELIZA ANN, born May 23, 1850.
1796. WILLIAM S., born April 20, 1854.
1797. JOSEPH, born December 27, 1861.

934. ABRAHAM, Fall River, Mass.

1798. ELIZA GIBBS, born February 13, 1822.
1799. JOHN JAY, born May 5, 1824.
1800. HORATIO NELSON, born January 10, 1828.

THE BORDEN FAMILY.

935. MAJOR, Fall River, Mass.

1801. SARAH E., born September 5, 1829; married Peleg Brightman, January 26, 1852.

1802. EMELINE A., born January 2, 1831; married Pardon Macomber, May 1, 1850.

1803. MARY P., born March 5, 1834.

1804. ARTHUR R., born April 20, 1836; married Sarah Jane Gunn, November 15, 1860.

1805. ASHIEL M., born June 8, 1838; married Maria J. White, August 19, 1862.

1806. CAROLINE H., born July 20, 1841; married David Waring, January 1, 1863.

1807. LOUISA M., born April 9, 1846; died in infancy.

938. LE FAVOUR, Fall River, Mass.

1808. MARY ANN, born July 26, 1832; married James Waring.

1809. ROGER W., born August 20, 1834.

1810. JOHN D. SCOTT, born December 31, 1835; died young.

1811. EUEL CHANNING, born February 12, 1838.

1812. DAVID PERRY.

939. ISRAEL, Fall River, Mass.

1813. HOLDER, born October 24, 1845.

940. ABEL.

1814. AMANDA M., born August 28, 1839.

1815. JULIANN, born April 17, 1844; died young.

1816. LUCY, born October 5, 1845.

1817. JULIA CLARA, born November 29, 1850.

946. EMELINE.

1818. MARY COOK LE VALLEY, born December 17, 1827; married Emer Smith Lovell, August 13, 1849.

1819. ARTHUR FENNER LE VALLEY.

1820. ADELAIDE REMINGTON LE VALLEY,

1821. WILLIAM PARKER LE VALLEY,

1822. BENJAMIN WARREN LE VALLEY,

1823. SARAH FRANCES LE VALLEY,

1824. JAMES GARLAND LE VALLEY,

EIGHTH GENERATION.

1825. EMELINE BORDEN LE VALLEY,

967. ARNOLD, Goldsboro, N. C.

1826. HARRIET MARIA, born October 20, 1827; married at Goldsboro, N. C., Dr. Charles Francis Dewey, brother of the father of Admiral George Dewey of Manila fame,; born September 12, 1825; died October, 1866.

1827. JAMES COLE, born June 18, 1829; died June, 1885; he was a captain in the Confederate army, Co. H, First North Carolina State troops; he married Mary Elizabeth Caruthers in 1852; she was born December 24, 1833.

1828. EDWIN BROWNRIGG, born July 5, 1831; married (1) Georgia Caroline Whitfield, born July 7, 1833; died December 7, 1872; after her death he married again.

1829. LOUISA, born 1835; died 1884; married Frank W. Kornegay.

1830. SARAH LAVINIA, born July 10, 1838; married Dr. John Miller, September 21, 1863.

1831. EUGENE ARNOLD, died in infancy.

1832. WILLIAM HENRY, born April 22, 1841, in Goldsboro, N. C.; he married (1) Sue Edmundson, October 15, 1863, daughter of William and Julia (Pipkin) Edmundson; (2) Julia Edmundson, October 23, 1873; after her death he married Hattie Kennedy, November 3, 1885.

981. THOMAS HENRY, Santa Ana, Cal.

1833 EMMA ADELIA, born October 9, 1859; married A. C. Black at Santa Ana, Cal.

1834. RICHMOND, born April 18, 1861.

1835. RUPERT, born April 18, 1861.

1836. ANNIE ELIZABETH, born May 21, 1863.

1837. WILLIAM HARRY, born February 10, 1867.

1838. CHARLES WEST, born September 22, 1871.

1839. ADA MAY, born April 4, 1879.

985. CHARLES HENRY, Cornwallis, N. S.

1840. MARY ARABELLA, born May 8, 1855.

1841. AUBREY NELSON, born June 30, 1856.

1842. ALICE ADELIA, born June 30, 1858.

1843. FRANK NOBLE, born August 25, 1858.

1844. EDGAR A., born February 23, 1861.

1845. WILLIAM LAWN, born February 3, 1864.

1846. FREDERICK CHARLES, born December, 1865.
1847. CLARA MAY, born June 3, 1866.
1848. JOHN ALFRED, born October 3, 1867.
1849. MINNIE ETTA, born January 27, 1872.
1850. OTTO EMERSON, born August 23, 1874.
1851. ETHEL ELENA, born December 4, 1878
1852. BLANCHE, born December 4, 1878.

989. SIDNEY D., Nova Scotia.

1853. HENRY F., born March 20, 1869.
1854. FREDDY W., born March 18, 1871.
1855. NAOMI M., born January 24, 1874.

994. EARL, Jr., Fall River, Mass.

1856. HENRY RUSSELL, born August 22, 1825.
1857. SAMUEL, born July 2, 1827.
1858. PHILANDER, born September 2, 1829; died young.
1859. MARSHALL W., born September 13, 1831.
1860. PHILANDER, born June 16, 1834.
1861. JOSEPH E., born April 12, 1840.

995. LAZARUS, Fall River, Mass.

1862. ABRAHAM E., born November 22, 1843; married Mary M Slocum.
1863. HANNAH LOUISA, born November 22, 1843; married Andrew Clinton Chace, born October 26, 1843; died June 11, 1879.
1864. ARNOLD B., born June 1, 1846; died young.
1865. CHARLES SUMNER.
1866. HOLDER, born December 12, 1850; died January 9, 1884.
1867. FRANKLIN BROWN, born May 14, 1853; married in Fall River, Mass., and has one son, Frank Nelson, born October, 1874.
1868. ABBY ARNOLD, born September 6, 1855.
1869. WILLIAM NELSON, born July 21, 1858.
1870. LAZARUS, born April, 1860; died in infancy.
1871. CLARA ELSBREE, born June 4, 1862.

1002. LABAN, Fall River, Mass.

1872. JOHN EARL, born July 20, 1824; married Sarah A. Harmer, May 15, 1846.

EIGHTH GENERATION.

1873. SIMEON, born May 2, 1833; married Eliza M. Lawton, November 4, 1856.

1874. EDWARD W. (REV.), married Jane Nichols.

1875. MALVINA, married Henry Hazlehurst.

1876. GILBERT, born ——; died 1863; married Elsie Velona Stanard, died April 12, 1843.

1877. AMANDA,

1878. ABBY, married —— Little of Fall River, Mass.

1879. JOSEPH D., married Anna Strickland.

1880. JAMES M., went to sea and was never heard from.

1881. REBECCA,

1882. HENRY, died young.

1883. ARIEL, died in infancy.

1003. MELVIN, Fall River, Mass.

1884. CAROLINE A., born July 20, 1828; married Simeon Nash.

1885. PHILANDER WILLIAM, born September 16, 1831.

1886. DANIEL, born July 25, 1833; died in infancy.

1887. ABBY D., born December 4, 1837.

1004. PHILANDER, Fall River, Mass.

1888. AMANDA M., born October 6, 1833; died in infancy.

1889. JIRUH S., born November 2,1835; m arried Sarah A. Washburn, October 23, 1862.

1890. EDWIN, born August 15, 1837; married Susan T. Macomber, April 3, 1862.

1891. EMILY V., born March 15, 1840; married Edward T. Gay, November 5, 1863.

1892. EMERSON F., born July 3, 1842.

1893. ARDELIA A., born September 7, 1845.

1894. EUGENE A., born April 20, 1847.

1005. LEANDER, Fall River, Mass.

1895. EVELYN, born October 7, 1836; died May 21, 1882.

1896. DELIA EDSON, born October 27, 1839; died August 28, 1840.

1897. THOMAS LAWRENCE, born July 9, 1841; died July 11, 1865.

1898. INEZ ELLA, born June 6, 1844.

258   THE BORDEN FAMILY.

1899. ERIC WARREN, born December 26, 1848; married Caroline C. Davis, October 22, 1873.

1900. OMER ELTON, born June 28, 1854; married Miriam Louise Slade of New York City, September 22, 1888.

1010. JEROME B., Somerset, Mass.

1901. GEORGE G., born 1844; married Ella Voorhees, March 30, 1866.

1902. FRANCIS COOK, born February 11, 1846.

1903. FLAVIUS JOSEPHUS, born September 13, 1850; married Ida E. Frahm at Vallejo, Cal., June 6, 1878.

1904. LIZZIE SEAVER, born November 13, 1859; married Charles E Chace, March 22, 1890.

1020. ISAAC, Fall River, Mass.

1905. JAMES A., born February 22, 1831.

1906. WILLIAM H., born September 10, 1834.

1025. WILLIAM N., Fall River, Mass.

1907. MARTHA M., born August, 1843; married John Laurence, January 15, 1860.

1908. ADONIRAM, born June 31, 1844.

1909. CELIA, born November, 1847.

1030. JOB, Fall River, Mass.

1910. CHARLOTTE, born August 10, 1851; died young.

1911. EMERETTA, born July 19, 1853.

1024. ABRAHAM E., Tiverton, R. I.

1912. ALBERT HENRY, born July 12, 1856.

1913. ALICE EVELINA, born July 28, 1860.

1061. OLNEY, Fairfield, N. Y.

1914. ISRAEL,

1915. TIMOTHY,

1916. VAN BUREN,

1917. OLNEY, born March 11, 1831, in Fairfield, N. Y.; married; resides at Jefferson Park, Colo.

1065. THOMAS F., Cazenovia, N. Y.

1918. LEVI B., born in Cazenovia, N. Y., May 29, 1835; died when 15 years of age.

1919. JOSEPH M., born in Cazenovia, N. Y., November 8, 1838; married Mary E. Mills, November, 1865, at Dundee, Ill.

EIGHTH GENERATION.

1920. THOMAS CLARKSON, born in Cazenovia, N. Y., December 7, 1842.

1921. HARRIET W., born September 8, 1844; died in infancy.

1922. MARY EMELINE LANGDON, born in Cazenovia, N. Y., October 5, 1846; married John H. Wells, son of John Hoxie Wells and Sarah, his wife.

1923. MINA L., born in Cazenovia, N. Y., June 2, 1849; died July 15, 1869.

1924. SEYMOUR SKIFF, born July 14, 1851.

1075. WILLIAM DEAN, Norwich, N. Y.

1925. FRANKLIN P., born December 7, 1853.

1926. ELMER D., born July, 1855.

1927. WILLIAM L., born February, 1857; killed by the railroad July 1, 1892.

1928. MARY M., born March, 1860; married Mr. Reynolds of Oakland, Cal.

1929. HENRY J., born February, 1862.

1930. GRANT S., born June, 1864; died in infancy.

1931. OWEN D., born June 20, 1865, at Norwich, N. Y.; married Frances Helen Comstock, daughter of Samuel Austin Comstock, May 26, 1886.

1932. PEARL A., born January 3, 1869; married Willard D. Ball, June 20, 1895. Mr. Ball is secretary of the Y. M. C. A. at Los Angeles, Cal.

1933. HATTIE A., born December 13, 1871.

1934. NORMAN T., born November 11, 1873.

1077. SAREITA, Tonica, Ill.

1935. GRACE L. HOXIE, born March 30, 1866.

1936. J. BENJAMIN HOXIE, born June 7, 1862; married Ella Velura Barker, September 4, 1894.

1050. THOMAS H., Fall River, Mass.

1937. JOHN, of the Wood Finishing Company.

1938. SARAH E., married R. Earl Bennett.

1939. ADA M.

1940. WILLIAM HERBERT, married in Boston, Mass.

1941. JUDITH A., married Oatis Gray.

1942. ANNIE L.

1943. NATHANIEL,

THE BORDEN FAMILY.

1081. JOHN, Chicago, Ill.

1944. WILLIAM, born August 6, 1850; married, and lives in Chicago.

1945. JOHN,

1946. JOSEPHINE,

1098. MARIA WEBB,

1947. WILLIAM HULL TELFAIR, born July, 1855; died 1884.

1948. JOHN S. TELFAIR, born August, 1857.

1949. HELEN H. TELFAIR, born November, 1858; married H. W. Clark.

1950. KATE J. TELFAIR, born April, 1860; died 1884; married W. Hallowes.

1951. ELIZA V. TELFAIR, born April, 1862.

1101. JOSEPH, Borden, Fresno county, Cal.

1952. MARGARET, born 1852; married Lovic Benjamin Pearce, October, 1882.

1953. RINNIE LEE, born 1857; married her cousin, Nathan Lane Borden.

1954. JOSEPH RUFFIN, born 1861.

1955. HARTWELL COCKE, born 1864.

1956. MARY GRAY, born 1866.

1957. CRAWFORD CLARK, born 1871.

1958. BESSIE, born 1874.

1959. THOMAS, born 1876.

1102. THOMAS JAMES, Pascagoula, Miss.

1960. NATHANIEL BARNETT. The following clipping is taken from the San Augustine News of June ——, 1896:

"Mr. N. B. Borden was born in Mississippi little more than thirty years ago, and possesses the strong characteristics of the Southern gentleman, together with the culture of a world-wide traveler and the learning of the university graduate. It was just after the ravages of the epidemic of 1888 that Mr. Borden appeared prominently in Florida. Fernandina was selected by him as his place of residence, and there he became actively engaged in the lumber business, which has proved such a successful venture. For many reasons he is one of the best-known men in the State, and has the honor of representing as consul half a dozen of the great nations, including Great Britain and Germany. Mr. Borden was credited by the Spanish government with having instigated the present revolution in Cuba—the only one, by the

EIGHTH GENERATION.

way, that has ever given Spain any uneasiness. This allegation is denied by Mr. Borden and his friends, but causes them no end of amusement. Mr. Borden is a great club man, and is a member of some of the leading clubs in the country. Yachting and fishing are among his favorite pastimes, and he enjoys a generous income by entertaining friends on board his yacht or at his bungalow, at the southmost end of Amelia Island. During the present season Mr. Borden has favored St. Agustine with half a dozen visits and luxuriates in the best entertainment the Ponce de Leon can afford."

1961. RUFFIN GRAY, married —— at Moss Point, Miss.

1962. BESSIE BYRN,

1963. THOMAS CRAWFORD, born at Moss Point, Miss., June 6, 1871; he married Elizabeth Jeffreys, daughter of Mr. and Mrs. W. O. Jeffreys, of Fernandina, Fla.

Thomas Borden was appointed Consul for the Argentine Republic June 6, 1893, at that time the youngest consul in the United States. He took part with his brother, N. B. Borden, in furnishing the first filibustering expedition to the Cuban insurgents. The expedition was stopped by the United States authorities at Fernandina, Fla., before the boats were able to get out.

    1103. MIRANDA, Eutaw Springs, Ala.

1964. JAMES CLAY CLARK, born July 12, 1864; married Clara David of St. Louis.

1965. BENJAMIN BORDEN CLARK, born December 18, 1866; died January 1, 1867.

1966. JOSEPH WALLACE CLARK, born July 3, 1868.

1967. THOMAS JAMES CLARK, born January 18, 1871.

    1104. WILLIAM ALFRED, New Bern, Ala.

1968. THOMAS CHENEY, born June 28, 1872.

1969. BENJAMIN RUFFIN, born June 29, 1878.

1970. ALICE MOORE, born June 17, 1884.

    1105. MARY ESTHER, Eutaw, Green county, Ala.

1971. WILLIAM THOMAS CHENEY, born February 26, 1872; married J. H. McGiffert.

1972. EDWARD FENWICK CAMPBELL, born August 17, 1881, died July 29, 1898, in Eutaw, Ala.

1973. BORDEN MONTGOMERY CAMPBELL, born September 3, 1883.

1974. MARTHA C. CAMPBELL, born October 18, 1886.

    1078. JAMES PENNINGTON, Greensboro, Ala.

1975. PARHAM BENJAMIN, born May 20, 1882.

THE BORDEN FAMILY.

1976. JAMES PENNINGTON, born July 28, 1886.

1977. ALFRED RUTLEDGE, born October 28, 1891.

1108. ELIZABETH GRAHAM, New Bern, Ala.

1978. LOVIC PEARCE, born ——; married his cousin, Margaret Borden, daughter of Joseph.

1979. MARY HOPE PEARCE

1980. PENELOPE PEARCE.

1109. MARY JAMES, New Berne, Ala.

1981. DAVID BORDEN GRACE.

1110. HANNAH WARD, Montgomery, Ala.

1982. GEORGIE PEARCE, born September 18, 1858; died January 15, 1866.

1115. NATHAN LANE, Borden, Fresno county, Cal.

1983. RHODES, born 1885.

1116. SHELDON, Los Angeles, Cal.

1984. CECIL ALEXANDER, born April 10, 1887.

1985. HARRY INNIS, born May 10, 1889.

1986. JULIET RHODES, born December 3, 1891.

1117. IVEY LEWIS, Oakland, Cal.

1987. HESTER BELL, born January 13, 1891.

1121. JAMES WALLACE, Alabama.

1988. FREDERICK WALLACE,

1989. AUGUSTA NELD.

1122. SARAH COART, Talladega, Ala.

1990. LYDIA HUGHSON BURR.

1991. ANN AARONA BURR, married Alexander Blackburn.

1992. ZAIDEE LOUISE BURR, married Samuel Howard Henderson.

1993. ESTHER WALLACE BURR, married John B. Robinson.

1994. WILLIE MILTON BURR.

1995. LYDIA ANTOINETTE BURR, married James M. Hicks.

1996. BORDEN HUGHSON BURR, born in Talladega, Ala. Although quite a young man he is one of the most prominent orators of this state. He delivered a stirring address in his native town on Decoration Day, 1899; and greater things are expected of him.

EIGHTH GENERATION.

1124. BENJAMIN, New Bern, Ala.

1997. THOMAS RICHARDSON.

1998. MARGARETA ROSA, married April 27, 1898, George Peabody Ide.

1999. ANNIE LEWIS, married Shepard McGhee, October, 1897.

1125. THOMAS SYDENHAM, Alabama.

2000. WILLIAM ALFRED.

2001. MARY ESTHER,

2002. CAROLINE LUCILE,

2003. FANNIE SHELDON,

2004. WALLACE JONES.

2005. SARAH CORDELIA,

2006. ELIZABETH McGEE.

1126. JOSEPH LANE, New Bern, Ala.

2007. THOMAS SHEPPARD, lieutenant in the United States navy, stationed on the United Staes war ship Brooklyn. He was given the rank of first lieutenant by brevet in the Marine Corps for distinguished services in the battle of Santiago, on the third day of July, 1898. The report of Capt. Murphy, U. S. M. C., contains the following: "I cannot speak too highly of the conduct of Lieut. Borden. His courage and excellent service proved him a valuable officer." Lieut. Borden was also warmly commended by Capt. Cook of the Brooklyn. He has since been promoted in the "personnel bill" to the rank of captain. November 17, 1897, he married Frances Caroline Wheatly, the daughter of Hon. Samuel E. Wheatly, Commissioner of the District of Columbia during President Cleveland's first administration.

2008. ALICE ASHLEY,

2009. SHIRLEY MOSS,

2010. LANE BEAL,

2011. JOSEPH LANE,

2012. JULIA CUNNINGHAM.

1132. WILLIAM JAMES, New York City.

2013. JAMES WALLACE, born December 3, 1867; died October 9, 1881.

2014. ANNA HOPKINS, born July 27, 1870.

2015. WILLIAM TEFFT, born April 19, 1880; died February 1, 1881.

1133. DAVID HENRY, New York.

2016. FRANCIS NELSON, born May 6, 1891, in North Dakota.

2017. ESTHER WALLACE, born March 14, 1893, in Ohio.

1136. MARY FRANCES SHELDON, Stamford, Ct.

2018. KATHERINE WALTON WOOLSEY, born July 16, 1856; died August 26, 1888; married Frederick Hastings Hamilton, October 16, 1878.

2019. WOOLSEY R. HOPKINS, born November 19, 1867; married on October 8, 1891, Helen Birdsall, daughter of James Birdsall and Mariana Townsend, his wife.

2020. SHELDON HOPKINS, born May 29, 1870; married Ella Scribner, daughter of James W. Scribner and Margaret E. Miller, his wife.

1137. CATHERINE JOSEPHINE SHELDON, New York City.

2021. KATHERINE BORDEN FRANKLIN, born October 23, 1877.

2022. SHELDON FRANKLIN, born June 21, 1879.

2023. ROSE CLAIRE FRANKLIN,

1155. HON. WILLIAM JOSEPH, Oxford, Fla.

2024. EDWIN GOSSON, born August, 1858; married Caroline Moench, January 26, 1884.

2025. WILLIE C., born August, 1859; married Mr. Treadaway, of Newnan, Ga.

2026. MALBERT TROUP, born June 3, 1861; married Mildred A. Harris of Lynchburg, Va., December 16, 1890, at Rockmark, Polk county, Ga. She was the daughter of James M. Harris and Mildred A. McCulloch, his wife. He lives in Cedartown, Ga. He has made some inventions in textile machinery, such as improvements on machines that wind yarns, threads, twines, etc., that has almost revolutionized this particular line of the textile business, and has also made several improvements in the modern bicycle, while not yet on the market, the patents have been allowed and will soon issue from the Patent Office, when the same will be the property of the public. Also an invention to facilitate the handling of mails (mail bags) that in the opinion of men versed in this branch of the government service will save the government millions of dollars in the one item referred to, besides saving at least nine-tenths of the time now required in handling this part of the service. This invention is now (1899) in the Patent Office waiting the action of the department.

2027. PELHAM, born 1864; married Mr. Harper of Corsicana, Texas.

2028. NANNIE, born 1867; married Mr. Frey of Newnan, Ga.

2029. IRMIE BONNIE, born 1869; married Mr. Martin of Newnan, Ga.

EIGHTH GENERATION.

2030. JOSEPH, born 1871: died in infancy.

2931. FRANK, born 1871;died in infancy.

2032. BENJAMIN F., born November 3, 1875.

1157. ANDREW CAMPBELL, Dallas, Texas.

2033. LYDIA CATHERINE, born April 26, 1858; died March 12, 1859.

2034. HENRY ALLEN, born October 25, 1860; married Martha Buckingham.

2035. NANCY LORENA, born February 9, 1864; died May 5, 1865.

2036. LULU ELLEN, born March 19, 1866; died June 9, 1867.

2037. CHARLES LOUIS, born June 21, 1868.

2038. DORA LOU, born January 29, 1871; married in Italy, Texas, Mr. J. M. D. Trammel.

2039. EPHIE TATE, born October 24, 1873.

2040. JOHN PICKENS, born August 30, 1876.

1169. MARY J. VAN METER Bowling Green, Ky.

2041. JOHN J. COOKE, born October 30, 1839; married in Rushton, Minn., in 1873, Lulu Peray; he died at Devil's Lake, Dakota, March 12, 1896.

2042. MARTHA A. COOKE, born February 11, 1842; married November 23, 1864 to Capt. Daniel Heaney, born in Ireland about 1834.

2043. CHARLES L., born in Bowling Green, Ky., April 29, 1845; married Ellen Dahl at Grand Forks, North Dakota, January, 1895.

2044. SAMUEL C., born in Bowling Green, Ky., November 26, 1848; died in Evansville, Ind., January 17, 1854.

1170. JULIA A. VAN METER, Bowling Green, Ky.

2045. LUCY USHER HOBSON, born October 2, 1837; died March 19, 1838.

2046. MARY ELIZA HOBSON, born March 31, 1842; died March 15, 1843.

2047. ELLEN FRANCES HOBSON, born March 31, 1842; died March 15, 1843.

2048. WILLIAM EDWARD HOBSON, born January 8, 1844, in Bowling Green, Ky. He served with distinction in the Federal army, in the Civil War, as colonel of the Thirteenth Kentucky Regiment of Volunteers. After the war he was in the Internal Revenue service. He later edited the Bowling Green Republican. He was postmaster at Bowling Green, Ky., during President Hayes administration. He is now engaged in farming near Bowling Green. On the 5th of March,

THE BORDEN FAMILY.

1773, he married Ida Thunar, youngest daughter of Judge H. K. Thunar. They have six children.

2049. JONATHAN HOBSON, born December 8, 1844.

2050. JOSEPH VAN METER HOBSON, born May 8, 1848.

2051. GEORGE ATWOOD HOBSON, born September 23, 1864.

1171. CAROLINE EVE VAN METER, Bowling Green, Ky.

2052. WILLIAM USHER ADAMS, born January 30, 1843. Married in Bowling Green, Ky., Martha Clarkson, born in 1846.

2053. MARY LELAND ADAMS, born July 28, 1843; died March 2, 1893. She married February 1, 1865, John Jacob Hilburn of Bowling Green; he died October 29, 1877.

2054. SAMUEL TYLER ADAMS, born July 12, 1846; married November 5, 1878, Sallie Porter, who was born September 5, 1858; he was many years marshal of the city of Bowling Green, Ky. He died December 17, 1893.

2055. JULIA WOODBURY ADAMS, born January 2, 1849; married December 11, 1884, William R. Carson of Bowling Green, Ky., born January 18, 1843; died March 9, 1885.

2056. CHARLES JOSEPH ADAMS, born October 25, 1851; married December 14, 1873, Mary Z. Harrison, who was born February 13, 1853.

2057. GEORGE BRADLEY ADAMS, born October 7, 1853; married November 16, 1882, Fannie P. Allen, who was born January 4, 1856.

1172. SAMUEL KIRK VAN METER, Bowling Green, Ky.

2058. MARTHA USHER VAN METER, born February 28, 1861; died in 1862.

2059. CHARLES CLINTON VAN METER, born September, 1862.

2060. MARY USHER VAN METER, born October 1, 1865; married December, 1884, Eugene Miller of Kansas City, Mo.

2061. WILLIAM SHARP VAN METER, born September, 1867; married Ella McGinnis, September 2, 1891.

1174. SARAH FRANCES VAN METER, Bowling Green, Ky.

2062. JAMES VAN METER CLARKSON, born April 14, 1858; married Nannie Clarkson.

2063. CLINTON CLAY CLARKSON, born December 18, 1859; died March 18, 1864.

1183. ELIZABETH ANN SWEETLAND, Bowling Green, Ky.

2064. WILLIAM ALEXANDER OBENCHAIN, born April 27, 1841, in Buchanan, Botetourt county Va. After attending a preparatory

PRES. W. A. OBENCHAIN.
2064.

school in his native town, he entered the Virginia Military Institute as a cadet, and graduated there in 1861, at the head of a class of thirty-eight members. He entered the Confederate service as a second lieutenant of artillery, C. S. A., at the beginning of the Civil War, but was soon after assigned to duty with a corps of engineers, in which he remained throughout the war. He was promoted captain for meritorious service in 1864. Professor of mathematics and military engineering in the Hillsboro, N. C., Military Academy, 1866-68; professor of mathematics and commandant of cadets in the Western Military Academy, Newcastle, Ky., under Gen. E. Kirby Smith, 1868-70; professor of French and German, and commandant of cadets in the University of Nashville, Tenn., 1870-73; elected professor of mathematics in Ogden College, Bowling Green, Ky., in 1878, and president of the college in August, 1883. A member of the National Committee of the Body of Councillors, and associate member of the faculty of the American Institute of Civics; member of the National University Committee of One Hundred; for some time member of the American Academy of Political and Social Science, and of the British Economic Association; member of the vestry, and junior warden, Christ Church Parish, Bowling Green, Ky., and lay reader; secretary and one of the organizers of the XV Club of Bowling Green, Ky., an active and influential literary society, composed of some of the ablest and most prominent men of Bowling Green; and an honorary member of the American Whig Society, of Princeton, N. J. He was also one of the honorary vice-presidents of the Department Congress of Higher Education of the International Congress of Education of the World's Columbian Exposition in 1893. He married, July 8, 1885, Eliza Calvert of Bowling Green, where she was born, February 11, 1856. She has won considerable reputation as a writer of both poetry and prose, and is a prominent contributor to some of the leading papers and magazines generally over the nom de plume of "Eliza Calvert Hall," but sometimes, especially in newspaper and political articles, over her real name.

2065. FRANCIS GARDNER OBENCHAIN, born February 15, 1843, in Buchanan, Botetourt county, Va. He was a Confederate soldier throughout the Civil War, serving in a battery of light artillery. He was promoted for gallant and meritorious conduct in the battle of Port Gibson, near Vicksburg, in 1863; was in Vicksburg during the siege of that place, and was taken prisoner when it surrendered. At the close of the war he was in command of his battery, the Botetourt Artillery. Since the war he has been engaged in the grocery and flour commission business. On November 25, 1875, he married Anna L. Brown, youngest daughter of Col. A. S. Brown, of Memphis, Tenn. He is now residing at Cocoanut Grove, Fla.

2066. MARY MARTHA OBENCHAIN, born March, 1845; died March 11, 1846.

2067. SALLIE SWEETLAND OBENCHAIN, born February 12, 1847.

2068. JAMES THOMAS OBENCHAIN, born March 8, 1849, in Buchanan, Botetourt county, Va. On account of the Civil War, which began when he was only about twelve years of age, his educational advantages were limited, but he had good native ability. He had also a pleasant address, and was a good talker. A few years before his death he was the promoter and one of the three organizers of an association of the retail clerks, known as the Retail Clerks' Union of Nashville, Tenn., that led to their emancipation from long hours of service at night. His services were recognized by his election to the vice-presidency of the union, which place he held from the organization to the time of his death. He married Frances Lou Collins in Nashville, Tenn., May 4, 1884; and died in that city of Bright's disease, June 23, 1893. He left no children.

2069. LETITIA ANN OBENCHAIN, born June 27, 1851.

2070. CAROLINE OBENCHAIN, born October 14, 1853.

2071. MARGARET OBENCHAIN, born October 14, 1853.

2072. ALICE VIRGINIA OBENCHAIN, born March 9, 1856.

2073. LURA BELL OBENCHAIN, born August 13, 1860.

2074. FLORENCE MOFFITT OBENCHAIN, born October 31, 1864; died in Mytheville, Va., May 31, 1881.

1184. MARY HESTER SWEETLAND, Virginia.

2075. WILLIAM ELEAZER STRICKLAND, born in Botetourt county, Va. He went West with his father in early youth, entered the Confederate service in 1861 as a member of the Second Missouri Cavalry. He married, August 23, 1863, Margaret Rebecca Chinn, who was born April 10, 1842.

2076. SALLIE ANN WOODVILLE STRICKLAND, born May 4, 1842; married Peter B. Riffe, June 5, 1862.

2077. MARY GEORGE STRICTLAND, born 1843; she married Nash J. Evans, October 28, 1869; died April 17, 1870.

1186. SAMUEL McFERRAN SWEETLAND, Memphis, Tenn.

2078. VENITIA SWEETLAND, born August 19, 1849; died January 27, 1850.

2079. SAMUEL ROBERT SWEETLAND, born September 15, 1856, in Memphis, Tenn. He married December 18, 1879, Maggie Lowe of Giles county, Tenn.

1188. ISAAC VAN METER SWEETLAND, Hamblin, W. Va.

2080. JOHN SAMUEL SWEETLAND, born April 20, 1846, in Pattonsburg, Va. He was Sheriff of Lincoln county, Va., for years.

EIGHTH GENERATION.

2081. MARY HESTER SWEETLAND, born June 30, 1847, in Battonsburg, Virginia; married March 18, 1868, Theodosius Adolphus Love.

2082. ANNIE HENRY SWEETLAND, born July 11, 1848; married March, 1870, William C. Wiley.

2083. CHARLES RUSSELL SWEETLAND, born January 13, 1850; died March 17, 1850.

2084. ELIZABETH OBENCHAIN SWEETLAND, born March 21, 1851; married James Hill in 1873.

2085. MAGGIE POAQUE SWEETLAND, born October 22, 1855; married February, 1874, Thomas J. Haile, M.D.

2086. CARRIE VAN METER SWEETLAND, born June 3, 1857; died September, 1871.

2087. LEWIS ROLFFE SWEETLAND, born February 14, 1859.

2088. MARTHA WALKER SWEETLAND, born January 11, 1861; married in December, 1886, Silas W. Oxley, M.D.

2089. VIRGINIA WATSON SWEETLAND, born August 3, 1862; married February, 1883, Longimus M. Sanford.

2090. SALLIE REBECCA, born July 27, 1867.

1191. SALLIE E. SWEETLAND, Greenup county, Ky.

2091. HENRY POWELL, born January 21, 1862.

2092. MARY POWELL, born February, 1864; married a Mr. Foster.

2093. LUCY, born January 11, 1866.

2094. LILLY, born November 5, 1867.

2095. LUKE, born May 15, 1873.

1192. JAMES O. SWEETLAND, Nevada county, Cal.

2096. JEFFERSON DAVIS SWEETLAND, born February 11, 1857; died June 16, 1857.

2097. LURA VIRGINIA SWEETLAND, born March 17, 1858; married October 31, 1875, Stephen R. Heath.

2098. GEORGE LEE SWEETLAND, born March 10, 1861.

2099. CARRIE VIRGINIA SWEETLAND, born November 20, 1864; died February 8, 1870.

2100. WILLIAM ALBERT, born June 13, 1866.

2101. HENRY PRESTON, born June 13, 1866; died March 5, 1885.

2102. LAURENCE GALE, born July 9, 1871.

THE BORDEN FAMILY.

1193. CAROLINE SWEETLAND, Greenup county, Ky.

2103. WILLIAM LUKE WOLCOTT, born December 26, 1860.

2104. ALBERT SWEETLAND WOLCOTT, born September 6, 1863; died May 14, 1875.

2105. VIOLA LEE WOLCOTT, born January 29, 1865; married February 18, 1890, Martin Butler Wilson.

2106. MATTIE WALKER WOLCOTT, born September 15, 1867; married January 18, 1885, Emory Walden Foreman; she died June 29, 1890.

2107. LURA DELINDA WOLCOTT, born December 17, 1869; died February 25, 1870.

2108. ALANSON HOWE WOLCOTT, born July 26, 1872.

2109. ELLA VAN METER WOLCOTT, born May 27, 1875.

1194. MARGARET SWEETLAND, Greenup county, Ky.

2110. CHARLES EDWARD POWELL, born December 5, 1864.

2111. ANNE FOSTER, born July 2, 1866; married in Galatin, Tenn., January 15, 1899, Ernest E. Chrisman.

2112. MINNIE BELLE, born September 30, 1868.

2113. CARRIE EVELYN, born August 5, 1871.

1197. JOSEPH CARPER, Virginia.

2114. ELIZABETH CARPER, married —— Echols.

2115. ROBERT CARPER, a Confederate soldier, killed near Williamsburg, Va., in 1862.

2116. WYNDHAM CARPER,

2117. ARIANA WILLIAMSON CARPER, married Wm. B. Bean, Sept. 11, 1866.

1198. DANIEL,

2118. LEAH, born August 15, 1807; married Dr. William Shaw, December 17, 1826.

2119. REBECCA, born February 5, 1809.

2120. JOHN, born June 30, 1801.

2121. MARY ANN, born May 5, 1812.

2122. BETSEY, born April 15, 1814; died April 29, 1814.

2123. DANIEL J., born February 27, 1815; died July 19, 1887; married (1) Irene Ely Babcock; (2) Mary Louise Cline, May 6, 1852.

2124. LYDIA, born April 20, 1817; married William Nelson, died February 26, 1871.

EIGHTH GENERATION.

2125. ELEANOR, born November 25, 1819; married William Woodard, September 17, 1845.

2126. HARRIET, born July 24, 1821; married December 31, 1846, Joseph Brown.

2127. NANCY, born July 2, 1823; married, January 2, 1842, Chester W. Clark.

2128. JULIA, born August 11, 1824; died August 17, 1824.

2129. CATHERINE, born April 8, 1830; married William A. Wilmot, December 19, 1847.

2130. ELIZABETH, born December 3, 1831; married James Nellis; died in Chicago, Ill., April 23, 1898.

2131. NELLIS, born August 10, 1837; died May 50, 1884, at Louisville, Kentucky.

1210. COL. THOMAS JEFFERSON COX, Zanesville, Ohio.

2132. JULIET COX, born March 16, 1847; died June 25, 1851.

2133. EDNA COX, born March 19, 1848; died June 22, 1851.

2134. JANE BUCKINGHAM COX, born July 31, 1856; died October 19, 1879.

2135. WILLIAM VAN ZANDT COX, born June 12, 1852; married October 27, 1886, Juliet H., daughter of Hon. M. G. Emery of Washington, D. C.; she was born January 4, 1858.

2136. ELIZABETH DENNISSON COX, born February 3, 1854; married January 4, 1879, Theodore G. Sullivan.

NINTH GENERATION

NINTH GENERATION.

1220. ELVIRA, Philadelphia, Pa.

2137. JOSEPH HAMILTON LINNARD, born September 27, 1860.

2138. THEODORE BORDEN LINNARD, born December 3, 1862; married Edna Clara Thiesing, daughter of Fredrick H. Thiesing and Madeline Zier, his wife.

2139. LOUIS TAUEZIN LINNARD, born February 6, 1866; died October 10, 1893.

1221. THEODORE, Philadelphia, Pa.

2140. ANNA ELVIRA, born in Philadelphia, May 2, 1870; married David J. Meyers, September 25, 1884.

2141. HELEN PAGE, born November 25, 1873; married Frank J. Burns, October 13, 1898.

2142. HATTIE JANES, born March 30, 1875; died September 25, 1877.

1222. JOHN, Philadelphia, Pa.

2143. WILLIAM NOSTON, born January 22, 1859; married Eleanor Caldwell Wilson, March 5, 1885; she was the daughter of Joseph M. Caldwell and Eleanor Morrow.

2144. THEODORE JOHN, born February 6, 1861, in Philadelphia. He is the inventor of the Perfection Shaking Grate; married Sarah Powell, May 22, 1883.

2145. MARY MORROW, born April 18, 1863; married Frank Sheppard Harris, April 26, 1887.

2146. EMMA PAGE, born September 17, 1868; married Charles Pittman Sutton.

2147. ELVIRA LINNARD, born February 3, 1872; married Howard M. Kain, February 18, 1893.

2148. ELIZABETH NOSTON, born April 17, 1874; married John Dunlap, February 19, 1895.

1224. HAMILTON, Chicago, Ill.

2149. SARAH ARMSTRONG, born in Portland, Or., February 15, 1871.

2150. FRANCES COLBURN, born in Philadelphia, September 26, 1873; married George Carr Purdy of Middletown, N. J., January 17, 1899.

2151. JOHN HAMILTON, born in Chicago, Ill., September 23, 1878.

2152. ANNA PAGE, born in Chicago, Ill., September 3, 1880; died April 20, 1881.

THE BORDEN FAMILY.

1232. JOB, Burlington county, N. J.

2153. SHERMAN,

2154. HAMILTON,

2155. GEORGE WILLIAM, born October 27, 1867, in Burlington county, N. J.; married Emma Sutton, March 18, 1892.

1233. STEPHEN C. D., Edna, Kan.

2156. PETER, died in infancy.

2157. NELLIE, born October, 1876.

2158. LILLIE, born October, 1876; married Oscar Chaffee.

2159. THEODORE, born October, 1879.

2160. FRANK, born May, 1882.

2161. GEORGE, born April, 1884.

1235. CHARLOTTE, Bordentown, N. J.

2162. HOWARD E. CARR, born 1863; died March 10, 1899, in Bordentown, N. J.

1249. JOHN W., Manasquan, N. J.

2163. JOHN CURTIS, born December 20, 1874; died March 31, 1899. About two years ago while a student in Princeton College he fell a victim to that dread disease, consumption, but by prompt medical attendance and an extended sojourn in the South, it was thought, and appearances seemed to indicate, that the malady had been mastered. But such proved not to be the case, and in a short time it was seen by those about him that the symptoms were alarming. All that the best medical skill could do or suggest was done to ward off the disease, and loving hearts constantly provided everything most abundantly that could be thought of to afford him relief, hoping to conquer the disease that sought to destroy the one they loved so much. But it was to no purpose, and he passed away most peacefully to his rest, surrounded by relatives and friends who loved him dearly in life for his many manly and sterling qualities, and who bitterly mourned the close of a life so full of bright prospects for usefulness and helpfulness for those about him. The deceased had been a student at Princeton for a number of years, first in the preparatory school and afterward in the college. At the time of being taken sick he was in his third year at the college. He was a close student and was working hard to fit himself for the pulpit and the lecture platform. His early death, just at the seeming consummation of all he had hoped for and worked so hard to obtain, seems most inexplicable to mortal view, but no murmur escaped the lips of the young man during his long illness, and those who now sit in the deeper shadow which is over the home he has left, are conscious that it was the will of a kind and all-wise Father,

NINTH GENERATION.

and are still putting their trust In Him, knowing His way must be best."

1253. PHOEBE ANN, Emleytown, N. J.

2164. IDA D. EMLEY, born February 9, 1856; married Capt. Edward Godfrey, U. S. A. Cavalry.

2165. THOMAS N. EMELEY, born March 11, 1858.

1254. JOSIAH, New Jersey.

2166. WILLIAM H., born July 26, 1863; died December 17, 1865.

2167. SAMUEL A., born ——; married June 6, 1895, Etta Poinsett.

1255. MARY R., Emleytown, N. J.

2168. EUGENE EMLEY, born June 4, 1855; married October 8, 1890, Anna Augusta Van Pelt.

1256. BEULAH, Allentown, N. J.

2169. ELIZA CORLIS BULLOCK, born March 8, 1862.

2170. HENRY EMLEY BULLOCK, born July 16, 1867.

1260. WALTER E., Jacobstown, N. J.

2171. BLANCHE E., born July 8, 1875; died February 15, 1876.

2172. JAMES B., born January 18, 1877.

2173. LIZZIE A., born December 12, 1887.

1263. JOSIAH BORDEN TILTON, Emleytown, N. J.

2174. JOSEPH H. TILTON, born November 7, 1864; died July 30, 1890.

2175. MARY BORDEN TILTON, born October 17, 1866; married William Kester, January 30, 1890.

2176. JOHN B. TILTON, born November 26, 1868; married Emma Hunsinger, September 9, 1891.

2177. ELLA B. TILTON, born March 26, 1872; married Harry Borden born August 25, 1866.

2178. ABBIE B., born September 18, 1881.

2179. BETH, born October 9, 1885.

1264. ELIZABETH TILTON, Emleytown, N. J.

2180. JOHN T. WAINWRIGHT, born May 10, 1868; died June 6, 1873.

2181. MARY FRANCES WAINWRIGHT, born October 4, 1869; died July 20, 1870.

2182. EMMA L. WAINWRIGHT, born February 22, 1871; married Alexander Buck June 20, 1890.

THE BORDEN FAMILY.

2183. AARON B. WAINWRIGHT, born March 31, 1876.

2184. ABBIE B. WAINWRIGHT, born March 1, 1878; died July 15, 1878.

2185. ANNA R., born November 13, 1872; died July 5, 1873.

1266. ANN POPE, New Jersey.

2186. WILLIAM S. COLLIER, born 1857; died March 26, 1878.

1267. GEORGE WASHINGTON, New Jersey.

2187. CLIFFORD, born September 13, 1864; married Emily Parker, December 23, 1891; she was born November 12, 1869.

2188. CLARA, born September 13, 1864; twin with Clifford; she died March 13, 1865.

2189. JOSIAH PARKER, born November 30, 1868.

2190. HARRY, born August 23, 1866; married Ella Borden Tilton, March 5, 1890.

2191. HOWARD, born February 5, 1871.

2192. ANNIE P., born March 18, 1874; married Wright Longstreet, January, 1891.

2193. GEORGE, born April 12, 1878.

2194. CARRIE, born November 14, 1881.

1269. CHARLES WESLEY, Jacobstown, N. J.

2195. STIRLING, born October 12, 1864; died October 2, 1865.

2196. CHARLES C., born August 23, 1866; married Mary C. Jordan.

2197. FRANK CONOVER, born October 4, 1870; died —

2198. FANNIE CONOVER, born October 4, 1870.

2199. JOHN LAIRD, born August 20, 1872.

2200. W. ROBERT, born October 26, 1879.

2201. J. A. GARFIELD, born April 22, 1881.

2202. M. AUGUSTA, born April 9, 1885.

1272. MARY E., New Jersey.

2203. THEODORE SCOTT, born July 15, 1859.

1273. MARTHA, New Jersey.

2204. BERTHA PAINTER, born November 10, 1873; married —

2205. GRANT HOFFMAN, born December 4, 1892.

1274. WILLIAM, New Jersey.

2206. APOLLO WOODWARD, born July 24, 1870; married Stella Wright, January 16, 1896.

NINTH GENERATION.

2207. FREDERICK WILLIAM, born February 24, 1876.

1275. SARAH, New Jersey.

2208. EMMA T. BUSSOM, born November 20, 1868; married Oscar Nutt.
2209. RICHARD BUSSOM, born December 3, 1870.
2210. JOEL B. BUSSOM, born July 27, 1875.
2211. LIZZIE W., born December 6, 1884; died December 6, 1884.
2212. ELMER W., born August, 1888.

1276. EDWARD, New Jersey.

2213. CHARLES S., born February 12, 1870; married (1) Gertrude Middleton, September 10, 1891; she died September 9, 1894; (2) Bertha Gaskell, August 1, 1897.

2214. GEORGE L., born March 5, 1873; married Edna Middleton, February 27, 1895.

2215. FRANK R., born May 21, 1874; married December 9, 1894, Bessie Dilatash.

2216. BERTHA, born August 23, 1875; married Fred Jones, January 9, 1894.

2217. AMY, born August 23, 1877.
2218. ELIZABETH T., born October 7, 1879.
2219. JAMES E., born February 9, 1882.

1279. HELEN, Ardmour, Pa.

2220. WILLIAM HENRY LOYD, born August 14, 1870.

2221. JOHN STRAWBRIDGE LOYD, born March 26, 1872; married November 16, 1898, Edith M. Parker, daughter of Mr. and Mrs. Robert M. Parker of Chicago, Ill.

1280. FRANCES S., Ardmour, Pa.

2222. FRANCIS REEVE, born February 16, 1870; died September 16, 1876.

2223. JULIA STRAWBRIDGE, born July 9, 1878.

1281. HENRY, New York.

2224. CAROLINE,

1283. BEULAH W. EMLEY, New Jersey.

2225. OLIVER E. LOGAN, born August, 1873.
2226. SCOTT LOGAN, JR.
2227. ELIZABETH G. LOGAN,
2228. RENA W. LOGAN.

THE BORDEN FAMILY.

1285. LAVINIA, Philadelphia, Pa.

2229. GUSTAVUS WYNNE COOK, born December 12, 1869; married Nannie Mumford Bright.

1288. JOSEPH EMLEY BORDEN, Philadelphia.

2230. RICHARD T. COOK BORDEN, born August 25 ,1892; died August 25, 1892.

LAVINIA COOK BORDEN, born April 25, 1894.

1289. JOHN, New Jersey.

2231. LAVINIA.

1303. ASHER, Allentown, Pa.

2232. WALTER AUGUSTUS, born April 28, 1863; married Anna F. Bockius, daughter of Dr. S. A. Bockius of Philadelphia, Pa.

2233. CAROLINE T., married —— Brown of Windsor, New Jersey.

2234. AMANDA, married —— Brown of Windsor, New Jersey.

1304. JOHN HANCE, Allentown, N. J.

2235. PETER R., born in Philadelphia, Pa., August 27, 1847; married (1) Eva Miller of Wilkesbarre, Pa., April 24, 1874; (2) Matie K. Spaulding of Forkston, Pa., May 1, 1883.

2236. THEODORE, born in Allentown, N. J., September 13, 1849.

2237. SIDNEY PARKER, born in Trenton, N. J., September 14, 1855.

1314. RANDALL, New Jersey.

2238. CHARLES.

2239. GEORGE.

2240. RANDALL.

1316. JOSIAH, New Jersey.

2241. JACOB

2242. RICHARD.

2243. ASHER.

1321. FRANCES HOPKINS, Chicago, Ill.

2244. BERTHA BORDEN PARKER, born June 6, 1873; died March 4, 1879.

2245. EDITH MERCER PARKER, born January 4, 1875; married John Strawbridge Loyd, November 16, 1898; son of Helen Borden and William Loyd of Ardmore, Pa.

2246. IRENE BORDEN PARKER, born October 3, 1882.

2247. ETHEL FRANCES PARKER, born April 21, 1884.

NINTH GENERATION.

1327. RICHARD, Red Bank, N. J.

2248. ANNIE MORRIS, born 1861; married —— Tilton.
2249. MINNIE FRANCES, born 1863; married —— Hill.
2250. SAMUEL MORRIS, born 1865.
2251. LAURA, born 1871; married —— Allaire.

1328. JOHN WHITE, Little Silver, N. J.

2252. JOHN FREDERIC, born 1864; he married in 1890.

1329. CAROLINE, Monmouth county, N. J.

2253. WILLIAM DYKES PONTIN, born 1859.
2254. GEORGIANA ALLERTON PONTIN, born 1861; married —— Field.
2255. CARRIE ELIZA PONTIN, born 1864; married —— Pagenstecher.
2256. SARAH EMILY PONTIN, born 1866; married —— Bates.
2257. ALFARETTA PONTIN, born 1875; married —— Ramage.

1330. GEORGE EDWIN, Monmouth county, N. J.

2258. EDWIN FLOYD, born 1897.

1333. JOHN, Sharpstown, N. J.

2259. GEORGE APPLE, born in Sharptown, N. J., September 8, 1851; died in Lakeview, Cal., November 9, 1894.
2260. WILLIAM HENRY, born in Sharptown, N. J., August 21, 1854.
2261. MARY JANE, born in Sharptown, N. J., October 20, 1856; married —— Smyth of Pensauken, Camden county, N. J.
2262. MORRIS PEAK, born near Sweedsboro, Gloucester county, N. J., May 30, 1859; married Olive Murphy, November 15, 1882, in Chicago, Ill.
2263. JOSIAH BRICK BORDEN, born in Woodstown, N. J., April 22, 1861.
2264. ELLEN HOWELL, born in Woodstown, N. J., February 24, 1865.
2265. CORNELIUS SELLERS, died at the age of 13 years.
2266. JOHN HANCE, born in Woodstown, N. J., April 8, 1870.

1335. JOHN FRANCIS, Meriden, Conn.

2267. FRANK EDWARD, born October 7, 1869; married Bertha R. Meyer, May 25, 1890. She is the daughter of Dr. John H. Meyer of New York city.
2268. CATHERINE AGNES, born at Meriden, Ct., November 19, 1872.
2269. JOHN FRANCIS, born July 8, 1874.

THE BORDEN FAMILY.

2270. LORETTA JANE FRANCES, born March 15, 1878.
2271. LILLIAN MAY, born May 21, 1881.
2272. MARGARET ELINE, born June 9, 1883.
2273. FREDERICK, born July 11, 1885.
2274. WALTER CHARLES, born July 23, 1889.

1336. JENNIE, New York City.
2275. LORETTA V. CAFFREY.

1340. HANNAH CHAMBERS, Rochester, N. Y.
2276. ALONZO BENJAMIN PALMER, born July 24, 1864; died September 13, 1867.
2277. MARGARET JESSIE PALMER, born January 31, 1868; died November 26, 1880.
2278. SYLVIA JOSIE PALMER, born January 16, 1870; died March, 1874.
2279. EDMUND LAYTON PALMER, born September 21, 1872; died August 4, 1873.

1341. ALMIRA CLAYTON, Clio, Mich.
2280. EMMA ELBERTINE RICE, born July 5, 1869; married William Schofield, November 21, 1894.
2281. MARY MARGARET RICE, born June 30, 1872; married William H. Parry, November 9, 1892; died February 5, 1894.
2282. CHARLES CLAYTON RICE, born January 7, 1876; died October 21, 1880.
2283. GEORGE CARLTON RICE, born July 17, 1882.

1345. GEORGE W., Auburn, Neb.
2284. ELSPETH BELL, born May 22, 1885.
2285. WILLIAM DUNCAN, born April 8, 1890.
2286. AUGUSTA HOPPER, born October 14, 1891, in Britton, Mich.
2287. HELEN ROSS, born June 1, 1893; in Gladwin, Mich.
2288. GEORGE WEBSTER, born September 22, 1895, in Auburn, Neb.

1348. MALVINA A., Shelton, Conn.
2289. HENRIETTA MALVINA SEELEY, born October 13, 1869.
2290. GENEVEVE AUGUSTA SEELEY, born June 17, 1871.
2291. GRACE MAUDE, born March 12, 1873; died May 17, 1875.
2292. SILVANUS, born April 21, 1874; died August 22, 1874.

NINTH GENERATION.

1352. PERCIVAL ERNEST, Shelton, Conn.

2293. CARLETON GRANVILLE, born August 15, 1892.
2294. SYLVANUS SEELEY, born March 10, 1894.
2295. PERCIVAL ERNEST, born November 18, 1896.
2296. ROBERT HENRY, born May 17, 1898.

1374. SETH, Fall River, Mass.

CAROLINE T., married James Robinson, March 20, 1854.
JULIA A., married William Robinson, 1858.
HENRIETTA
      1375. BAILEY.
EMMA C. H.
      1376. ISAAC, Fall River, Mass.

2297. ISAAC H., born February 21, 1829; married Louisa Burns January 4, 1852.

2298. AMANDA C., born April 9, 1835; married George Hill December 23, 1853.

2299. DAVID B., born April 9, 1835; married Ruth Hambley, daughter of Benjamin Hambley.

2300. HENRY CLAY, born May 8, 1840; married Emma Noros.

2301. JAMES CLARENCE, born October 13, 1844.

1379. DEACON JOSEPH C., Fall River, Mass.

2302. MARY MARIA D., born August 31, 1835; married James W. Martin November 2, 1851.

2303. STEPHEN B., born September 3, 1838; married Ellen Eastwood, June 26, 1860, at Fall River.

2304. ANGENETTE, born June 2, 1841; died in infancy.

2305. JOSEPH F., born August 4, 1843; married Marianna Hussey, May 16, 1866.

2306. HANNAH G. J., born February 18, 1846; married Charles M. Horton, February 18, 1867.

2307. EMMA C., born February 18, 1849; married George Daves.

2308. JAMES W. M., born January 16, 1851; died September 1, 1898; married Carrie Shaw, in Fall River, Mass.

2309. CHARLES F., born September 24, 1854; married Annie L. Remington June 8, 1880; she died July 3, 1895.

2310. SETH A., born November 15, 1857; married Susan Gibbs of Fall River.

THE BORDEN FAMILY.

1387. ANDREW, Fall River, Mass.

2311. ANN ELIZA, born September 15, 1836.

2312. ADELAIDE, born June 26, 1839; married Stephen Paine of Providence, R. I., May 29, 1861.

2313. ANDREW, born July 30, 1843; married Ellen A. Bliss.

1388. PHILIP D., Fall River, Mass.

2314. ABBY D., born March 7, 1841; married George Borden son of James, November 19, 1865.

2315. WILLIAM, born August 15, 1843; died May 19, 1864.

2316. THOMAS S., born June 9, 1848.

2317. PHILIP D., born December 23, 1850; married Adelaide E. Scoville, November 24, 1875.

2318. FRANK, born November 13, 1853; married Elizabeth S. Pierce, July 15, 1875.

1395. IRENE LUTHER, New York City.

2319. CORNELIUS N. BLISS of New York city.

"He was born in Fall River, Mass., sixty-six years ago. Since boyhood he has been identified with cotton manufacturies. In 1866 he went to New York city, and has since been one of the most highly esteemed leaders in mercantile circles. Until he entered President McKinley's Cabinet as Secretary of the Interior he had held no political office, although other Presidents had offered him Cabinet positions, both because of his fitness and because of his prominence in national Republican conventions. He is universally trusted and esteemed. He is the head of the great wholesale dry goods house of Bliss, Fabyan & Co., and is a director in several banks, insurance companies, trust companies, etc.

1396. SARAH BORDEN LUTHER.

2320. CATHERINE WHITE NEWHALL,

2321. STEDMAN NEWHALL,

2322. HENRY BORDEN NEWHALL,

1398. THOMAS J., Fall River, Mass.

2323. HARRIET MINERVA, born June 15, 1856; married Rufus Waterman Bassett, September 13, 1882.

2324. ANNA HOWLAND, born March 30, 1858.

2325. RICHARD, born January 16, 1860; died August 25, 1861.

2326. CARRIE LINDLEY, born September 17, 1864.

NINTH GENERATION.

1402. MATTHEW C. D., New York City.

2327. WILLIE OWEN, born February 12, 1867; died March 4, 1868.
2328. BERTRAM H., born October 3, 1868.
2329. HARRY E., born December 19, 1870; died January 1, 1872.
2330. MATTHEW,
2331. HOWARD SEYMOUR.

1410. E. CORINNA, Philadelphia, Pa.

2332. CORINNA KEEN, born November 4, 1867; married Dr. Walter J. Freeman November 3, 1892.
2333. FLORENCE KEEN, born December 8, 1868.
2334. DORA KEEN, born June 24, 1870.
2335. MARGARET KEEN, born March 5, 1878.

1399. RICHARD B., Fall River, Mass.

CHARLES N.
RICHARD P.
LILLIE M.
NELLIE L.

1400. EDWARD P., Fall River, Mass.

E. SHIRLEY,

1401. WILLIAM H. H., Fall River, Mass.

MARY I.

1407. ELLEN, Fall River, Mass.

2336. ELLEN CORINNA PAINE, born August 10, 1851, at Fall River, Mass.; died September 12, 1896, in East Greenwich, R. I. She was married to Prof. Ray Green Huling, head master of the English High School of Cambridge, Mass. He is a graduate of Brown University, 1869, and holds a degree of A. M. from Harvard University, and of S. C. D. from Brown. He was editor of "School and College" in 1892, and "Lecturer," in Harvard University in 1898.

1411. JEFFERSON, Fall River, Mass.

2337. JEFFERSON, born November 16, 1869.
2338. JOHN WESTALL, born September 28, 1871.
2339. ELIZABETH, born February 9, 1873; married —— Nichols.

1413. SPENCER, Fall River, Mass.

2340. SPENCER, born September 8, 1872.
2341. LEONORA, born May 2, 1874.

THE BORDEN FAMILY.

2342. ALFRED, born June 25, 1875.

2343. BROOKS, born August 5, 1878; killed January 31, 1894, on a grade crossing of the N. Y., N. H. & H. R. R.

2344. FLORENCE, born December 31, 1882.

   1416. ANDREW J., Fall River, Mass.

2345. EMMA LEONORA, born March 1, 1851.

2346. ALICE ESTHER, born May 3, 1853.

2347. LIZZIE DREW, born July 19, 1861.

   1421. CHARLES WILLIAM, Fall River, Mass.

2348. HERBERT G.

2349. HIRAM F.

   1428. JEROME COOK, Fall River, Mass.

2350. MARY EMMA, born May 15, 1871; married Charles D. Pierce of Fall River.

2351. BESSIE, born April 19, 1874.

2352. FANNIE, born November 8, 1876.

2353. BERTHA, born April 25, 1882.

2354. THEODORA, born October 27, 1885.

   1434. CHRISTOPHER, Fall River, Mass.

2355. JONATHAN, born May 15, 1841; married Mary M. Estes.

2356. ALICE, born November 16, 1842; married George H. Hicks, January 1, 1862.

2357. MARY E., born December 7, 1844; married Isaac W. Howland of Little Compton.

2358. OTHNEIL, born August 24, 1850.

2359. EDWIN, born June 26, 1850.

2360. PHOEBE SARAH, born April 14, 1858.

   1438. ALANSON, New Bedford, Mass.

2361. WILLIAM A.

2362. LAURA EVELINA,

   1444. ISAAC N., Fort Pierre, South Dakota.

2363. MERT A., born in Ogle county, Ill. ,December 26, 1861; he is married, and lives at Fort Pierre, South Dakota.

   1465. SIMEON, Fall River.

2364. SIMEON,

2365. SARAH,

NINTH GENERATION.

1468. NATHANIEL B., Fall River, Mass.

2366. NATHANIEL BRIGGS, born March 4, 1871; married Annie R. Smith, June 4, 1895.
2367. ANNIE BROWN, born December 4, 1877.
2368. ARNOLD BUFFUM, born March 19, 1882.
2369. LOUISE GOULD, born October 11, 1883.

1473. WILLIAM G., Fall River, Mass.
2370. HELEN.
2371. WILLIAM ARTHUR,

1488. CYRUS H., Fall River, Mass.
2372. ANNIE LAURIE,
2373. HARRIET L., born August 19, 1851.
2374. ALFRED M., born May 11, 1853.
2375. CHARLES E., born October 25, 1858.
2376. CYRUS E., born January 30, 1860.
2377. ANNA M., born December 9, 1861.

1499. BENJAMIN F., Fall River, Mass.
2378. ABRAHAM O., born April 30, 1851.
2379. WILLIAM H., born July 26, 1855.
2380. EDWARD E., born November 27, 1864; died in infancy.

1522. DR. FREDERICK W. Ottawa, Canada.
2381. ELIZABETH M., born October 3, 1874.
2382. HAROLD LOTHROP, born May 23, 1876.
2383. J. MAUDE, born March 23, 1879.

1535. ANNIE MARIA. Port Williams, Nova Scotia.
2384. MARTHA ADELIA BORDEN, born October 6, 1864.
2385. SUSAN MARIA BORDEN, born February 25, 1866.
2386. MARGERY ELLEN BORDEN, born August 11, 1868.

1554. ROBERT ALLISON, Moncton, New Brunswick.
2387. LILA REEVE, born March 30, 1878.
2388. SARAH ALLISON, born December 19, 1879.
2389. MAUD JEAN, born February 27, 1881, died June 5, 1882.

1555. BYRON CRANE, Sackville, New Brunswick.
2390. ELAINE ALLISON, born December 6, 1887.

THE BORDEN FAMILY.

2391. GLADYS ALLISON, born December 6, 1887, in Sackville, Nova Scotia.

    1556. LILA LAVINIA.

2392. MARION LOCKWOOD SMITH, born January 23, 1889.

2393. DONALD BORDEN SMITH, born October 5, 1896.

    1583. ELIJAH C., Kingsport, Nova Scotia.

2394. BLANCHE T., born June 20, 1870.

2395. FRED L., born March, 1875.

2396. ARTHUR F., born July, 1878.

2397. ALICE L., born August, 1884.

    1606. WILLIAM C., Philipsburg, Canada.

2398. LAURA LOUISA, born September 27, 1849; married a Mr. Flemming of Philipsburg, Canada.

2399. MILTON LUTHER, born January 9, 1853.

2400. ASA HERBERT, born April 29, 1855.

2401. CHARLES EDGAR, born February 8, 1858; married Josephine T. Greene in New York city, October 5, 1891.

2402. JOHN FRANKLIN, born April 18, 1860.

2403. JOSEPHINE, born November 1, 1863; married a Mr. Gardner.

2404. FRED WILLIAM, born October 27, 1866; lives in Manchester, Ma.

2405. MABEL HATTIE, born May 27, 1875.

    1631. GEORGE, Canada.

2406. JENNIE.

2407. BESSIE.

    1633. NELSON, Boston, Mass.

2408. ARIETTA J., born December 16, 1882.

2409. WINDSOR M., born January 16, 1894.

    1643. HENRY C., Fall River, Mass.

2410. ELLA C., born September 4, 1856.

2411. KNIGHT H., born January ——

    1644. JAMES A., Fall River, Mass.

2412. ELLEN LINCOLN, born April 16, 1860.

2413. ELMA, born February 20, 1863.

    1645. GEORGE R., Fall River, Mass.

DR. MILBANK JOHNSON.
2424.

NINTH GENERATION. 289

2414. LILLIAN V., born August 18, 1863.

2415. MARY E. D., born March 15, 1865.

1677. HENRY LEE, Tonti, Ill.

2416. WILLIE CLAY, died September, 1872, at Borden, Colorado county, Texas.

2417. GERALD MARK BORDEN, born January 28, 1871; married Lucille Papin of St. Louis, Mo.

2418. LEWIS LAMBERT BORDEN, born June 20, 1875.

1679. PHILADELPHIA WHEELER, Houston, Tex.

2419. LEWIS CASS JOHNSON, born September 12, 1857; died October 9, 1869.

2420. GAIL BORDEN JOHNSON, born November 11, 1859; married Mary Willie Borden, daughter of James C. Borden of Galveston, Tex., and Palmyra Atkinson, his wife, November 29, 1880.

2421. WALDO PASCHAL JOHNSON, born January 6, 1863; married Mary Lois Brown, daughter of James C. and Theresa Brown of Elgin, Ill., March 4, 1885.

2422. VIRGINIA LEE JOHNSON, born January 13, 1865; married Isaac Milbank of New York. He is vice-president of the Borden Condensed Milk Company.

2423. EMMA ORA JOHNSON, born February 17, 1868; married Thomas Paschal Borden, son of John Rolden Borden and Mary Jane McKee, his wife, in Yonkers, N. Y., March 16, 1887.

2424. MILBANK JOHNSON, born October 13, 1871, in Columbus, Tex.; married Louiez Leavitt Lothrop, daughter of Franklin B. Lothrop and Cornelia Norris, his wife. He is a graduate of Chicago Medical College, and has taken a post graduate course at Johns Hopkins University. Dr. Johnson was the original promoter of this work on Borden Genealogy, but his devotion to his studies and the demands of his increasing practice made it impossible for him to keep up the work.

2425. CARRIE BORDEN JOHNSON, born July 27, 1874, in Columbus, Texas; married John Barnes Miller, son of J. E. and Sarah (Barnes) Miller of Port Huron, Mich., in Yonkers, N. Y., April 17, 1895.

1681. MARY JANE, Hartford, Conn.

2426. ETTIE CHARLOTTE MUNSILL, born June 10, 1867; married John Ulrich, October 27, 1887, in Hartford, Conn.

2427. GAIL BORDEN MUNSILL, born January 6, 1864; married Ruth Holmes of Winsted, Conn., January 16, 1895.

2428. MARCUS MUNSILL, born October 12, 1868; died in Hartford, Connecticut, August 25, 1894; married Eva May Graves of Hartford, Conn., December 7, 1892.

2429. CLIFFORD LEE MUNSILL, born, April 28, 1874; died in Winchester, March, 1876.

1682. JOHN GAIL, Wallkill, N. Y.

2430. PENELOPE ADALINE, born April 3, 1868; married Louis Alford Hamilton of New York city, January 15, 1896.

2431. GAIL, born in Brewsters, N. Y., May 12, 1872; married December 5, 1893, Helen May Valk, daughter of L. Bolton Valk and Ellen Leonard, his wife, of Brooklyn, N. Y.

2432. BEATRICE, born November 6, 1875; died in Brewsters, N. Y. 1876.

2433. LEWIS MERCER, born in Brewsters, N. Y., June 22, 1879.

2434. MARION, born in Wallkill, N. Y., March 9, 1883.

1683. JOHN ROLDEN, Texana, Texas.

2435. THOMAS PASCHAL, born January 6, 1856; married March 16, 1887, Emma Ora Johnson, daughter of J. W. Johnson and Philadelphia W. Borden, his wife. (See children under No. 2423.)

2436. JAMES MC KEE, born July 25, 1858; married (1) Alice Christine Lauck, June 14, 1887; after her death he married (2) Mai Niernsee, October 24, 1894.

2437. NANNIE, born November 3, 1861; died November, 1862.

2438. ROLDEN, born October 11, 1864; married Russell Low.

1684. JAMES COCHRAN, Galveston, Texas.

2439. HATTIE L., born October 7, 1859; married (1) November 9, 1881, Alexander C. Harcourt, born April 13, 1856; died in Beaumont, Texas, November 19, 1882; he was the son of Judge John T. Harcourt of Galveston, Texas; (2) on February 11, 1886, to William Clayton Weld of Elgin, Ill. They removed to Los Angeles, Cal., in 1887.

2440. MARY WILLIE, born November 3, 1861; married in Galveston, Texas, November 29, 1880, Gail Borden Johnson, son of Philadelphia W. Borden. (See No. 2420 for children).

2441. FLORA PARKER, born May 25, 1865, in Jackson county Texas; married in Galveston, Texas, September 24, 1884, Robert James Davis, son of Joseph Augustus Davis and Martha Finnie, his wife.

2442. ROBERT STAFFORD, born April 11, 1884.

2443. MAIE PHILA, born December 24, 1885.

2444. JAMES GAIL, born May 21, 1886; died November 9, 1886.

2445. ADINE DEMIS, born December 26, 1888.

NINTH GENERATION. 291

    1685. MILAM, Columbia, Texas.
2446. MILAM UNDERWOOD,
    1686. GUY, San Antonio, Texas.
2447. NELLIE HAMILTON, born June 21, 1884.
2448. GUY, born October 20, 1886.
2449. PASCHAL PAVOLO, born December 3, 1890.
    1687. WILLIAM JOSEPH, Galveston, Texas.
2450. FLORENCE LOVISA, born in Galveston, Texas, March 7, 1881; died May 25, 1881.
2451. EMMA ELIZABETH, born in Hico, Texas, August 25, 1885.
    1689. SIDNEY G., Sharpsburg, Texas.
2452. HENRY LEE, born at Sharpsburg, Texas, January 6, 1879.
2453. JOSEPHINE, born at Sharpsburg, Texas, March 10, 1881.
2454. ALICE, born at Sharpsburg, Texas, September 7, 1882.
2455. SIDNEY GAIL, born at Sharpsburg, Texas, October 27, 1884.
2456. ARTHUR, born at Sharpsburg, Texas, June 21, 1886; died June, 1896.
2457. LORETTA, born at Sharpsburg, Texas, August 15, 1888.
2458. LUCILE, born at Sharpsburg, Texas, January 15, 1891.
    1692. FIDELIA, Fort Worth, Texas.
2459. WILLIE C. DUFFEL, born December, 1871.
2460. JOHN P. DUFFEL, born March, 1873; died in infancy.
2461. CARRIE L. DUFFEL, born 1875; married Marshall Thompson.
2462. MARY LILLIE DUFFEL, born 1877; married John Cobb.
2463. JOHN P. DUFFEL, born 1879; died in infancy.
2464. HATTIE DUFFEL, born 1880; died in 1886.
2465. LESLIE B. DUFFEL, born July, 1882.
2466. OLA B. DUFFEL, born January, 1885.
2467. L. MAY DUFFEL, born November, 1887.
2468. SAM B. DUFFEL, born June, 1890.
2469. SIDNEY G. DUFFEL, born July, 1892.
2470. HORACE E. DUFFEL, born January 3, 1896.
    1696. LEE DE WITT, Galveston, Texas.
2471. EDWARD RAY, born in Galveston, Texas, January 17, 1893.
2472. MINNIE LEE, born in Galveston, Texas, August 2, 1895.

THE BORDEN FAMILY.

1698. IDA MINETTA, Minneapolis, Minn.

2473. GRACE BLEETHEN, born May 16, 1885; died 1898.

1699. ADMIRAL FARRAGUT, Los Angeles, Cal.

2474. MARIE GERTRUDE, born March 18, 1899; died March 19, 1899.

1700. EVA GERTRUDE, Pomona, Cal.

2475. EDITH GERTRUDE MARSHALL, born June 22, 1892.

2476. HORATIO VARNUM MARSHALL, born December 25, 1894.

1707. DANIEL, Fall River, Mass.

2477. WILLIAM C., born February 7, 1837; married Alice Tobey, October 16, 1865.

2478. GILBERT H., born August 27, 1830; died in infancy.

2479. ANNIE E. L., born November 8, 1842.

1710. BENJAMIN, Fall River, Mass.

2480. DANIEL B., born February 14, 1843.

2481. HENRY N., born March 31, 1845.

2482. FRANKLIN S., born 1858.

2483. LEONARD I., born November, 1861.

2484. AURETIA, born October, 1863.

1713. BRADFORD D., Fall River, Mass.

2485. HANNAH, died young.

2486. SARAH A.

2487. LUCY,

2488. CASSANDRA,

2489. LOUISA,

2490. MARY, died in infancy.

1722. STEPHEN, Fall River, Mass.

2491. LUCY, married Benjamin Parsons.

2492. BENJAMIN,

2493. THOMAS LEONARD,

2494. JAMES,

2495. NANCY,

2496. HARRIET MATILDA,

2497. REASON,

2498. SARAH,

NINTH GENERATION.

1723. SAMUEL, Fall River, Mass.

2499. BRADFORD, married Sarah Holt.
2500. MARY MARIA,
2501. ALFRED PARDY,
2502. JOSEPH, married Alvisa Mackey.
2503. ABRAHAM, died unmarried.
2504. SAMUEL,
2505. SARAH JOSEPHINE,
2506. LUCY,
2507. THOMAS,
2508. ABIGAIL,
2509. JAMES FORD,
2510. SANFORD,
2511. BENJAMIN,
2512. ABRAHAM,
2513. SAMUEL,

1730. JOSEPH, Fall River, Mass.

2514. WILLIAM, born January 26, 1835; married Louisa Parkhurst, April 6, 1860.
2515. CHARLES JOSEPH, born December 15, 1837; married Harriet C. Hills, October 13, 1858.

1737. ROBERT E., Fall River, Mass.

2516. PHOEBE E., born September 28, 1834; married George Busby.
2517. ANN F., born December 11, 1835; married George H. Davis December 25, 1859.
2518. JOHN R., born July 9, 1839.
2519. ROBERT T., born 1841, accidentally killed at sixteen years of age.

1740. THOMAS E., Fall River, Mass.

2520. THOMAS E., born May 31, 1844.
2521. FANNIE M., born February 12, 1848, died in infancy.
2522. HATTIE J., born July 20, 1854.
2523. ARTHUR T., born May 31, 1856.

1753. JOHN H., Fall River, Mass.

2524. SARAH JANE, born October 15, 1843.
2525. MARY VALENTINE, born October 29, 1845.

THE BORDEN FAMILY.

2526. HARRIET ELIZA, born December 27, 1847.

1756. REV. THOMAS, Fall River, Mass.

2527. GERTRUDE.

2528. GEORGE.

1762. ISAIAH, Fall River, Mass.

2529. BETSY JANE, born May 15, 1862, married Hiram C. Borden, son of Richard.

1764. JOHN L., Fall River, Mass.

2530. JOHN FRANCIS, born February 4, 1847.

2531. RUTH ANN, born November 27, 1848, married George H. C. Walter, July 3, 1867.

2532. CHARLES H., born 1850.

2533. WILLIAM H., born November, 1851.

2534. RICHARD D., born November 23, 1855.

2535. ALBERT H., born December 4, 1857.

2536. CHARLES A., born February, 1861.

2537. MARY MATILDA, born September 12, 1863.

1765. STEPHEN, Fall River, Mass.

2538. SYLVIA, born July 3, 1846.

2539. LOUISA M., born January 7, 1847.

2540. STEPHEN A., born October 14, 1850.

2541. SARAH A., born August 15, 1853.

2542. CHARLES E., born November 15, 1856.

2543. IDA MAY, born July 14, 1862.

1776. JOHN A., Fall River, Mass.

2544. HIRAM.

1783. LEANDER O., Fall River, Mass.

2545. ABBY JANE, born November 20, 1855.

2546. GEORGE A., born June 2, 1858, died March 2, 1895.

2547. ELLEN R., born October 6, 1863.

1793. CHARLES A., Fall River, Mass.

2548. JAMES EDWARD, born February 12, 1864.

1818. MARY COOK LE VALLEY, Fall River, Mass.

2549. HENRY LOVEL, born ——, died in infancy.

2550. MARY EMELINE LOVEL, born 1852, February 6, married Rev. M. R. Minnich, October 1, 1846.

NINTH GENERATION.

2551. GEORGE ELEAZER LOVEL, died in infancy.

1826. HARRIET MARIA, Goldsboro, North Carolina.

2552. CHARLES DEWEY, born December, 1851, in Goldsboro, North Carolina; married (1) Mary Alice Steele, December 16, 1874; she died in 1893, and Charles Dewey married (2) Mrs. Annie Snow Konnegay, February 16, 1898.

2553. MARY WEBBER DEWEY, born May, 1853, married Charles Slocum.

2554. ANN MARIA DEWEY, born March, 1855, married Junius Slocum.

2555. GEORGE WOOD DEWEY, born January, 1857, lives in New York City.

2556. THOMAS WEBBER DEWEY, born march 27, 1859, married Eliza Sledge Mial, May 20, 1886.

2557. ERNEST BROWNRIGG DEWEY, born January, 1861, married Sallie B. Arrington, June 10, 1884.

2558. HARRIET M. DEWEY, born July 1, 1865.

1827. JAMES COLE, Wilmington, North Carolina.

2559. EDWIN BORDEN, born April 16, 1853, married (1) October 10, 1878, Nepple Wallace. She died February, 1884; (2) Octavia Stockton Wallace, September 29, 1888.

2560. MARIA LOUISE, born November 13, 1854, married Isaac H. McLeod, December 31, 1874.

2561. JAMES COLE, Jr., born October 26, 1856, in Goldsboro, North Carolina, married in 1881, Belle Abel.

2562-. MINNIE BELLE, born February 5, 1859, married Adolphus P. Lynch, December 7, 1881.

2563. CHARLES E., born April 23, 1861, married (1) Annie L. Whitehead, April 30, 1884; she died October 2, 1892, and Mr. Borden married (2) Hattie Taylor, October 23, 1892.

2564. HERBERT LEIGH, born November 28, 1863, married Mary R. Moffitt, April 8, 1896.

2565. ALICE WALLACE, born June 30, 1866, died September 22, 1867.

2566. HUGH D., born October 3, 1868, died June 4, 1869.

2567. KATE COLE, born July 7, 1871, married December, 1893, Edward R. Wotten.

2568. ANNIE MEAD, born April 2, 1875, died July 4, 1897. She was married to John M. Banner, February 23, 1895.

1828. EDWIN BROWNRIGG, Goldsboro, North Carolina.

2569. ARNOLD, born 1854, married Eunice Hemphill, November 5, 1879.

2570. ANNIE B., born 1856, married Matthew L. Lee, December 5, 1882.

2571. FRANK KORNEGAY, born July 12, 1857, married Sallie Jones, December 4, 1887.

2572. CARRIE W., born 1859, married Frank Daniel, December 22, 1885.

2573. LOUISA, born August, 1861, married Adam C. Davis, February 26, 1887.

2574. MARY C., born 1865.

2575. EDWIN BROWNRIGG, Jr., born 1868.

2576. JOHN LEMUEL, born 1863, married Mary Rowena Micks.

2577. GEORGIA L., born March 23, 1874, died September 29, 1879.

2578. MURRAY, born 1878.

2579. MABEL MORAN, born November 6, 1876.

2580. MINNIE DEEMS, born November 6, 1876, died June 9, 1877.

2581. PAUL, born 1882.

2582. ELIZABETH, born 1880.

1830. SARAH LAVINIA, Goldsboro, North Carolina.

2583. JOHN C. MILLER, born September 10, 1865, married Mabel Fisher of Athens, Ohio, June 20, 1899.

2584. CHARLES BORDEN MILLER, born December 13, 1866, married Anna Burwell, October 11, 1893.

2585. LOUISE MILLER, born September 28, 1868, married Randolph Macon Michaux, April 20, 1892.

2586. HUGH LEE MILLER, born October 25, 1870.

2587. MARY BROWNRIGG MILLER, born December 17, 1872, married Leslie Southerland.

2588. FRANK MARVIN MILLER born August 1, 1874.

2589. ROBERT BASCOM MILLER, born September 3, 1876.

2590. BESSIE WEST MILLER, born April 24, 1881.

1832. WILLIAM H., Goldsboro, North Carolina.

2591. JULIA, born July 5, 1865, died September 9, 1889; married Beverly Sydnor Jerman, December 5, 1888.

2592. WALTER EUGENE, born November 30, 1867, married Mattie Curtis Fuller, daughter of Judge Thomas C. Fuller of Raleigh, North Carolina, February 14, 1894.

NINTH GENERATION.

2593. GEORGIA, born November 5, 1870, married Dr. W. H. Cobb, December 15, 1893.

2594. SUE, born July 7, 1872, married William Douglass of Charleston, South Carolina, October 26, 1898.

2595. SALLIE born August 1, 1874.

2596. WILLIAM HENRY, born December 5, 1877.

2597. RACHEL MOYE. born August 15, 1886.

2598. SYDNOR JERMAN, born February 23, 1894.

1862. ABRAHAM EARL, Providence, R. I.

2599. HARRIET SOPHRONIA, born in Tiverton R. I. married June 9, 1897, William Judson Brown. She died in Providence, R. I., March 22, 1898.

2600. ALICE BROWNELL, born in Tiverton, R. I.

2601. SARAH LOUISA.

1863. HANNAH LOUISA, Fall River, Mass.

2602. LOUISA SUMNER CHACE, born in Fall River, February 6, 1868.

2603. HATTIE CLINTON CHACE, born January 9, 1870, died September 8, 1872.

1876. GILBERT, Illinois.

2604. NETTIE, born August 5, 1861, married Mr. Baldwin of Chicago, Ill.

2605. LABAN GILBERT, born June 1, 1863, married Maude Peet, October 28, 1891, at Crystal Lake, Illinois.

1899. ERIC WARREN, Fall River, Mass.

2606. LAURENCE LEANDER, born September 13, 1874.

2607. RAYMOND DAVIS, born September 3, 1877.

2608. ALDEN EDSON, born September 9, 1886.

1900. OMER ELTON, Fall River, Mass.

2609. ELSIE, born July 21, 1889.

2610. HELEN, born November 8, 1890.

2611. HAROLD, born December 29, 1895.

1901. GEORGE G., Somerset, Mass.

2612. AMANDA PIERCE, born December 11, 1867.

2613. WILLARD PEACHY, born July 29, 1869.

2614. LINDON COOK, born April 4, 1885.

THE BORDEN FAMILY.

1903. FLAVIUS JOSEPHUS, San Francisco, Cal.

2515. FREDERICK RUDOLPH, born April 6, 1879.

1919. JOSEPH M., Dundee, Illinois.

2616. CHARLES J., born August 6, 1866, died February 20, 1869.
2617. BERTHA M., born December 24, 1868, died March 6, 1886.
2618. STELLA L., born March 7, 1870, married March 4, 1896, Harold Hemb of Dundee, Illinois.
2619. NINA SUSAN PRENTICE, born November 4, 1871.
2620. JOSEPHINE C., born September 23, 1873.
2621. CLARA B., born March 4, 1876.

1922. MARY E. L., Elgin, Illinois.

2622. GEORGE L. WELLS, born May 31, 1871.

1931. OWEN D., New Berlin, N. Y.

2623. WILLIAM AUSTIN, born February 26, 1887.

1936. J. BENJAMIN HOXIE, Tonica, Ill.

2624. GUY DELAMATER HOXIE, born March 2, 1896.

1971. WILLIAM THOMAS CHENEY, Eutaw, Ala.

2625. MARY ESTHER M'GIFFERT, born February 23, 1897.

1949. HELEN TELFAIR, Jacksonville, Florida.

2626. ARCHIE BORDEN CLARK, born in Jacksonville, Florida, in 1897. He has brothers and sisters, but their names have not been sent in time to insert them.

1963. THOMAS CRAWFORD, Fernandina, Florida.

2627. ELIZABETH JEFFREYS, born September 2, 1897.

ANNIE BYRN, born December 20, 1898.

1978. LOVIC B. PEARCE, Selma, Ala.

2628. JOSEPH BORDEN PEARCE, born November 7, 1885, died July 26, 1886.
2629. GEORGE LOVIC PEARCE, born November 29, 1887, died July 23, 1888.
2630. FANNIE GRAY PEARCE, born March 10, 1891, in Selma, Alabama.
2631. ELIZABETH GRAHAM PEARCE, born January 8, 1892, died May 2, 1893.
2632. LOVIC BENJAMIN PEARCE, born May 15, 1894, in Selma, Alabama.

NINTH GENERATION.

1991. ANN AARONA BURR, Talladega, Alabama.

2633. BURR BLACKBURN.

1992. ZAIDEE L. BURR, Talladega, Ala.

2634. NONA SHEPPARD HENDERSON.
2635. SAMUEL HOWARD HENDERSON.
2636. KATHERINE ALICE HENDERSON, born July, 1896.

1993. ESTHER W. BURR, Talladega, Ala.

2637. HENRY BURR ROBINSON.

1995. LYDIA ANTOINETTE BURR, Talladega, Ala.

2638. MARION ANTOINETTE HICKS, born August 2, 1898.

1999. ANNIE LEWIS, Talladega, Ala.

2639. MATTHEW BORDEN M'GHEE, born July, 1898.

2007. CAPT. THOMAS S., Washington, D. C.

2640. SAMUEL WHEATLY, born April 22, 1899.

2018. KATHERINE WALTON WOOLSEY, Stamford, Ct.

2641. CECIL WOOLSEY HAMILTON, born June 9, 1881.
2642. STUART HAMILTON, born May 2, 1888.

2019. WOOLSEY R. HOPKINS, Stamford, Ct.

2643. CATHERINE WOOLSEY HOPKINS, born January 5, 1893.
2644. WOOLSEY ROGERS HOPKINS, born December 7, 1895.

2024. EDWIN GOSSON, San Francisco, Cal.

2645. FREDERICK WILLIAM, born April 9, 1886.
2646. HENRY FORNEY, born January 18, 1888.
2647. EMMA CLAUDINE, born November 20, 1892.
2648. HARRIET MAY, born June 14, 1894.

2026. MALBERT TROOP, Cedartown, Ga.

2649. CHRISTINE, born October 7, 1891.

2034. HENRY ALLEN, Dallas, Texas.

2650. ADDIE LOU.
2651. BERTIE LAKE.
2652. HENRY GRADY.

2038. DORA LOU, Italy, Texas.

2653. CHESLEY TRAMMEL.

THE BORDEN FAMILY.

2042. MARTHA A. COOKE, Bowling Green, Ky.

2654. WILLIAM D. HEANEY, born in Rochester, Minnesota, November 28, 1865, married in January, 1896, to Clara Meyers of St. Paul, Minnesota.

2655. MATTIE I. HEANEY, born May 18, 1868, in Rochester, Minn.; married June 14, 1893, T. M. White, born in Topeka, Kansas.

2048. WILLIAM EDWARD HOBSON, Bowling Green, Ky.

2656. MOREHEAD HOBSON, born August 12, 1874.

2657. JULIA HOBSON, born February 13, 1876.

2658. JAMES THOMAS HOBSON, born March 13, 1879.

2659. EDWARD WILLIAM HOBSON, born March 3, 1881.

2660. MARY ATWOOD HOBSON, born August 15, 1884.

2661. MARGARET, born March 18, 1890.

2052. WILLIAM USHER ADAMS, Bowling Green, Ky.

2662. JAMES ADAMS, born 1868, died in infancy.

2663. CARRIE ADAMS, born August, 1871, married in 1886 to James Creider. He died in 1889, and she in July, 1897, leaving two children.

2053. MARY LELAND ADAMS, Bowling Green, Ky.

2664. CARRIE HILBURN, born November 15, 1865; died February 12, 1866.

2665. EVA ROMOLINE HILBURN, born June 6, 1867; married September 11, 1894, to Rev. James Bolton.

2666. FREDERIC S. HILBURN, born November 1, 1869.

2667. JULIETTE CLOTHILDE HILBURN, born February 3, 1872, married July 7, 1891, to John B. Preston, A.M., Professor of Ancient Languages and French in Ogden College, Bowling Green, Kentucky.

2668. JACOB VAN METER HILBURN, born April 8, 1874, married September, 1891, Ermine Morgan.

2054. SAMUEL TYLER ADAMS, Bowling Green, Ky.

2669. MARY LELAND ADAMS, born October 5, 1879.

2670. LOU MAULEY ADAMS, born November 9, 1881.

2671. GEORGE PORTER ADAMS, born December 18, 1883.

2672. SAMUEL TYLER ADAMS, born July 14, 1885.

2673. CAROLINE EVE ADAMS, born June 26, 1888.

2674. JULIAN PRESTON ADAMS, born November 11, 1891.

NINTH GENERATION.

2056. CHARLES JOSEPH ADAMS, Bowling Green, Ky.

2675. NANNIE ADAMS, born October 17, 1874, died January 12, 1875.

2676. DAISY ADAMS, born February 29, 1876, married November 11, 1897, to Charles Collier

2677. BENJAMIN ADAMS, born April 23, 1877, died September 16, 1877.

2678. WILLIAM CARSON ADAMS, born November 17, 1880.

2679. HARRISON ADAMS, born September 18, 1885.

2057. GEORGE BRADLEY ADAMS, Bowling Green, Ky.

2680. PATSY USHER ADAMS, born May 14, 1884.

2681. KIMBLE ADAMS, born September 6, 1885.

2060. MARY USHER VAN METER, Kansas City, Missouri.

2682. KIRK MILLER.

2683. WALTER MILLER.

2684. CESNA MILLER.

2061. WILLIAM SHARP VAN METER, Bowling Green, Ky.

2685. CLINTON C. VAN METER, born April, 1893.

2686. MARY BELL VAN METER, born 1895.

2062. JAMES VAN METER CLARKSON, Bowling Green, Ky.

2687. CLINTON CLARKSON.

2688. JAMES CLARKSON.

2064. WILLIAM A. OBENCHAIN, Bowling Green, Ky.

2689. MARGARY OBENCHAIN, born September 19, 1886.

2690. WILLIAM ALEXANDER OBENCHAIN, born September 21, 1888.

2691. THOMAS HALE OBENCHAIN, born November 2, 1891.

2692. CECELIA CALVERT OBENCHAIN, born March 25, 1895.

2065. FRANCIS G. OBENCHAIN, Cocoanut Grove, Fla.

2693. JENNETTE BROWN OBENCHAIN, born August 22, 1876, in Bowling Green, Ky.

2694. ELIZABETH SWEETLAND OBENCHAIN, born August 15, 1878, in Forrest City, Ark.

2695. FANNIE MAUDE OBENCHAIN, born January 16, 1881, in Grenada, Mississippi.

2696. ALEXANDRA OBENCHAIN, born April 28, 1885, in Chicago, Ill.

THE BORDEN FAMILY.

2075. WILLIAM E. STRICKLAND, Virginia.

2697. VIRGINIA ELIZABETH STRICKLAND, born December 16, 1865, married W. G. Bruce, February 22, 1885.

2698. SARAH HESTER STRICKLAND, born January 26, 1868; married E. H. Bruce, January 22, 1886.

2699. MARGARET REBECCA STRICKLAND, born January 1, 1870, married M. R. Eskridge, January 10, 1887.

2700. PANDORA ANN STRICKLAND, born November 12, 1871, married J. S. Bennett, April 15, 1887.

2701. LAURA LEONA STRICKLAND, born December 19, 1873, died August 15, 1876.

2102. WILLIE MARGIE STRICKLAND, born January 16, 1876, died July 14, 1878.

2703. NELLIE EWING STRICKLAND, born July 16, 1878, died July 17, 1878.

2704. LILLY MAY STRICKLAND, born July 20, 1881.

2705. JOHN WALTER ELEAZER STRICKLAND, born June 14, 1884.

2706. WILLIAM ANTHONY NEAL STRICKLAND, born June 14, 1884.

2076. SALLIE A. W. STRICKLAND, Virginia.

2707. WALTER SCOTT RIFFE, born January 20, 1865, married Martha Grind Staff, February 17, 1886.

2708. MARTHA HESTER RIFFE, born May 16, 1867, married John Duncan.

2079. SAMUEL R. SWEETLAND, Memphis, Tenn.

2709. SIGNAL WALTER SWEETLAND, born September 1, 1880, in Giles county, Tennessee.

2710. MOLLIE B. SWEETLAND, born July 21, 1882, died October 2, 1885.

2711. HICKMAN SWEETLAND, born April 12, 1884, died March 26, 1885.

2712. ROSS RICHARDSON SWEETLAND, born March 17, 1886.

2713. MEDORA C. SWEETLAND, born April 15, 1888.

2714. ESSIE HAGAN SWEETLAND, born March 23, 1890.

2081. MARY HESTER SWEETLAND, Pattonsburg, Va.

2715. ANNIE LOVE.

2716. CHARLES LOVE.

2717. MARY LOVE.

NINTH GENERATION.

2082. ANNIE E. SWEETLAND, Pattonsburg, Va.

2718. ROBERT WILEY.
2719. NETTIE WILEY.
2720. MATTIE WILEY.
2721. SALLIE WILEY.
2722. EDITH WILEY.
2723. FRANK WILEY.
2724. LAURA WILEY.
2725. WYATT WILEY.
2726. ELIZABETH WILEY.

2084. ELIABETH OBENCHAIN SWEETLAND, Virginia.

2727. MATTIE HILL.
2728. ANNIE HILL.
2729. ELLA HILL.
2730. BULAH HILL.
2731. JAMES HILL.
2732. WALKER HILL.
2733. JOHN HILL.

2085. MAGGIE P. SWEETLAND, Virginia.

2734. WILLIAM SWEETLAND HAILE.
2735. LAURA DRAKEFORD HAILE.

2097. LURA V. SWEETLAND, Nevada county, California.

2736. MAUDE VIRGINIA HEATH, born February 4, 1879.
2737. CHESTER SWEETLAND HEATH, born January 9, 1883.

2123. DANIEL J.

2738. JENNETTE MERAB, born Dec. 14, 1847; married William H. S. Empey, September 5, 1866, in Adams, New York.
2739. BUEL.
2740. WILLIAM CLINE, born May 19, 1858, married Jennie E. Adams, daughter of John Quincy Adams and Emely Lincoln, his wife, in Fulton, New York, October 27, 1883. Dr. W. C. Borden is a surgeon in United States Army
2741. ELIAS.

2135. WILLIAM VAN ZANDT COX, Washington, D. C.

2742. EMERY COX, born May 23, 1888.

THE BORDEN FAMILY.

2743. HAZEL VAN ZANDT COX, born February 14, 1890.
2744. THEODORE G. COX, born August 17, 1894.
      2136. ELIZABETH DENNISSON COX.
2745. ALGERNON SIDNEY SULLIVAN, born ——, died in infancy.
2746. ETHEL VAN ZANDT SULLIVAN, born June 28, 1884.
2747. ACLEEN SULLIVAN, born December 28, 1887, died April 24, 1894.
2748. THEODORE G. SULLIVAN, born April 19, 1895.

TENTH GENERATION

TENTH GENERATION.

2138. THEODORE B. LINNARD, Philadelphia, Pa.

2749. ELIZABETH CHARLOTTE LINNARD, born July 31, 1892.

2750. HELEN EMMA LINNARD, born September 15, 1987.

2143. WILLIAM NOSTON, Philadelphia, Pa.

2751. ELEANOR WILSON, born March 1, 1887, in Philadelphia, Pennsylvania.

2752. JOHN MORROW, born February 20, 1889.

2753. FRANK HARRIS, born April 3, 1893.

2754. JOSEPH WILSON, born July 3, 1895.

2144. THEODORE J., Philadelphia, Pa.

2755. JOHN ALFRED, born 1884, in Philadelphia, Pennsylvania.

2145. MARY MORROW, Philadelphia, Pa.

2756. ANNA ETHEL HARRIS, born March 14, 1888, in Phliaedlphia, Pennsylvania.

2757. GERTRUDE BORDEN HARRIS, born May 13, 1890.

2758. MARRION MORROW HARRIS, born June 1, 1892.

2146. EMMA PAGE, Philadelphia.

2759. BENJAMIN FRANKLIN SUTTON, born November 26, 1892, died February 19, 1893.

2158. LILLIE, Edna, Kansas.

2760. EZRA STEPHEN CHAFFEE, born 1896, died June, 1897.

2761. JOHN CHAFFEE, born December 25, 1897.

2147. ELVIRA LINNARD, Philadelphia, Pa.

2762. EDWARD SEALIEZE KAIN, born in Philadelphia, Pennsylvania, January 20, 1897.

2235. PETER R., Tunkhannock, Pa.

2763. GARRICK MALORY, born June 15, 1875. He is being educated at Cornell University, Ithica, N. Y.

2764. JOHN FRANCIS, born February 15, 1879.

2765. EVART P., born April 19, 1881.

2766. FANNIE ELVINA, born April 19, 1886.

2767. FRANK SPAULDING, born April 19, 1886.

THE BORDEN FAMILY.

2768. GRANVILLE S., born August 18, 1893.

    2232. WALTER AUGUSTUS, Philadelphia, Pa.

2769. WALTER AUGUSTUS, born July 14, 1888.

2770. HAROLD ATLEE, born September 15, 1891.

    2248. ANNIE MORRIS, Red Bank, N. J.

2771. HAZEL TILTON, born 1887.

2772. RICHARD BORDEN TILTON, born 1889.

    2249. MINNIE FRANCES, Red Bank, N. J.

2773. LESLIE BORDEN HILL, born 1891.

    2254. GEORGIANNA A. PONTIN, New Jersey.

2774. CARRIE MILLARD FIELD, born 1884.

2775. EDNA FRANCES FIELD, born 1887.

2776. MABEL WINSTON FIELD, born 1889.

    2255. CARRIE E. PONTIN, New Jersey.

2777. ALBERTINA AUGUSTA PAGENSTECHER, born 1890.

2778. FLORENCE MABEL PAGENSTECHER, born 1892.

2779. CHARLES FREDERIC RUDOLPH PAGENSTECHER, born 1893.

    2256. SARAH EMILY PONTIN, New Jersey.

2780. ESTHER LILLIAN BATES, born 1889.

    2252. JOHN FREDERICK, New Jersey.

2781. ORVILLE CARROLL, born 1892.

2782. HELEN LOUISE, born 1894.

2783. JOHN WHITE, born 1897.

2784. EDWARD ALLAIRE, born 1897.

    2262. MORRIS PEAK, Colorado Springs, Col.

2785. HERMAN FRANK, born January 6, 1884, died July 7, 1887.

2786. ARATHUSA ELLIOTT, born August 11, 1886, died August 6, 1891.

2787. MARJORIE VAN HORN, born April, 1888.

    2267. FRANK EDWARD, New York City.

2788. MARIAN, born in New York, August 11, 1891.

2789. JOHN FRANCIS, born in New York November 30, 1894.

    2297. ISAAC H., Fall River, Mass.

2790. MARY,

TENTH GENERATION.

2791. AMANDA E., born October 31, 1852.

2792. GEORGE E., born February 2, 1855.

2793. ERNEST A., born March 31, 1861; died November 27, 1864.

2314. ABBY, D. K., Fall River, Mass.

2794. JAMES ELGAR BORDEN, born October 2, 1867; married Helen A. E. Pearce, born March 19, 1866.

2795. FRANK HERBERT BORDEN, born January, 1869; married Josephine H. Trasher, September, 1892.

2796. SIDNEY FREEMAN BORDEN, born October 16, 1873.

2299. DAVID B., New Bedford, Mass.

2797. ARTHUR H., born ——, married Mattie ——.

2798. ELMER E., born June, 1882, died February 3, 1894.

2799. WILLIAM BRADFORD, born December 4, 1868, died at Fall River, February 20, 1896, married Fannie Maria Brown.

2309. CHARLES FREDERICK, Fall River, Mass.

2800. IDA EASTMAN, born September 15, 1881.

2801. ROBERT REMINGTON, born July 6, 1884.

2802. EDWARD, born July 6, 1886.

2803. CHARLES FREDERICK, born December 4, 1892.

2175. MARY BORDEN TILTON, New Jersey.

2804. JOSEPH TILTON KESTER, born May 3, 1894.

2805. EDITH KESTER, born September 30, 1898.

2176. JOHN B. TILTON, New Jersey.

2806. ELSIE MARIE TILTON, born November 21, 1891.

2807. MARTIE ROSE TILTON, born June 16, 1895.

2177. ELLA B. TILTON, New Jersey.

2808. PERCY BORDEN, born June 18, 1891.

2809. ETHEL BORDEN, born August 29, 1893.

2182. EMMA L. WAINWRIGHT, New Jersey.

2810. BESSIE W. BUCK, born June 20, 1891.

2811. DOROTHY C. BUCK, born January 11, 1893.

2812. EMILY T. BUCK, born October 15, 1897.

2213. CHARLES S., son of Edward, New Jersey.

2813. EDWARD, born October 24, 1892.

THE BORDEN FAMILY.

2814. VERNON, born October 14, 1897.

2229. GUSTAVUS WYNNE COOK, Philadelphia.

2815. NANNIE WYNNE COOK, born October 26, 1897.

2192. ANNIE P., New Jersey.

2816. LILLIAN LONGSTREET, born May 20, 1892.

2817. RUSSEL LONGSTREET, born May 9, 1894.

2818. GEORGE L. LONGSTREET, born July 15, 1897.

2196. CHARLES C., Jacobstown, N. J.

2819. ALLEN BRADFORD, born August 18, 1895.

2323. HARRIET MINERVA, Fall River, Mass.

THOMAS BORDEN BASSETT, born August 24, 1883.

FREDERIC WATERMAN BASSETT, born April 23, 1885.

MARGARET BASSETT, born January 26, 1888.

CHARLES FRENCH BASSETT, born April 5, 1891, died December 26, 1891.

CONSTANCE BASSETT, born January 19, 1896.

2332. CORINNA KEEN, Philadelphia, Pa.

2820. WILLIAM KEEN FREEMAN, born January, 1899.

2336. ELLEN CORINNA PAINE, Cambridge, Mass.

2821. ELLEN PAINE HULING, born July 24, 1880, in New Bedford, Mass.

2822. ELIZABETH LILLIBRIDGE HULING, born December 7, 1882, in New Bedford, Mass.

2823. ALICE BORDEN HULING, born September 2, 1885, in Fitchburg, New Bedford.

2824. CORINNE WILCOX HULING, born January 10, 1888.

2825. RAY GREEN HULING, born July 13, 1890.

2363. MERT A., Fort Pierre, South Dakota.

2826. RUTH, born July 14, 1896.

2366. NATHANIEL BRIGGS, Fall River.

Daughter, born January 8, 1899.

2350. MARY EMMA, Fall River, Mass.

2827. MADELINE PIERCE.

2401. CHARLES EDGAR, New York City.

2828. CLAUDIUS ROOME, born January 2, 1893.

TENTH GENERATION.

2420. GAIL BORDEN JOHNSON, Los Angeles, Cal.

2829. PHILA BORDEN JOHNSON, born in Houston, Tex., March 13, 1882.

2830. RAY LUCILE JOHNSON, born in Elgin, Illinois, January 19, 1885.

2831. VIRGINIA LESLIE JOHNSON, born in Elgin, Illinois, July 19, 1886.

2421. WALDO PASCHAL JOHNSON, Alhambra, Cal.

2832. PASCHAL BORDEN JOHNSON, born in Elgin, Illinois, July 28, 1886.

2833. COLVIN LEE JOHNSON, born Elgin, Ill., June 1, 1888.

2834. GAIL BROWN JOHNSON, born in Alhambra, California, October 23, 1889.

2835. RICHARD WALDO JOHNSON, born in Alhambra, California, April 27, 1891.

2422. VIRGINIA LEE JOHNSON, Yonkers, N. Y.

2836. PHILA NICHOLS MILBANK,, born November 17, 1887, in Yonkers, New York.

2837. LAURENCE MILBANK, born in Yonkers, New York, November 26, 1889.

2838. LEE BORDEN MILBANK, born in Yonkers, New York, April 6, 1897.

2423. EMMA ORA JOHNSON, Washington, D. C.

2839. ROLDEN LEE BORDEN, born March 15, 1888, in Washington, D. C.

2840. CARRIE M'KEE BORDEN, born October 11, 1890, in Washington, D. C.

2841. LEWIS COLONNA BORDEN, born January 11, 1892.

2424. DR. MILBANK JOHNSON, Los Angeles, California.

2842. LOUIEZ LESTER JOHNSON, born August 23, 1894, in Alhambra, California.

2843. EVELYN GAIL JOHNSON, born December 7, 1897, in Alhambra, California.

2425. CARRIE BORDEN JOHNSON, Pasadena, California.

2844. PHILADELPHIA BORDEN MILLER, born April 17, 1896, in Port Huron, Michigan.

2845. JOHN BORDEN MILLER, born August 4, 1897, in Pasadena, California.

THE BORDEN FAMILY.

2426. ESTHER CHARLOTTE MUNSILL, New Haven, Ct.

2846. LESLIE BORDEN ULRICH born in New Haven, Connecticut, August 30, 1889.

2847. ANNA MARGUERITE ULRICH, born in New Haven, Connecticut, October 27, 1892.

2848. JOHN MUNSILL ULRICH, born in New Haven, Connecticut, June, 1897.

2428. MARCUS MUNSILL, Hartford, Conn.

2849. MARCUS MILLS MUNSILL, born January, 1893.

2431. GAIL, Alhambra, California.

2850. ALEXINE RAMONA, born January 26, 1896, at Alhambra, California.

2436. JAMES M'KEE, New York City.

2851. MARY M'KEE, born July 5, 1898, in Washington, D. C.

2439. HATTIE LOVISA, Los Angeles, California.

2852. LESLIE LEE HARCOURT, born September 8, 1882, in Beaumont, Texas, died June 16, 1885, in Elgin, Illinois.

2853. GAIL JOHNSON WELD, born in Elgin, Illinois, November 28, 1886, died November 29, 1886.

2854. MILDRED BORDEN WELD, born in Alhambra, California, July 29, 1888, died in Weatherford, Texas, July 5, 1891. Is buried in Alhambra, California.

2855. JAMES ROMAINE WELD, born in Alhambra, California, November 2, 1890.

2856. ELEANOR MARION WELD, born in Alhambra, California, April 20, 1895.

2441. FLORA PARKER, Weatherford, Texas.

2857. NELLIE LEE DAVIS, born in Weatherford, Texas, March 22, 1886.

2858. WINNIE DAVIS, born in Weatherford, Texas, October 10, 1892.

2859. AUGUSTUS BORDEN DAVIS, born in Weatherford, Texas, December 12, 1894.

2546. GEORGE A., Fall River, Mass.

FREDERICK C., born in Fall River, is proprietor of Borden's Bakery.

2461. CARRIE L. DUFFEL, West, McLennan county, Texas.

2860. MARSHALL BORDEN THOMPSON, born December, 1897.

2462. M. LILLIE DUFFEL, McLennan county, Texas.

2861. HARVEY L. COBB, born February, 1894.

TENTH GENERATION. 313

2499. BRADFORD, Fall River, Mass.

2862. ROBERT HENRY, born September 26, 1860.
2863. SAMUEL NELSON, born March 19, 1862.
2864. WILLIAM EDWARD, born December 6, 1863.
2865. SARAH JEANETTE, born June 28, 1856.

2550. MARY EMELINE LOVEL, Philadelphia, Pa.

2866. EMER LOVEL MINNICH, born ——, died in infancy.
2867. GEORGE WESCOTT MINNICH, born March 1, 1876, died November 6, 1889.
2868. MICHAEL REED MINNICH, died in infancy.
2869. MARIE LE VALLEY MINNICH, born February 14, 1878.
2870. CLARA FRANK MINNICH, born August 19, 1882.
2871. SIDNEY LEE MINNICH, died in infancy.
2872. CHARLES HOWARD MINNICH, born October 19, 1887.

2552. CHARLES DEWEY, Goldsboro, North Carolina.

2873. CHARLES FRANCIS DEWEY, born August 31, 1875, died November 6, 1876.
2874. MARY ALICE WEBBER DEWEY, born November 24, 1876, died September 14, 1880.
2875. AUGUSTA DEWEY, born February 21, 1878, died October 8, 1880.
2876. GEORGE STEELE DEWEY, born August 19, 1881.
2877. FRANK KORNEGAY DEWEY, born June 6, 1883, died 1886.
2878. THOMAS AUGUSTUS DEWEY, born June 6, 1883.
2879. HANNAH DEWEY, born June 24, 1885.
2880. ERNEST MILLER DEWEY, born December 23, 1886.
2881. HARRIET MARIA DEWEY, born January 17, 1889.
2882. EDWIN STEELE DEWEY, born August 14, 1891, died January, 1892.

2554. ANN MARIA DEWEY, Goldsboro, N. C.

2883. CHARLES DEWEY SLOCUMB, born January 8, 1886.
2884. LOUISA KORNEGAY SLOCUMB, born November 30, 1889.

2556. THOMAS WEBBER DEWEY, New Bern, North Carolina.

2885. EDWIN MIAL DEWEY, born June, 1887.
2886. VICTORIA LE MAY DEWEY, born October, 1889.
2887. BROWNRIGG HEFFENON DEWEY, born August, 1892.

THE BORDEN FAMILY.

2559. EDWIN, Wilmington, N. C.

2888. MARY, born December 31, 1879.
2889. ALICE WALLACE, born April 13, 1883.
2890. EDWIN, born February 14, 1886, died 1886.
2891. DURALD STOCKTON, born January 28, 1891.

2560. MARIE L., North Carolina.

2892. GEORGIA HULSE M'LEOD, married M. L. Stover.
2893. MAUDE ALLISON M'LEOD.

2561. JAMES COLE, New Orleans, Louisiana.

2894. JAMES COLE, born December 21, 1881.
2903. CHARLES BORDEN LYNCH, born March 26, 1892.
2896. EDWIN H., born May 9, 1888.
2897. MARY S., born October 10, 1890.

2562. MINNIE B., Goldsboro, N. C.

2898. JAMES BORDEN LYNCH, born January 29, 1883.
2899. GEORGE GREEN LYNCH, born September 20, 1884.
2900. HERBERT ADOLPHUS LYNCH, born August 17, 1886.
2901. CARRIE BORDEN LYNCH, born 1888; died 1888.
2902. MARY BORDEN LYNCH, born 1890; died 1890.
2903. CARLES BORDEN LYNCH, born March 26, 1892.
2904. RUTH BRADFORD LYNCH, born April 4, 1895.
2905. ANNIE MEAD LYNCH, born October 20, 1897.

2567. KATE COLE, North Carolina.

2906. MARY N. WOOTTEN,
2907. BELLE C. WOOTTEN,

2568. ANNIE M., North Carolina.

2908. ALLAN C. BANNER, born January 12, 1896.

2569. ARNOLD, Goldsboro, N. C.

2909. ANNIE BELLE, born August 11, 1880.
2910. RACHAEL HEMPHILL, born March 12, 1884.
2911. EUNICE, born June 17, 1892.
2912. JAMES HEMPHILL, born December 27, 1894; died March 25, 1895.

2570. ANNIE B., North Carolina.

2913. EDWIN LEE, born February 19, 1884.
2914. GEORGIA LEE, born June 10, 1886.

TENTH GENERATION. 315

2571. FRANK KORNEGAY, Goldsboro, N. C.

2915. FRANK KINNON, born September 30, 1888, in Goldsboro, N. C.
2916. ARNOLD, born December 18, 1889; died August 29, 1890.
2917. JULIA, born August 13, 1892.
2918. MILDRED BROWNRIGG, born July 28, 1891, in Goldsboro, N. C.
2919. EDWIN BROWNRIGG, born February 6, 1895,
2920. SARAH ELIZABETH, born July 6, 1897.

2573. LOUISA, North Carolina.

2921. ANNIE LEE DAVIS, born February 26, 1888.
2922. ADAM CLARK DAVIS, born November 21, 1889.
2923. GEORGIA LAVINIA DAVIS, born October 2, 1891.
2924. EDWIN BORDEN DAVIS, born March 1, 1894.

2572. CARRIE W..

2925. FRANK BORDEN DANIELS, born March 25, 1887.
2926. GEORGIA SEABROOK DANIELS, born July 25, 1889.
2927. MARY CLEAVES DANIELS, born March 26, 1894.

2576. JOHN LEMUEL, Goldsboro, N. C.

2928. ROWENA HICKS, born July 27, 1892.
2929. MARY CARROW, born January 3, 1894.
2930. MARGARET DOANE, born March 30, 1895.
2931. VIRGINIA WHITFIELD, born August 7, 1897.
2932. CATHERINE WENTWORTH, born March 6, 1899.

2584. CHARLES BORDEN MILLER, Goldsboro, N. C.

2933. ETHEL BURWELL MILLER, born September 9, 1894.
2934. CHARLES BORDEN MILLER, Jr., born February 19, 1898.

2585. LOUISE MILLER, North Carolina.

2935. SARA BORDEN MICHAUX, born November 7, 1893.
2936. MARY LOUISE MICHAUX, born June 13, 1895.
2937. EDWIN RANDOLPH MICHAUX, born December 13, 1897.

2591. JULIA, Goldsboro, N. C.

2938. WILLIAM BORDEN JERMAN, born 1889, September 9.

2592. WALTER EUGENE, Goldsboro, N. C.

2939. WALTER ECCLES, born November 12, 1894.
2940. THOMAS FULLER, born November 20, 1897.

THE BORDEN FAMILY.

2593. GEORGIA, North Carolina.
2941. WILLIAM BORDEN COBB, born August, 1894.
2942. DONALD BROWNRIGG COBB, born July 8, 1898.
2605. LABAN GILBERT, Chicago, Ill.
2943. HAZEL MAUD, born August 6, 1892.
2944. HELEN ELSIE, born November 12, 1893.
2655. MATTIE J. HEANEY, Bowling Green, Ky.
2945. KATHERINE WHITE, born in Wilbur, Washington, August 14, 1895.
2663. CARRIE ADAMS, Bowling Green, Ky.
2946. LOUIS CREIDER, born 1887.
2947. CHARLES CREIDER, born 1889.
2667. JULIETTE C. HILBURN, Bowling Green, Ky.
2948. MARY ANNIS PRESTON, born in Bowling Green, Ky., November 6, 1896.
2668. JACOB VAN METER HILBURN, Bowling Green, Ky.
2949. MORGAN TYLER HILBURN, born June 9, 1897.
2697. VIRGINIA ELIZABETH STRICKLAND.
2950. WILLIAM BRUCE.
2698. SARAH HESTER STRICKLAND.
2951. MAUDE BRUCE, born 1887.
2699. MARGARET R. STRICKLAND.
2952. CHARLES WILLIAM ESKRIDGE, born July 27, 1898.
2707. WALTER S. RIFFE.
2953. PETER B. RIFFE, born December 16, 1886.
2708. MARTHA HESTER RIFFE.
2954. SUSAN DUNCAN.
2700. PANDORA A. STRICKLAND.
2955. ALLIE RANDOLF BARNETT, born October 13, 1888.
2738. JENNETTE MERAB, Clinton, Mo.
2956. EDWARD BORDEN EMPEY, born June 17, 1867.
2957. FLORENCE IRENA EMPEY, born May 31, 1881.
2740. DR. WILLIAM C., U. S. A.
2958. DANIEL LE RAY, born October 25, 1887.
2959. WILLIAM AYRES, born March 20, 1890.

ELEVENTH GENERATION.

ELEVENTH GENERATION.

2793. JAMES EDGAR, Fall River, Mass.
2960. EDNA LOUISE, born March 13, 1894.
2961. PHILIP PEARCE, born June 21, 1898.
    2794. FRANK HERBERT, Fall River, Mass.
2962. HILDRETH, born July 10, 1893.
2963. EARL KINSLEY, born February 5, 1895; died July 24, 1896.
    2797. ARTHUR H., New Bedford, Mass.
2964. ELMER ESTES, born January 17, 1899.
    2892. GEORGIA HULSE MC LEOD, North Carolina.
2965. MARY LOUISE STOVER, born June 30, 1897.

APPENDIX.

374. FRANCIS, New Jersey.

FRANCIS, his son, married Hannah Lambert Holmes and had eight children as follows:

1. WILLIAM LAMBERT, married C. Godett in 1871.
2. RANDOLPH; married Caroline Maxon in 1879.
3. ABRAHAM HOLMES; married Emily Bunn in 1886.
4. FRANCIS.
5. MARGARET.
6. JERUSIA L.
7. OLIVIA.
8. ANNIE.

1. WILLIAM LAMBERT.

9. HARRY; marired Sarah Shuts.
10. GRACE; married Edward ——
11. LOTTIE; married G. Brown.
12. WILLIAM.

2. RANDOLPH.

13. NELLIE.
14. MAY.
15. BESSIE.

3. ABRAHAM HOLMES.

16. CHARLES.
17. FRANCIS.

1440. AMASA, Fall River.

GEORGE A. BORDEN, son of No. 1440, Amasa of Fall River; married Patience W. Shaw, December 29, 1847. They had five children as follows:

1. WILLIAM H., born June 11, 1849, he married and had one child Lillian W., Fall River
2. GEORGETTA L., born July 21, 1856. Died February 25, 1863.
3. FRANK L., born February 10, 1861. He married and has three children Ernest Raymond and Robert.
4. FRED C., born June 17, 1864, Fall River.
5. FLORENCE V., born October 24, 1867; married——Johnston of Fall River they have four children Marion, George, Spencer and Dana Johnston. The last two are twins.

THE BORDEN FAMILY.

2117. ARIANA W. CARPER, Virginia.

1. MARY CLAUD BEAN, born March 5, 1868.
2. JOSEPH HERBERT BEAN, born March 4, 1871; married November 15, 1897; Adeline Peters; they have one child, Mary Bennett Bean, born September 20, 1898.
3. ROBERT BENNETT BEAN, born March 24, 1874.
4. ANNA KATHERINE BEAN, born January 19, 1877.
5. WYNDHAM RANDOLPH BEAN, born November 7, 1878.
6. ARTHUR PARSONS BEAN, born October 24, 1881.
7. GEORGE EDWARD BEAN, born April 28, 1885.
8. VIRGINIA HAMPTON BEAN, born May 2, 1889.

INDEX.

A

464.	Aaron	Fall River, Mass.
661.	Aaron	Monmouth, N. J.
663.	Aaron, R.	Emeleytown, N. J.
949.	Abbey, (Borden)	Fall River, Mass.
1179.	Abbie, A., (Stafford)	
2314.	Abby	Fall River, Mass.
1868.	Abby, Arnold	Fall River, Mass.
961.	Abby, (Carlton)	Slatersville.
1887.	Abby, D.	Fall River, Mass.
2545.	Abby, Jane	Fall River, Mass.
1878.	Abby, (Little)	Fall River, Mass.
940.	Abel	Fall River, Mass.
941.	Abel	Fall River, Mass.
514.	Abel	Fall River, Mass.
469.	Abel	Fall River, Mass.
200.	Abigail	Burlington, N. J.
2508.	Abigail	Fall River, Mass.
251.	Abigail, (Durfee)	Fall River, Mass.
257.	Abigail, (Durfee)	Gloucester, R. I.
164.	Abigail, (Pritchard)	Virginia.
472.	Abigail, (Wordell)	Fall River, Mass.
470.	Abner	Fall River, Mass.
1790.	Abner, D.	Fall River, Mass.
244.	Abraham	Fall River, Mass.
476.	Abraham	Fall River, Mass.
508.	Abraham	Fall River, Mass.
754.	Abraham	Fall River, Mass.
896.	Abraham	Fall River, Mass.
934.	Abraham	Fall River, Mass.
2512.	Abraham	Fall River, Mass.
310.	Abraham	Havana, Cuba.
33.	Abraham	Newport, R. I.
421.	Abraham	Nova Scotia.
124.	Abraham	Portsmouth, R. I.
193.	Abraham	Providence.
399.	Abraham	Tiverton.
532.	Abraham	Tiverton, R. I.
743.	Abraham	Westport.
733.	Abraham, B.	Fall River, Mass.
1862.	Abraham E.	
		Fall River, Mass.
1053.	Abraham, E.	Tiverton, R. I.
2378.	Abraham, O.	Fall River, Mass.
1356.	Ada	New Jersey.
1839.	Ada, M	California.
1939.	Ada, M	Mass.
463.	Adams	Fall River, Mass.
906.	Adams	Fall River, Mass.
2678.	Adams, Benjamin C.	Ky.
2673.	Adams, Caroline Eve	Ky.
2662.	Adams Carrie, (Creider)	Ky.
2056.	Adams Chas J	Kentucky.
2676.	Adams, Daisy, (Collier)	Ky.
2057.	Adams, George Bradley	Kentucky.
2671.	Adams, George Porter	Ky.
2679.	Adams, Harrison	Ky.
2055.	Adams, Julia W., (Carson)	Kentucky.
2674.	Adams, Julian P	Ky.
2681.	Adams, Kimble	Ky.
2670.	Adams, Lou McAnley	Ky.
2053.	Adams, Mary, (Hilburn)	Kentucky.
2669.	Adams, Mary Leland	Ky.
2680.	Adams, Patsy Usher	Ky.
2054.	Adams, Samuel, T	Kentucky.
2672.	Adam, Samuel Tyler	Ky.
2052.	Adams William, U	Kentucky.
2650.	Addie Lou	Dallas, Texas.
2312.	Adelaide, (Paine)	Providence, R. I.
2445.	Adine Demis	Galveston, Texas.
1699.	Admiral, F	Los Angeles, Cal.
780.	Adolphus, K	Nova Scotia.
1506.	Adrienne, J	Mass.
1383.	Adrienne, (Lovel)	Fall River, Mass.
1438.	Alanson	New Bedford.
1355.	Albert	New Jersey.
2535.	Albert, H	Fall River, Mass.

INDEX.

No.	Name	Location
1912.	Albert, H	Tiverton, R. I.
2608.	Alden Edson	Fall River, Mass.
924.	Alexander	Fall River, Mass.
1405.	Alexander	Fall River, Mass.
1306.	Alexander	New Jersey.
2850.	Alexine Romona	Alhambra, Cal.
2342.	Alfred	Fall River, Mass.
1603.	Alfred, C	Cornwallis, N. S.
2374.	Alfred, M	Fall River, Mass.
2501.	Alfred Pardy	Fall River, Mass.
1977.	Alfred, R	Greensboro, Ala.
154.	Alice	Beauford, N. C.
2454.	Alice	Sharpsburg, Texas.
2008.	Alice Ashley	Alabama.
2600.	Alice Brownell	Providence, R. I.
2346.	Alice Esther	Fall River, Mass.
1913.	Alice Evelina	Tiverton, R. I.
2356.	Alice, (Hicks)	Fall River, Mass.
2397.	Alice L	Nova Scotia.
528.	Alice, (Manchester)	Tiverton, R. I.
1970.	Alice Moore	New Berne, Ala.
2889.	Alice Wallace	Wilmington, N. C.
317.	Alice, (Ward)	Beaufort, N. C.
2895.	Allan, H	New Orleans.
2719.	Allen Bradford	New Jersey.
960.	Almira	Tiverton, R. I.
1341.	Almira, C., (Rice)	Michigan.
1069.	Almira, (Skiff)	Cazenovia, N. Y.
1007.	Alonso	Fall River, Mass.
1018.	Alphonso	Fall River, Mass.
1601.	Althea, M	Cornwallis N. S.
1814.	Amanada, M	Mass.
1792.	Amanada, M., (Taylor)	Mass.
1533.	Amanda	Nova Scotia.
2234.	Amanda, (Brown)	New Jersey.
2298.	Amanda, C., (Hill)	Mass.
793.	Amanda, (Caldwell)	N. S.
2791.	Amanda, E	Fall River, Mass.
1558.	Amanda, (Lockhart)	Mass.
2612.	Amanda Fierce, Somerset, Mass.	
1442.	Amantha, Tepper	Fall River, Mass.
1041.	Amasa	Fall River, Mass.
1440.	Amasa	Fall River, Mass.
1237.	Amelia, (Canes)	
68.	Amey	Bordentown, N. J.
38.	Amey, (Chase)	Tiverton, R. I.
1772.	Amey Maria	Fall River, Mass.
12.	Amey, (Richardson)	Flushing, L. I.
766.	Amey, (Rodie)	Nova Scotia.
894.	Amos	Fall River, Mass.
1626.	Amos	Mass.
355.	Amos	Middletown, N. J.
1252.	Amos	Red Bank, N. J.
1462.	Amy	Fall River, Mass.
108.	Amy	N. J.
2217.	Amy	New Jersey.
1509.	Amy	Nova Scotia.
102.	Amy, (Foy)	Shrewsbury, N. J.
231.	Amy, (Morris)	Manasquan, N. J.
186.	Amy, (Potts)	Bordentown, N. J.
408.	Amy, (Westgate)	Tiverton.
734.	Amy, (Wilcox)	Fall River, Mass.
1908.	Andoniram	Fall River, Mass.
1041.	Andrew	Fall River, Mass.
1387.	Andrew	Fall River, Mass.
2313.	Andrew	Fall River, Mass.
792.	Andrew	Grand Prie, N. S.
1507.	Andrew	Mass.
1205.	Andrew	Philadelphia, Pa.
1157.	Andrew C	Waco, Texas.
1416.	Andrew, J	Fall River, Mass.
1311.	Andrew, J	New Jersey.
1432.	Angenette	Fall River, Mass.
1242.	Ann	Burlington, N. J.
1593.	Ann	Cornwallis, N. S.
294.	Ann	Fall River, Mass.
560.	Ann	Fall River, Mass.
1042.	Ann	Fall River, Mass.
221.	Ann	Monmouth.
126.	Ann	Portsmouth, R. I.
951.	Ann, (Crary)	Fall River, Mass.
749.	Ann, (Chase)	Portsworth
250.	Ann, (Durfee)	Fall River, Mass.
2311.	Ann Eliza	Fall River, Mass.
1490.	Ann, Eliza, (Shearman)	Nova Scotia.
1308.	Ann Eliza, (Zelley)	New Jersey.
2517.	Ann F., (Davis)	Fall River, Mass.
1705.	Ann, (Fowler), (Marvel)	Mass.
179.	Ann, (Hance)	Freehold, N. J.
340.	Ann, (Hopkinson)	Philadelphia, Pa.
682.	Ann, (Kiby)	N. J.
387.	Ann L., (Hance)	N. J.
1646.	Ann, M., (Capwell)	Fall River, Mass.
677.	Ann, (Potts)	N. J.
30.	Ann, (Slocum)	Providence, R. I.
581.	Ann T	Beaufort, N. C.
844.	Anna	Scituate, R. I.
493.	Anna, (Cook)	Tiverton.
2140.	Anna F., (Meyers)	Philadelphia, Pa.
479.	Anna, (Fisher)	Fall River, Mass.
1139.	Anna, (Given)	Alabama.
1118.	Anna Helen	San Francisco, Cal.

2014.	Anna Hopkins..........New York.
2324.	Anna Howland, Fall River, Mass.
709.	Anna, L., (Montgomery)..........
2377.	Anna, M.......Fall River, Mass.
1666.	Anna, M..........Scituate, R. I.
1836.	Anne, E........................
	..................North Carolina.
859.	Anne, E., (Cameron)..............
	................Fall River, Mass.
274.	Anne, (Jameson).................
	..................Fall River, Mass.
1239.	Annie..........Burlington, N. J.
1152.	Annie, (Alexander).........Tenn.
2570.	Annie, B........Goldsboro, N. C.
2919.	Annie Belle......Goldsboro, N. C.
2367.	Annie Brown....Fall River, Mass.
2627.	Annie Byrn.....Fernandina, Fla.
1586.	Annie E........Cornwallis, N .S.
2479.	Annie, E. L....Fall River, Mass.
1100.	Annie, H., (Jackson....Alabama.
618.	Annie, (Johnson)................
	................Fall River, Mass.
1942.	Annie, L.....................Mass.
2372.	Annie Laurie....Fall River, Mass.
1999.	Annie Lewis, (McGhee)...........
	........................Alabama.
2568.	Annie, M., (Banner).............
2248.	Annie, M., (Tilton)..New Jersey.
1535.	Annie Maria, (Borden)............
2192.	Annie, P., (Longstreet)......N. J.
2206.	Apollo, W.......................
665.	Apollo, W....................N. J.
1045.	Ardelia.........Fall River, Mass.
1893.	Ardelia, A......Fall River, Mass.
794.	Ardelia, A., (Fisher).........N. S.
794.	Ardelia, A., (Fisher), ...
	..................Nova Scotia.
1377.	Ardelia, (Brow)..................
	................Fall River, Mass.
858.	Ardelia, (Scarle)................
	................Fall River, Mass.
2408.	Arietta, J..........Boston, Mass.
491.	Arnold ....Fall River, Mass.
967.	Arnold.........Goldsboro, N. C.
2569.	Arnold.........Goldsboro, N. C.
2368.	Arnold Buffum...................
	................Fall River, Mass.
1655.	Arthur..............Scituate, R. I.
2456.	Arthur........Sharpsburg, Texas.
1538.	Arthur, C...................Japan.
2396.	Arthur E..............Nova Scotia.
2797.	Arthur H..........New Bedford.
1553.	Arthur, H..........Nova Scotia.
1804.	Arthur, R.......Fall River, Mass.
2523.	Arthur, T......Fall River, Mass.
197.	Asa.............Burlington, N. J.

430.	Asa ....Canada.
559.	Asa............Fall River, Mass.
835.	Asa ....Quebec.
2400.	Asa Herbert...............Canada.
316.	Asher............Allentown, N. J.
689.	Asher................New Jersey.
1303.	Asher................New Jersey.
2243.	Asher................New Jersey.
1805.	Ashiel, M........Fall River, Mass.
1841.	Aubrey, N.........Nova Scotia.
1561.	Augusta..........Chelsea, Mass.
2286.	Augusta Hopper....Auburn Neb.
1595.	Augusta, M....Cornwallis, N. S.
1542.	Augusta Maria.....Nova Scotia.
1989.	Augusta, N..............Alabama.
843.	Auldis.............Scituate, R. I.
2484.	Auretia..........Fall River, Mass.
1009.	Avery ....
510.	Avis, (Bessey)..Fall River, Mass.
478.	Avis, (Hutchin's)..................
	..................Killingly, Conn.

B

2908.	Banner, Allen C..North Carolina.
580.	Barclay..........Beaufort, N. C.
2955.	Barnett Allie R........Kentucky.
1602.	Barry...........Cornwallis, N. S.
	Bassett, Charles T...............
	................Fall River, Mass.
	Bassett, Constance ....
	................Fall River, Mass.
	Bassett Frederick W.............
	................Fall River, Mass.
	Bassett, Margaret ....;
	................Fall River, Mass.
2819.	Bassett, Thomas B...............
	................Fall River, Mass.
1066.	Bateman........Cazenovia, N. Y.
2780.	Bates, Esther L.....New Jersey.
1375.	Bailey..........Fall River, Mass.
964.	Bailey, E........Providence, R. I.
334.	Beck, Adam ....Tennessee.
331.	Beck, Benjamin....Giles Co., Va.
338.	Beck, Hannah (Hohn)..Virginia.
337.	Beck, Hester (Van Meter).......
	.... Virginia.
332.	Beck, Jacob.........Augusta, Va.
333.	Beck, John ....Virginia.
336.	Beck, Joseph ....Virginia.
335.	Beck, Mary (Carper)....Virginia.
1086.	Benjah ....Fall River, Mass.
520.	Benjah ....Tiverton.
1256.	Beulah, (Bullock)....New Jersey.
1124.	Benjamin ....Alabama.
593.	Benjamin.........Beaufort, N. C.

INDEX.

160. Benjamin ....Botetourte, Va.
11. Benjamin...Burlington Co., N. J.
432. Benjamin ....Canada.
62. Benjamin....Cooper Creek, N. J.
814. Benjamin........Cornwallis, N. S.
99. Benjamin........Eversham, N. J.
293. Benjamin......Fall River, Mass.
407. Benjamin......Fall River, Mass.
1710. Benjamin......Fall River, Mass.
1724. Benjamin......Fall River, Mass.
2492. Benjamin......Fall River, Mass.
2511. Benjamin......Fall River, Mass.
206. Benjamin........Gloucester, N. J.
115. Benjamin......Manasquan, N. J.
64. Benjamin........Middleton, N. J.
386. Benjamin............New Jersey.
260. Benjamin..........Newport, R. I.
602. Benjamin..........North Carolina.
419. Benjamin ....Nova Scotia.
992. Benjamin............Nova Scotia.
129. Benjamin......Portsmouth, R. I.
35. Benjamin........Providence, R. I.
236. Benjamin..........Rumson, N. J.
142. Benjamin........Swansea, Mass.
249. Benjamin........Tiverton, R. I.
271. Benjamin........Tiverton, R. I.
443. Benjamin ....Tiverton.
889. Benjamin..........Tiverton, R. I.
954. Benjamin..........Tiverton, R. I.
1732. Benjamin..........Tiverton, R. I.
44. Benjamin ....Virginia.
318. Benjamin..........Beaufort, N. C.
325. Benjamin ....Virginia.
1106. Benjamin, C.............Alabama.
1476. Benjamin C.....Fall River, Mass.
1029. Benjamin, D...Fall River, Mass.
1495. Benjamin F....Fall River, Mass.
2032. Benjamin, F.........Oxford, Fla.
1584. Benjamin, H...Cornwallis, N. S.
1969. Benjamin, R............New Bern.
1487. Bernice C (Bassett)...............
2353. Bertha.........Fall River, Mass.
2216. Bertha, (Jones) ......New Jersey.
2651. Bertie, Lake.........Dallas, Tex.
2328. Bertram H............New York.
2351. Bessie.........Fall River, Mass.
1958. Bessie ....California.
1962. Bessie, Byrn.........Mississippi.
1559. Bessie, (Crocker) ....
........................Providence, R. I.
872. Betsey, (Durfee)..Tiverton, R. I.
1059. Betsey, (Hanchett)....New York.
676. Betsey, (Hayes) ....Ohio.

2529. Betsey J., (Borden)...............
........................Fall River, Mass.
1665. Betsey, P.........Scituate, R. I.
396. Betsey, (Valentine) ....
........................Fall River, Mass.
2633. Blackburn, Burr........Alabama.
2171. Blanche E......Jacobstown, N. J.
2394. Blanche T............Nova Scotia.
2319. Bliss, Cornelius N....New York.
897. Bradford........Fall River, Mass.
2499. Bradford.......Fall River, Mass.
1712. Bradford, D...Fall River, Mass.
860. Bradford, Marcy (Turpin)........
........................Providence, R. I.
861. Bradford, Ruth.Providence, R. I.
1519. Brenton ....Nova Scotia.
2951. Bruce, Maude ....Kentucky.
2950. Bruce, William........Kentucky.
2810. Buck, Bessie......New Jersey.
2811. Buck, Dorothy......New Jersey.
2812. Buck, Emily T......New Jersey.
2729. Buel ....Missouri.
2169. Bullock, Eliza C.....New Jersey.
2170. Bullock, Henry E...New Jersey.
1991. Burr, Ann A (Blackburn)........
........................Alabama.
1996. Burr, Borden H..........Alabama.
1993. Burr, Esther W., (Robinson)....
........................Alabama.
1995. Burr, Lydia A., (Hicks).........
........................Alabama.
1990. Burr, Lydia H..........Alabama.
1994. Burr, Willie Milton.....Alabama.
1992. Burr, Zaidee L., (Henderson)....
........................Alabama.
2212. Bussom, Elmer W...New Jersey.
2208. Bussom, Emma T., (Nutt)........
........................New Jersey.
2210. Bussom, Joel......New Jersey.
2209. Bussom, Richard....New Jersey.
503. Byard..............Nova Scotia.
986. Byard..............Nova Scotia.
1555. Byron, C.........Sackville, N. B.

C

2275. Caffrey, Loretta V.....New York.
1062. Caleb.................New York.
1973. Campbell, Borden M....Alabama.
1972. Campbell, Edward F....Alabama.
1974. Campbell, Martha F....Alabama.
2293. Carleton, G........Shelton, Conn.
1397. Caroline.......Fall River, Mass.
1329. Caroline.............New Jersey.
2224. Caroline.............New York.

INDEX.   325

1806.	Caroline H., (Waring)............Massachusetts.	1909.	Celia............Fall River, Mass.
2003.	Caroline L...............Alabama.	1454.	Chace, Alfred C.....Portsmouth.
2003.	Caroline, L............Alabama.	1454.	Chace, Alfred C............Portsmouth, R. I.
944.	Caroline M., (Oswell) (Brown).	1447.	Chace, Amy, A., (Almy)............Portsmouth.
1884.	Caroline, (Nash)..Massachusetts.		
1551.	Caroline O...........Nova Scotia.	1447.	Chace, Amy A., (Almy)............Portsmouth, R. I.
1119.	Caroline Sneed.........Alabama.		
2233.	Caroline, T., (Brown)............New Jersey	1448.	Chace, Borden......Portsmouth.
		1448.	Chace, Borden ....Portsmouth, R. I.
2296.	Caroline, T.,(Robinson)............Fall River, Mass.	1452.	Chace, Charles......Portsmouth.
2117.	Carper, Ariana W., (Bean)............Virginia.	1452.	Chace, Charles ....Portsmouth, R. I.
2114.	Carper, Elizabeth, (Echols)............Virginia.	1451.	Chace, Eliza, (Fowler)............Brooklyn, N. Y.
1196.	Carper, George.........Virginia.	2603.	Chace, Hattie " . ...Fall River, Mass.
1195.	Carper, James............Virginia.	2602.	Chace, Louisa S............Fall River, Mass.
1197.	Carper, Joseph..........Virginia.		
2115.	Carper, Robert ....Virginia.	1453.	Chace, Nathaniel B..Portsmouth.
2116.	Carper, Wyndham......Virginia.	1453.	Chace, Nathaniel B............Portsmouth, R. I.
2162.	Carr, Howard E....Bordentown.		
2194.	Carrie..............New Jersey.	1449.	Chace, Philip........Portsmouth.
2326.	Carrie, L........Fall River, Mass	1449.	Chace, Philip ....Portsmouth, R. I.
2840.	Carrie, McKee.Washington, D. C.		
2572.	Carrie W., (Daniels)............Wilmington, N. C.	1450.	Chace, Sarah, (Davol)............Brooklyn, N. Y.
2117.	Casper, Ariana W, (Bean)............Virginia.	1450.	Chace, Sarah, (Daval)............Portsmouth, R. I.
2114.	Casper, Elizabeth, (Echols)............Virginia.	1446.	Chace, Simon B.....Portsmouth.
2115.	Casper, Robert..........Virginia.	1446.	Chace, Simon B............Portsmouth, R. I.
2116.	Casper, Wyndham......Virginia.		
1717.	Cassandra......Fall River, Mass.	2761.	Chaffee, John........Edna, Kan.
2488.	Cassandra......Fall River, Mass.	657.	Charles........Burlington, N. J.
1549.	Cassie, B., (Gilmore)............Nova Scotia.	1240.	Charles........Burlington, N. J.
		972.	Charles......Fall River, Mass.
156.	Catherine ....Beaufort, N. C.	1245.	Charles............New Jersey.
1513.	Catherine ....Nova Scotia.	1357.	Charles............New Jersey.
1145.	Catherine ....Tennessee.	2238.	Charles............New Jersey.
2268.	Catherine, A..........New York.	1733.	Charles............Tiverton, R. I.
803.	Catherine A., (Dawson)..........	2536.	Charles A......Fall River, Mass.
181.	Catherine, (Britton) ....Bordentown, N. J.	1793.	Charles A............Farnham, Mass.
827.	Catherine, Cornwallis.......N. S	1541.	Charles A......Nova Scotia.
1771.	Catherine E., (Hambley)............Massachusetts.	1574.	Charles A......Nova Scotia.
		2196.	Charles, C....Jacobstown, N. J.
1099.	Catherine H., (Brown)............City Point, Fla.	1331.	Charles C............New Jersey.
		915.	Charles E......Fall River, Mass.
1088.	Catherine T., (Mott)............Fall River, Mass.	2542.	Charles E......Fall River, Mass.
422.	Catherine, (Turner) ....Tiverton, R. I.	2375.	Charles E......Fall River, Mass.
		701.	Charles E............New York.
2932.	Catherine W....Goldsboro, N. C.	1579.	Charles E ....Nova Scotia
2129.	Catherine, (Wilmot)	2563.	Charles E......Wilmington, N. C
2129.	Catherine, (Wilmot)....Virginia.	2401.	Charles, Edgar ....Canada
1984.	Cecil A........Los Angeles, Cal.	2309.	Charles F.....Fall River, Mass

INDEX.

2803.	Charles, Frederick .... Fall River, Mass.	2662.	Clarkson, James V....Kentucky.
929.	Charles H......Fall River, Mass.	2828.	Claudius R......New York, N. Y.
2532.	Charles H.....Fall River, Mass.	1334.	Clement A............New Jersey.
985.	Charles H...........Nova Scotia.	1597.	Clement...........Nova Scotia.
1250.	Charles H........Red Bank, N. J.	2187.	Clifford..............New Jersey.
2616.	Charles J............Dundee, Ill.	1596.	Clifford ....Nova Scotia.
2515.	Charles, Joseph.Fall River, Mass.	1605.	Clifford A............Nova Scotia.
2037.	Charles L............Dallas, Tex	2042.	Cobb, Donald B.................. Goldsboro, N. C.
2037.	Charles, L..........Dallas, Texas.	2861.	Cobb, Harvey L..........Texas.
925.	Charles L......Fall River, Mass.	2941.	Cobb, William B.................. Goldsboro, N. C.
1789.	Charles Lott ......Massachusetts.	654.	Collan ....New Jersey.
2335.	Charles, N......Fall River, Mass.	2186.	Collier, William S....New Jersey.
1433.	Charles R......Fall River, Mass.	738.	Cook...........Fall River, Mass.
1433.	Charles, R......Fall River, Mass.	85.	Cook, Amey ....
1415.	Charles S......Fall River, Mass.	83.	Cook, Deborah ....
1415.	Charles, S......Fall River, Mass.	77.	Cook, Elizabeth ....
667.	Charles S......Jacobstown, N. J.	2229.	Cook, Gustavus W................ Philadelphia, Pa.
2213.	Charles, S...........New Jersey.	80.	Cook, Hannah ....
1236.	Charles T......Burlington, N. J.	79.	Cook, John....Portsmouth, R. I.
1865.	Charles T......Fall River, Mass.	81.	Cook, Joseph ....
1838.	Charles W.............California.	82.	Cook, Martha ....
1421.	Charles W.....Fall River, Mass.	76.	Cook, Mary....Portsmouth, R. I.
1421.	Charles, W......Fall River, Mass.	1497.	Cook, Mary A.................... Fall River, Mass.
1277.	Charles W............New Jersey.	2815.	Cook, Nannie W................. Philadelphia, Pa.
1235.	Charlotte, (Carr) (Fillman)......	86.	Cook, Samuel..Portsmouth, R. I.
1543.	Charlotte, (Cox)....Nova Scotia.	78.	Cook, Sarah ....
767.	Charlotte, (Harrington) .... Nova Scotia.	84.	Cook, Thomas..Portsmouth, R. I.
1971.	Cheney, T. W., (McGiffert)...... Alabama.	2043.	Cooke, Charles L.....N. Dakota.
2649.	Christine ....Cedartown, Ga.	2043.	Cooke, Charles L..North Dakota.
1434.	Christopher.....Fall River, Mass.	2041.	Cooke, John J..........Kentucky.
1434.	Christopher....Fall River, Mass.	2041.	Cooke, John J.........Kentucky.
239.	Christopher........Tiverton, R. I.	2042.	Cooke, Martha, (Heaney)......... Kentucky.
862.	Church, Alfred B......Elgin, Ill.	2042.	Cooke, Martha, (Heaney)......... Kentucky.
862.	Church, Samuel M................ Brewsters, N. Y.	1213.	Cox, Alexander S................
2188.	Clara................New Jersey.	640.	Cox, Amy, (Vandevoie).......... Zanesville, O.
2621.	Clara B.............Dundee, Ill.	1216.	Cox, Angeline S., (Sites).........
1871.	Clara Elsbree.....Massachusetts.	1217.	Cox, Augustus C................
1664.	Clara J..........Scituate, R. I.	644.	Cox, David J.....Zanesville, O.
1604.	Clarence H.........Nova Scotia.	2136.	Cox, Elizabeth D., (Sullivan).... Washington.
1023.	Clarissa, (Atwood)	2136.	Cox, Elizabth D., (Sullivan)..... Washington.
1068.	Clarissa, (Knowlton)..New York.	1218.	Cox, Elizabeth, (Taylor)........
1752.	Clarissa, (Winslow) .... Massachusetts.	2742.	Cox, Emery....Washington, D. C.
2626.	Clark, Archie B................. Jacksonville, Fla.	645.	Cox, Ezekiel T.....Zanesville, O.
1964.	Clark, James C....St. Louis, Mo.	1214.	Cox, Ezekiel Taylor..............
1966.	Clark, Joseph W......Alabama.	2743.	Cox, Hazel....Washington, D. C.
1967.	Clark, Thomas J......Alabama.	647.	Cox, Horatio J.....Zanesville, O.
2687.	Clarkson, Clinton......Kentucky.		
2688.	Clarkson, James......Kentucky.		
2662.	Clarkson, James V....Kentucky.		

638. Cox, James.........Zanesville, O.
642. Cox, Jonathan......Zanesville, O.
636. Cox, Joseph.........Zanesville, O.
1212. Cox, Lavinia (Sledgewick)........
637. Cox, Lewis........Zanesville, O.
1215. Cox, Maria M., (Van Renselear).
639. Cox, Mary, (Bateman)...........
................Zanesville, O.
1219. Cox, Mary S., (Spangler)....Ohio.
646. Cox, Morgan R....Zanesville, O.
634. Cox, Richard......Zanesville, O.
643. Cox, Samuel........Zanesville, O.
1211. Cox, Samuel Sullivan.............
2744. Cox, Theodore ....
................Washington, D. C.
641. Cox, Thomas.....Zanesville, O.
1210. Cox, Thomas J.....Zanesville, O.
635. Cox, William......Zanesville, O.
2135. Cox, Willilam V.....Washington.
2135. Cox, William V.....Washington.
1957. Crawford C............California.
2947. Crieder, Charles ....
............Bowling Green, Ky.
2946. Crieder, Louis ....
............Bowling Green, Ky.
2376. Cyrus, E........Fall River, Mass.
1488. Cyrus H.........Fall River, Mass.

D

360. Daniel........Emleytown, N. J.
442. Daniel........Fall River, Mass.
1707. Daniel........Fall River, Mass.
1141. Daniel ....Kentucky
1198. Daniel ....New York.
2480. Daniel B.......Fall River, Mass.
2123. Daniel J........New York, N. Y.
2958. Daniel, Le Ray................
................New York, N. Y.
1302. Daniel S.............Burlington.
504. Daniel S........Cornwallis, N. S.
1247. Daniel S........Red Bank, N. J.
662. Daniel S........Redbank, N. J.
1445. Daniel W.....Fall River, Mass.
700. Daniel W............New Jersey.
2925. Daniels, Frank B................
................North Carolina.
2926. Daniels, Georgia S..............
................North Carolina.
2927. Daniels, Mary C................
................North Carolina.
71. David ....Bordentown, N. J.
911. David............Fall River, Mass.
471. David............Fall River, Mass.
2299. David B....New Bedford, Mass.
1135. David, Henry. ....Borden, Cal.

777. David, Henry.......Nova Scotia
414. David, Horton......Nova Scotia.
1812. David, Perry....Fall River, Mass.
603. David W.........New Bern, Ala.
2859. Davis, A. Borden................
................Weatherford, Tex.
2922. Davis, Adam C..North Carolina
2921. Davis, Annie L..North Carolina.
2944. Davis, Edwin B..North Carolina.
2923. Davis, Georgia L................
.... North Carolina.
2857. Davis, Nellie Lee................
................Weatherford, Tex.
2858. Davis, Winnie ....
................Weatherford, Tex.
240. Deborah, (Brayton) ....
................Tiverton, R. I.
166. Deborah, (Hendley).....Virginia.
718. De Grauw, Chas. E. New Jersey.
719. De Grauw, Mary, (Sherman)....
................Clairmont, N. J.
717. De Grauw, W. N................
900. Delano, (Durfee) ....
................Fall River, Mass.
1503. Delia W., (Manchester).........
................Fall River, Mass.
2554. Dewey, Ann M., (Slocum).......
.... North Carolina.
2887. Dewey, Brownrigg H............
................North Carolina.
2552. Dewey, Charles..North Carolina.
2885. Dewey, Edward M...............
................North Carolina.
2557. Dewey, Ernest Brownrigg......
.... North Carolina.
2880. Dewey, Ernest M................
................North Carolina.
2876. Dewey, George S................
................North Carolina.
2555. Dewey, George Wood............
.... North Carolina.
2879. Dewey, Hannah..North Carolina.
2558. Dewey, Harriet M ....
.... North Carolina.
2881. Dewey, Harriet M...............
................North Carolina.
2553. Dewey, Mary W., (Slocum)......
.... North Carolina.
2878. Dewey, Thomas A..............
................North Carolina.
2556. Dewey, Thomas Webber........
.... North Carolina.
2886. Dewey, Victoria L...............
................North Carolina.
57. Dinah ....
16. Dinah...........Providence, R. I.
9331. Dollie, Church.Fall River, Mass.
2038. Dora, Lou, (Trammel)....Texas.

INDEX.

2461.	Duffel, Carrie, (Thompson)...... Texas.		2580.	Edward E. .... Fall River.
2464.	Duffel, Hattie ....Texas.		666.	Edward N. J......................
2470.	Duffel, Horace E..........Texas.		1400.	Edward P. ....Fall River.
2467.	Duffel, L. May.............Texas.		686.	Edward P. .... Philadelphia.
2465.	Duffel, Leslie B..........Texas.		1324.	Edward P. .... Washington.
2462.	Duffel, Mary L., (Cobb)...Texas.		1547.	Edward Ferry ...... Nova Scotia.
2466.	Duffel, Ola B.............Texas.		2472.	Edward Ray .... Galveston, Tex.
2468.	Duffel, Sam B.............Texas.		674.	Edward T. .... New Jersey.
2469.	Duffel, Sidney G..........Texas.		1296.	Edward T. .... New Jersey.
2459.	Duffel, Willie C. ....Texas.		1890.	Edwin .... Fall River.
2954.	Duncan, Susan .... Kentucky.		2359.	Edwin .... Fall River.
2891.	Durald, S.... Wilmington, N. C.		2559.	Edwin .... Wilmington, N. C.
1456.	Durfee, Amy.. Fall River, Mass.		2919.	Edwin B. .... Goldsboro, N. C.
1461.	Durfee, Julia M.................... Fall River, Mass.		1828.	Edwin Brownrigg................. North Carolina.
1455.	Durfee, Lucy, (Borden)........... Fall River, Mass.		2575.	Edwin Brownrigg .... North Carolina
1458.	Durfee, Mary .... Fall River, Mass.		2258.	Edwin Floyd ...... New Jersey.
			2024.	Edwin Gosson ...... Oxford, Fla.
1460.	Durfee, Sarah A.................. Fall River, Mass.		2896.	Edwin H. .... New Orleans.
			2390.	Elaine Allison.. New Brunswick.
1459.	Durfee, Simeon B................. Fall River, Mass.		1204.	Eleanor. ....
			1251.	Eleanor .... New Jersey.
1457.	Durfee, Thomas .... Fall River, Mass.		558.	Eleanor, (Thomas) ....
			2751.	Eleanor W. ...... Philadelphia.
	E		2125.	Eleanor, (Woodward) ....
			1560.	Elfrida, (Darling) Massachusetts.
1410.	E. Corinna, (Keen).. Philadelphia.		614.	Eli .... Tennessee.
2335.	E. Shirley .... Fall River.		1159.	Eli .... Texas.
507.	Earl .... Fall River.		1621.	Elias .... Hartford, Ct.
947.	Earl .... Fall River.		2741.	Elias .... New York.
974.	Earl .... Fall River, Mass.		522.	Elihu.............Fall River.
848.	Earl D. .... Scituate, R. I.		285.	Elijah .... Pompey, N. Y.
2963.	Earl Kinsley .... Fall River.		1583.	Elijah C. ...... Cornwallis N. S.
429.	Ebenezer .... Canada.		1623.	Eliza .... Canada.
258.	Ebenezer ...... Gloucester, R. I.		1795.	Eliza Ann .... Fall River.
1620.	Edgar .... Michigan.		952.	Eliza, (Brightman) (Westgate.)
1844.	Edgar A. .... Nova Scotia.		1738.	Eliza D., (Hearsy) ...... Grafton
1610.	Editha, A. M., (Lind), Wisconsin		809.	Eliza, (Ells)........ Nova Scotia.
1343.	Edmund J. .... Michigan.		761.	Eliza, (Gardiner) .... Fall River.
694.	Edmund W. .... Michigan.		1798.	Eliza Gibbs .... Fall River.
2960.	Edna Louise .... Fall River.		1408.	Eliza O., (Durfee) ... Fall River.
252.	Edward .... Fall River.		987.	Eliza, (Welton) ..... Nova Scotia.
1874.	Edward .... Fall River.		389.	Elizabeth .... Fall River.
2802.	Edward .... Fall River.		466.	Elizabeth .... Fall River.
365.	Edward .... New Jersey.		104.	Elizabeth ...... Imlaystown, N. J.
1276.	Edward .... New Jersey.		1070.	Elizabeth .... Madison Co., N. Y.
2813.	Edward .... New Jersey.		121.	Elizabeth .... Portsmouth, R. I.
420.	Edward .... Nova Scotia.		130.	Elizabeth ...... Portsmouth, R. I.
782.	Edward .... Nova Scotia.		190.	Elizabeth .... Providence.
1516.	Edward .... Nova Scotia.		1153.	Elizabeth .... Tennessee.
126.	Edward .... Westport, R. I.		1788.	Elizabeth A........ Fall River.
2784.	Edward A. .... New Jersey.		1540.	Elizabeth A. ...... Nova Scotia.

670.	Elizabeth A., (Emley) .... New Jersey.	2412.	Ellen Lincoln .... Fall River.
1259.	Elizabeth A., (Warren)	1407.	Ellen, (Paine) .... Fall River.
303.	Elizabeth, (Borden)..Fall River.	2547.	Ellen R. .... Fall River.
916.	Elizabeth, (Borden) ..Fall River.	2413.	Elma .... Fall River.
705.	Elizabeth C., (Berger) .... Eatontown, N. J.	1926.	Elmer D........... ....
232.	Elizabeth, (Chandler) .... Manasquan, N. J.	2964.	Elmer Estes.......... Fall River.
756.	Elizabeth, (Collins) ..New York.	2609.	Elsie .... Fall River.
303.	Elizabeth, (Cook).......Tiverton.	2284.	Elspeth Bell ...... Auburn, Neb.
553.	Elizabeth, (Cook) ......Tiverton.	1663.	Elvira .... Fall River.
220.	Elizabeth, (Corlies) .... Monmouth Co.	2147.	Elvira L., (Kain) ..Philadelphia.
1033.	Elizabeth, (Davis))Lawton.)....	1220.	Elvira, (Linnard) ..Philadelpnia.
185.	Elizabeth, (Douglass) .... Bordentown.	1802.	Emeline E., ...Macomber, Mass.
1243.	Elizabeth, (Ellis) ....New Jersey.	1134.	Emeline G., (Hargous).New York.
520.	Elizabeth, (Gifford) ...Westport.	946.	Emeline, (Le Valley) .Fall River.
2581.	Elizabeth Goldsboro .... North Carolina.	1911.	Emeretta .... Fall River.
764.	Elizabeth, (Hance) .... Massachusetts	1892.	Emerson F. .... Fall River.
831.	Elizabeth, (Haskell) .... Rhode Island.	710.	Emily T., (Monell) ....
633.	Elizabeth, (Hunt) .... Bordentown, N. J.	1891.	Emily V., (Guy) ......Fall River.
2627.	Elizabeth Jeffreys .... Fernandina, Fla.	1283.	Emley Buelah W., (Logan.)......
90.	Elizabeth, (Latham) .... Johnston, R. I.	2168.	Emley Eugene ...... New Jersey.
2381.	Elizabeth M. ....Ottawa, Canada.	2164.	Emley Ida, (Godfrey) .... New Jersey.
2006.	Elizabeth McGhee .....Alabama.	1284.	Emley Mary E. ......New Jersey.
313.	Elizabeth (Monroe)......Bristol.	1300.	Emley S., (Phillips)..New Jersey.
2148.	Elizabeth N., (Dunlap) .... Philadelphia.	2165.	Emley Thomas N...New Jersey.
2130.	Elizabeth, (Nellis) ......Chicago.	1282.	Emley Walter S. ....New Jersey.
781.	Elizabeth, (Newcomb) .... Nova Scotia.	1833.	Emma A., (Black) ....California.
2339.	Elizabeth, (Nichols) .... Massachusetts.	1140.	Emma, (Brown) ....Alabama.
168.	Elizabeth, (Patton) ..Tennessee.	2307.	Emma C., (Daves) .... Massachusetts.
878.	Elizabeth, (Simmons) .... Cape Cod, Mass.	2297.	Emma C. H. .... Fall River.
2218.	Elizabeth T. .... New Jersey.	2647.	Emma Claudine, San Francisco.
1108.	Elizabeth T., (Pearce), Alabama.	2451.	Emma E. .... Hico, Tex.
748.	Elizabeth, (Tasker) ....Tiverton.	2956.	Empey Edward B., Clinton, Mo.
363.	Elizabeth, (Tooley)..New Jersey.	2957.	Empey Florence I., Clinton, Mo.
406.	Elizabeth, (Wales) ....Boston. .... Newburgh, N. Y	2345.	Emma Leonora ......Fall River.
140.	Elizabeth, (Wardwell) .... Swansea, R. I.	2146.	Emma P., (Sutton) Philadelphia.
1278.	Ella .... New Jersey.	2039.	Ephie T.............Dallas, Tex.
1552.	Ella A., (Woodman.)..............	1008.	Erastus ...... Fall River, Mass.
2410.	Ella C. ....Fall River.	1899.	Eric Warren .... Fall River.
1619.	Ellen, (Beek) ..Bedford, Quebec.	2793.	Ernest A. ....Fall River.
2264.	Ellen Howell .... New Jersey.	2952.	Eskridge Chas. W....Kentucky.
		1128.	Esther A., (Aylesworth) Troy, O.
		2017.	Esther W. ....Alabama.
		2809.	Ethel .... New Jersey.
		1661.	Eudora..........Scituate, R. I.
		1409.	Eudora S., (Dean)....Fall River.
		1894.	Eugene A. .... Fall River.
		1323.	Eugenia D. .... Washington.
		820.	Eunice, Cornwallis, Nova Scotia.
		2911.	Eunice ....Goldsboro, N. C.
		1154.	Euphemia .... Tennessee.
		1700.	Eva G., (Marshall) ....California
		2765.	Evart P. .... Tunkhannock.

INDEX.

No.	Name	Location
927.	Eveline, (Read)	Fall River.
20.	Experience	Portsmouth, R. I.

F

No.	Name	Location
2352.	Fannie	Fall River.
2198.	Fannie	Jacobstown, N. J.
1478.	Fannie A. (Durfee)	Tiverton.
2766.	Fannie E.	Tunkhannock.
2003.	Fannie S.	Alabama.
870.	Fanny (Howard)	Warren, R. I.
1092.	Ferdinand	Fall River.
1692.	Fidelia Duffel	Texas.
902.	Fidelia (Durfee)	Fall River.
2774.	Field Carrie M.	New Jersey.
2775.	Field Edna F.	New Jersey.
2776.	Field, Mabel, W.	New Jersey.
1082.	Finis G., (Macomber)	Fall River.
1903.	Flavius J.	San Francisco.
1316.	Flora E., (Exelby)	Michigan.
2441.	Flora Parker	Weatherford, Tex.
2344.	Florence	Fall River.
1290.	Florence	New Jersey.
2150.	Frances C., (Purdy)	New Jersey.
1321.	Frances H., (Parker)	Chicago.
1280.	Frances S	Philadelphia.
683.	Francis	Allentown, N. J.
58.	Francis	Coopers Creek, N. J.
1594.	Francis	Cornwallis, N. S.
1598.	Francis	Cornwallis, N. S.
211.	Francis	Imlaystown, N. J.
170.	Francis	Mansfield, N. J.
106.	Francis	Nottingham, N. J.
364.	Francis	Philadelphia, Pa.
5.	Francis	Shrewsbury, N. J.
23.	Francis	Shrewsbury, N. J.
374.	Francis	Shrewsbury, N. J.
1902.	Francis Cook	Massachusetts.
2017.	Francis Nelson	Alabama.
2222.	Francis Reeve	Ardmore, Pa.
2318.	Frank	Fall River.
2197.	Frank	Jacobstown, N. J.
1315.	Frank	New Jersey.
2160.	Frank Edna	Kansas.
2267.	Frank Edward	New York.
2795.	Frank H.	Fall River.
2753.	Frank Harris	Philadelphia.
2915.	Frank K.	Goldsboro, N. C.
2571.	Frank Kornegay	North Carolina.
1843.	Frank N.	Nova Scotia.
2215.	Frank R.	New Jersey.
2767.	Frank S.	Tunkhannock.
1118.	Franklin	Denton Co., Tex.
1867.	Franklin B.	Fall River.
1675.	Franklin J. A.	
2021.	Franklin Katherine B.	New York.
1925.	Franklin P.	New York.
2023.	Franklin Rose Claire	New York.
2482.	Franklin S.	Fall River.
2022.	Franklin Sheldon	New York.
2395.	Fred L.	Nova Scotia.
1851.	Fred W.	Nova Scotia.
2404.	Fred William	Manchester Mass.
1091.	Frederick	Fall River.
2273.	Frederick	New York.
1439.	Frederick A.	Fall River.
1123.	Frederick Asa	Alabama.
1846.	Frederick C.	Nova Scotia.
1587.	Frederick P.	Cornwallis, N. S.
1760.	Frederick P.	Fall River.
2615.	Frederick R.	San Francisco.
1988.	Frederick W.	Alabama.
2207.	Frederick W.	New Jersey.
1522.	Frederick W.	Ottawa, Canada.
2645.	Frederick William	San Francisco.
991.	Freedom (Bishop)	Nova Scotia.
506.	Freedom O., (Brompton.)	
2820.	Freeman Wm. Keen	Philadelphia.

G

No.	Name	Location
2431.	Gail	Alhambra, Cal.
862.	Gail	Elgin, Ill.
256.	Gail	Gloucester, R. I.
439.	Gail	Norwich, N. Y.
877.	Gardner	Fall River.
2763.	Garrick M.	Tunkhannock.
910.	General	Fall River.
681.	George	Allentown, N. J.
1161.	George	Calhoun Co., Ala.
2161.	George	Edna, Kan.
277.	George	Fall River.
904.	George	Fall River.
1037.	George	Fall River.
2528.	George	Fall River.
147.	George	Fall River, Mass.
457.	George	Fall River, Mass.
655.	George	New Jersey.
2193.	George	New Jersey.
2239.	George	New Jersey.
1575.	George	Nova Scotia.
524.	George	Tiverton.
266.	George	Tiverton.
2546.	George A.	Fall River.
2259.	George Apple	New Jersey.
2792.	George E.	Fall River.

INDEX. 331

No.	Name	Location
1330.	George E.	New Jersey.
703.	George F.	Red Bank.
1546.	George F.	Nova Scotia.
515.	George G.	Fall River.
1901.	George G.	Somerset, Mass.
1770.	George H.	Fall River.
1787.	George H.	Fall River.
1151.	George H.	Palestine, Tex.
2214.	George L.	New Jersey.
1133.	George P., Capt.	U.S.A.
1645.	George R.	Fall River.
2155.	George W.	Burlington, N. J.
1345.	George W.	Nebraska.
668.	George W.	New Jersey.
798.	George W.	Nova Scotia.
993.	George W.	Nova Scotia.
2288.	George Webster	Auburn, Neb.
977.	George Wellington	Nova Scotia.
1271.	Georgeanna, (Parr.)	
2593.	Georgia, (Cobb).	Goldsboro, N. C.
2417.	Gerald Mark	Chicago.
1317.	Garrett	New Jersey.
2527.	Gertrude	Fall River.
278.	Gideon	Fall River.
1876.	Gilbert	Illinois.
1477.	Gilbert B.	Fall River.
2391.	Gladys Allison	New Brunswick.
1360.	Glentworth D.	New York.
1981.	Grace, David B.	Birmingham, Ala.
2768.	Granville S.	Tunkhannock.
1686.	Guy	San Antonio, Tex.
2448.	Guy	San Antonio, Tex.

H

1529.	H. C.	Halifax.
2735.	Haile, Laura D.	Virginia.
2734.	Halle, W. S.	Virginia.
1933.	Hallie A.	California.
2154.	Hamilton	Burlington, N. J.
1224.	Hamilton	Chicago, Ill.
2641.	Hamilton Cecil W.	Stamford, Ct.
2642.	Hamilton Stuart H.	Stamford, Ct.
716.	Hance Benjamin	New Jersey.
1371.	Hance, Elizabeth	Illinois.
1373.	Hance Howard	Illinois.
1372.	Hance Susan, (Haviland).	Illinois.
207.	Hannah	Gloucester.
223.	Hannah	Monmouth, N. J.
697.	Hannah	New York.
742.	Hannah, (Babbitt)	Tiverton.

876.	Hannah, (Borden)	Fall River.
264.	Hannah, (Borden)	Tiverton.
735.	Hannah, (Cook)	Fall River.
100.	Hannah, (Coxe)	Shrewsbury, N. J.
1703.	Hannah, (Crocker)	Fall River.
448.	Hannah, (Earl)	Tiverton.
933.	Hannah, (Elsbree)	Fall River.
2306.	Hannah G J.	Horton, Mass.
1504.	Hannah H., (Gifford).	Fall River.
216.	Hannah, (Hawkins)	New Jersey.
1863.	Hannah L., (Chace)	Massachusetts.
184.	Hannah, (Lawrence)	Bordentown, N. J.
159.	Hannah, (Mace)	North Carolina.
1340.	Hannah, (Palmer)	Buffalo.
161.	Hannah, (Rogos)	Virginia.
1746.	Hannah, (Smith)	Bristol.
384.	Hannah T., (Hartshorn)	N. J.
707.	Hannah T., (Hartshorn)	New Jersey.
1475.	Hannah W., (Chace).	Fall River.
1110.	Hannah W., (Pearce)	Wallace, Ala.
853.	Harley P.	Scituate, R. I.
2611.	Harold	Fall River.
2770.	Harold, A.	Philadelphia.
2382.	Harold Lathrop	Ottawa, Can.
1207.	Harriet	
693.	Harriet	New Jersey.
2126.	Harriet, (Brown.)	
1127.	Harriet C.	Alabama.
783.	Harriet, (Childs)	Nova Scotia.
2525.	Harriet Eliza	Fall River
2373.	Harriet L.	Fall River.
713.	Harriet, (Lawson)	Newburgh, N. Y.
2323.	Harriet M., (Bassett).	Fall River.
1826.	Harriet M., (Dewey)	North Carolina.
2496.	Harriet Matilda	Fall River.
2648.	Harriet May	San Francisco.
966.	Harriet, (Peckham)	Fall River.
2599.	Harriet Sophronia, (Brown)	Massachusetts.
1747.	Harriet, (Talbot)	Fall River.
2756.	Harris Anna E.	Philadelphia.
2757.	Harris, Gertrude B.	Philadelphia.
2758.	Harris, Marian M.	Philadelphia.
2190.	Harry	New Jersey.
1985.	Harry Innis	Los Angeles.
1370.	Hartshorn, Margaret	New Jersey.
1369.	Hartshorn Wm.	New Jersey.

1955.	Hartwell C............ California.	1856.	Henry R. .... Fall River.
594.	Hatch Alice......North Carolina.	2348.	Herbert G........... Fall River.
596.	Hatch Asa.......North Carolina.	1550.	Herbert H. ....Nova Scotia.
599.	Hatch George...North Carolina.	2564.	Herbert Leigh.Wilmington, N. C.
597.	Hatch Helley....North Carolina.	307.	Hervey ....Portsmouth
598.	Hatch Hope......North Carolina.	1987.	Hester Bell ...... Oakland, Cal.
600.	Hatch Mary......North Carolina.	226.	Hester, (Lippincott) ....
595.	Hatch William...North Carolina.		.... Toronto, Canada.
2522.	Hattie J............. Fall River.	771.	Hezekiah ....Nova Scotia.
2439.	Hattie L., (Weld)...Los Angeles	2638.	Hicks Marion H........Alabama.
615.	Hawkins .... Alabama.	2962.	Hildreth .... Fall River.
1149.	Hawkins .... Alabama.	2665.	Hilburn Frederick S..Kentucky.
2943.	Hazel, Maude ....Chicago.	2668.	Hilburn Jacob V...............
2737.	Heath, Chester S......California.		.... Bowling Green, Ky.
2736.	Heath, Maude V......California.	2667.	Hilburn Julia C., (Preston)......
2655.	Heaney Mattie I., (White)	444.	John .... Tiverton.
	.... Kentucky.	531.	John .... Tiverton,
2654.	Heaney William D....Kentucky.		.... Kentucky.
2370.	Helen.................Fall River.	2949.	Hilburn, Morgan T....Kentucky.
2610.	Helen .... Fall River.	2728.	Hill Annie .... Virginia.
2944.	Helen. Elsie............ Chicago.	2730.	Hill, Bulah ....Virginia.
2782.	Helen L............ New Jersey.	2729.	Hill Ella .... Virginia.
1279.	Helen, (Loyd.)	2731.	Hill, James .... Virginia.
2141.	Helen Page, (Burns)............	2733.	Hill, John ....Virginia.
	.... Philadelphia.	2773.	Hill, Leslie B........New Jersey.
2287.	Helen Ross ....Auburn, Neb.	2727.	Hill, Mattie .... .Virginia.
2636.	Henderson Katherine A.Alabama.	2732.	Hill Walker ....Virginia.
2634.	Henderson Nona S......Alabama.	2544.	Hiram .... Fall River.
2635.	Henderson Samuel H..Alabama.	1063.	Hiram .... New York.
2296.	Henrietta ....Fall River.	1420.	Hiram C. .... Fall River.
804.	Henrietta, (Buckley).Nova Scotia.	2349.	Hiram F..............Fall River.
1518.	Henry .... Boston.	2659.	Hobson Edward W....Kentucky.
874.	Henry .... Fall River.	2051.	Hobson George A.....Kentucky.
907.	Henry ....Fall River.	2658.	Hobson James T......Kentucky.
1728.	Henry ....Fall River.	2049.	Hobson Jonathan ....Kentucky.
1638.	Henry ....Montreal, Canada.	2050.	Hobson ..Joseph Van M..........
1281.	Henry .... New York.		.... Kentucky.
498.	Henry .... Nova Scotia.	2657.	Hobson Julia H......Kentucky.
577.	Henry .... Portsmouth.	2661.	Hobson Margaret......Kentucky.
1095.	Henry .... Portsmouth.	2660.	Hobson Mary A.......Kentucky.
2034.	Henry A............. Dallas, Tex.	2656.	Hobson Morehead.....Kentucky.
1578.	Henry A. .... Nova Scotia.	2048.	Hobson William E....Kentucky.
1643.	Henry C. .... Fall River.	899.	Holder ....Fall River.
2300.	Henry Clay ....Fall River.	18133.	Holder .... Fall River.
1853.	Henry F. .... Nova Scotia.	1485.	Holder B............. Fall River.
2646.	Henry Forney....San Francisco.	53.	Holmes Catherine ....
2652.	Henry Grady........ Dallas, Tex.		.... Portsmouth, R. I.
1600.	Henry H......Cornwallis, N. S.	49.	Holmes Jonathan ....
1929.	Henry J..............New York.		.... Portsmouth, R. I.
1677.	Henry Lee ....Chicago.	56.	Holmes Joseph ....
2452.	Henry Lee ....Sharpsburg, Tex.		.... Portsmouth, R. I.
685.	Henry M. .... Philadelphia.	55.	Holmes Lydia ....
2481.	Henry N............ Fall River.		.... Portsmouth, R. I.
1571.	Henry Pope ...... Nova Scotia.	54.	Holmes Martha ....
			.... Portsmouth, R. I.

INDEX. 333

52.	Holmes Mary .... Portsmouth, R. I.	747.	Isaac .... Fall River.
48.	Holmes Obediah .... Portsmouth, R. I.	843.	Isaac .... Fall River.
50.	Holmes Samuel .... Portsmouth, R. I.	1020.	Isaac .... Fall River.
		1040.	Isaac .... Fall River.
		1376.	Isaac .... Fall River.
51.	Holmes Sarah .... Portsmouth, R. I.	1393.	Isaac .... Fall River.
		1047.	Isaac ....Tiverton, R. I.
157.	Hope .... Beaufort, N. C.	437.	Isaac .... Scituate.
42.	Hope, (Almy) ....Tiverton, R. I.	1495.	Isaac, A............... Fall River.
290.	Hope, (Brownell) ....Fall River.	2297.	Isaac, H............... Fall River.
726.	Hope, (Cook) ....Tiverton.	1444.	Isaac, N........Cayugo Co., N. Y.
410.	Hope, (Durfee) ....Tiverton.	608.	Isaac P. .... Alabama.
395.	Hope, (Graves) ......Fall River.	694.	Isaac P. .... New Jersey.
319.	Hope, (Hatch)..Jones Co., N. C.	1292.	Isabella, (Ommie)...New Jersey.
881.	Hope, (Turner)........Fall River.	1662.	Isadora F............ Fall River.
2643.	Hopkins Catherine W............ Stamford, Ct.	453.	Isaiah .... Fall River.
		1762.	Isaiah .... Fall River.
2020.	Hopkins Sheldon ....Connecticut.	477.	Israel ....Fall River.
2019.	Hopkins Woolsey R.Cinnecticut.	939.	Israel .... Fall River.
2644.	Hopkins Woolsey Rogers....... Stamford, Ct.	1914.	Israel .... New York.
		1047.	Ivey, Lewis........ Oakland, Cal.
632.	Hopkinson, (Joseph). .... Philadelphia.	1117.	Ivey, Lewis........Oakland, Cal.
1622.	Horatio ....Quebec, Canada.		**J**
1800.	Horatio Nelson ......Fall River.		
2191.	Howard ....New Jersey.	2201.	J. A. Garfield..Jacobstown, N. J.
2331.	Howard Seymour ....New York.	2383.	J. Maude ......Ottawa, Canada.
1935.	Hoxie Grace L........Tonica, Ill.	196.	Jacob .... Burlington, N. J.
2624.	Hoxie Guy Delamatu .... Tonica, Ill.	2241.	Jacob .... New Jersey.
		1325.	Jacob, P............ New Jersey.
1936.	Hoxie J. Benj......... Tonica, Ill.	1206.	James ....
2566.	Hugh D........Wilmington, N. C.	787.	James .... Avonport, N. S.
347.	Huldah, (Sprague) .... Johnston, R. I.	198.	James .... Burlington, N. J.
		838.	James .... Canada.
2823.	Huling, Alice B......Cambridge.	59.	James......Cooper Creek, N. J.
2824.	Huling, Corinne W..Cambridge.	96.	James .... Eversham, N. J.
2822.	Huling, Elizabeth L..Cambridge.	919.	James .... Fall River.
2821.	Huling, Ellen P......Cambridge.	1791.	James .... Fall River.
2825.	Huling, Ray Green..Cambridge.	2494.	James .... Fall River.
		65.	James............Freehold, N. J.
	I	111.	James .... Imlaystown, N. J.
1358.	Ida .... New Jersey.	1164.	James .... Mississippi
2799.	Ida Eastman ....Fall River.	688.	James .... New Jersey.
1698.	Ida M., (Blithen).....Minnesota.	212.	James ....Nottingham, N. J.
2543.	Ida, May.............. Fall River.	1057.	James .... Pompey, N. Y.
1899.	Inez, Ella........ Massachusetts.	368.	James ....Shrewsbury, N J.
727.	Irene, (Butler.)	1644.	James, A............. Fall River.
1483.	Irene, P............... Fall River.	1905.	James, A............. Fall River.
2029.	Irmie, B., (Martin)..Newnan, Ga.	1248.	James, A....Monmouth Co., N. J.
391.	Isaac .... Fall River.	2172.	James, B...... Jacobstown, N. J.
513.	Isaac .... Fall River.	2301.	James Clarence......Fall River.
556.	Isaac .... Fall River.	1684.	James, Cochran..Galveston, Tex.
723.	Isaac .... Fall River.	2894.	James Cole........ New Orleans.
758.	Isaac .... Fall River.	1827.	James Cole......North Carolina.

2561.	James Cole ...Wilmington, N. C.	300.	John	Fall River.
706.	James E...... Eatontown, N. J.	730.	John	Fall River.
2219.	James, E............. New Jersey	760.	John	Fall River.
2794.	James Edgar ..... Fall River.	973.	John	Fall River.
2548.	James Edward ..... Fall River.	1085.	John	Fall River.
2509.	James Ford .... Fall River.	1937.	John	Fall River.
1474.	James H............. Fall River.	134.	John .... Fall River, Mass.
1591.	James, J........ Cornwallis, N. S.	326.	John .... Hardeman Co.. Tenn.
1548.	James Martin...... Nova Scotia	555.	John .... Indiana.
2436.	James McKee........ New York.	1616.	John .... Iowa.
818.	James, Newton ... Nova Scotia.	230.	John.......... Manasquan, N. J.
1976.	James, P...... Greensboro, Ala.	687.	John.............. New Jersey.
1107.	James, P......... Hale Co., Ala.	691.	John .... New Jersey.
1121.	James, W.............. Alabama.	1289.	John .... New Jersey.
606.	James W...... Fort Wayne, Ind	1319.	John .... New Jersey.
2308.	James W. M........ Fall River.	519.	John .... New York.
105.	Jane .... Imlaystown, N. J.	591.	John .... North Carolina.
222.	Jean ...... Monmouth Co., N. J.	500.	John .... Nova Scotia.
869.	Jane, (Howard) ....Warren, R. I.	1521.	John .... Nova Scotia.
982.	Jane R............. Nova Scotia.	1151.	John .... Palestine, Tex.
1074.	Jeannette...... Madison Co., N. Y.	1222.	John .... Philadelphia.
731.	Jefferson .... Fall River.	1607.	J hn...... Philipsburg, Canada.
1411.	Jefferson .... Fall River.	7.	John........ Portsmouth, R. I.
2337.	Jefferson .... Fall River.	150.	John..........Portsmouth, R. I.
2738.	Jennette M., (Empey)..Missouri.	34.	John .... Providence, R. I.
1336.	Jennie, (Caffrey)..... New York.	435.	John .... Scituate.
1310.	Jennie, (Conover).. New Jersey.	141.	John .... Scituate, R. I.
903.	Jeremiah .... Fall River.	254.	John .... Scituate, R. I.
112.	Jeremiah....... Manasquan. N. J.	1656.	John .... Scituate, R. I.
376.	Jeremiah ..... Shrewsbury, N. J.	107.	John .... Shrewsbury, N. J.
2938.	Jermain William B........N. C.	217.	John......... Shrewsbury, N. J.
1010.	Jerome B....... Somerset. Mass.	371.	John .... Shrewsbury, N. J.
1429.	Jerome, Cook...... Fall River.	37.	John .... Swansea, R. I.
516.	Joanna, (Hoskins) .... Newport.	616.	John .... Tennessee.
487.	Joanna, (Orswell) .... Fall River.	280.	John .... Tiverton.
1232.	Job .... Burlington, N. J.	891.	John .... Tiverton, R. I.
279.	Job .... Fall River.	1147.	John .... Walker Co., Ala.
518.	Job .... Fall River.	95.	John .... Wilmington, N. C.
1030.	Job .... Fall River.	1776.	John, A............. Fall River.
847.	Job, W........... Scituate, R. I.	990.	John A........... Nova Scotia.
1889.	Joch, S............... Fall River.	1848.	John, A........... Nova Scotia.
617.	Joel .... Alabama	808.	John, Alexander .. Nova Scotia.
219.	Joel .... Monmouth Co., N. J	2755.	John Alford .... Philadelphia.
1158.	Joel, Eli .... Hope, Ark.	648.	John Allen .... Philadelphia.
1200.	John..........	1076.	John, B...... Madison Co., N. Y.
619.	John .... Arkansas.	2.	John, brother of Richard......
1122.	John ...... Borden Springs, Ala.	569.	John C................ Fall River.
1246.	John .... Burlington, N. J.	1093.	John, C............. Fall River.
427.	John .... Canada.	1051.	John, C.............. Tiverton.
1081.	John .... Chicago.	2163.	John, Curtis .. Manasquan, N. J.
1945.	John .... Chicago.	828.	John, D........ Cornwallis, N. S.
60.	John .... Cooper's Creek, N. J.	1257.	John, E...... Emleytown, N. J.
163.	John .... Covington, Ky.	1872.	John, Earl........... Fall River.

INDEX. 335

1335.	John F.............	Connecticut.
2764.	John F.............	Tunkhannock.
1405.	John, Francis........	Fall River.
2530.	John Francis........	Fall River.
2269.	John Francis........	New York.
2789.	John Francis........	New York.
2402.	John Franklin............	Canada.
2252.	John Frederick....	New Jersey.
1682.	John, Gail........	Walkill, N. Y.
1753.	John, H.............	Fall River.
671.	John H.............	New Jersey.
1304.	John, H.............	New Jersey.
845.	John, H.........	Scituate, R. I.
2157.	John, Hamilton........	Chicago.
2266.	John Hance........	Philadelphia.
1359.	John Harvey........	New York.
1496.	John J.............	Fall River.
1799.	John, Jay............	Fall River
1764.	John, L.............	Fall River.
380.	John L.........	Manasquan, N. J.
2199.	John, Laird....	Jacobstown, N. J.
2576.	John Lemuel....	Goldsboro, N. C.
1089.	John, Levi.....	Portsmouth, R. I.
2752.	John Morrow......	Philadelphia.
1570.	John, N.............	Nova Scotia.
1577.	John, Nelson......	Nova Scotia.
1.	John of Kent Co.......	England.
865.	John P.............	Borden, Tex.
2040.	John, Pickens......	Dallas, Tex.
2518.	John R.............	Fall River.
1683.	John, Rolden...........	Texas.
1294.	John, S.............	New Jersey.
1333.	John, (Sharpstown).	New Jersey.
1249.	John, W.......	Manasquan, N. J.
715.	John W.............	New Jersey.
1328.	John, W.............	New Jersey.
2783.	John W.............	New Jersey.
812.	John, Wells........	Nova Scotia.
2338.	John Westall........	Fall River.
1527.	John William..	Ottawa, Canada.
2833.	Johnson Calvin L.	Alhambra, Cal.
2425.	Johnson Carrie B (Miller).......	Pasadena, Cal.
2423.	Johnson Emma O., (Borden)....	Washington.
2843.	Johnson Evelyn G...	Los Angeles.
2420.	Johnson Gail B....	Los Angeles.
2834.	Johnson Gail B..	Alhambra, Cal.
2842.	Johnson Louise L..	Los Angeles.
2424.	Johnson Milbank...	Los Angeles.
2832.	Johnson Paschal B.............	Alhambra, Cal.
2829.	Johnson Phila B....	Los Angeles.
2830.	Johnson Ray L....	Los Angeles.

2835.	Johnson Richard W..............	Alhambra, Cal.
2831.	Johnson Virginia L..	Los Angeles.
2422.	Johnson Virginia L., (Milbank)..	New York.
2421.	Johnson Waldo P..	Alhambra, Cal.
70.	Jonathan....	Bordentown, N. J.
2355.	Jonathan...........	Fall River.
98.	Jonathan........	Gloucester, N. J.
205.	Jonathan.......	Gloucester, N. J.
785.	Jonathan.........	Horton, N. S.
790.	Jonathan.........	Horton, N. S.
415.	Jonathan.........	Nova Scotia.
1569.	Jonathan...........	Nova Scotia.
398.	Jonathan............	Tiverton.
620.	Joseph.............	Arkansas.
8.	Joseph........	Barbadoes, W. I.
320.	Joseph........	Beauford, N. C.
69.	Joseph........	Bordentown, N. J.
182.	Joseph......	Bordentown, N. J.
341.	Joseph......	Bordentown, N. J.
1067.	Joseph........	Cazenovia, N. Y.
61.	Joseph....	Cooper's Creek, N. J.
247.	Joseph.............	Fall River.
305.	Joseph.............	Fall River.
467.	Joseph.............	Fall River.
720.	Joseph.............	Fall River.
759.	Joseph.............	Fall River.
922.	Joseph.............	Fall River.
1730.	Joseph.............	Fall River.
1777.	Joseph.............	Fall River.
1797.	Joseph.............	Fall River.
2502.	Joseph.............	Fall River.
39.	Joseph........	Fall River, Mass.
137.	Joseph........	Fall River, Mass.
604.	Joseph.............	Fresno, Cal.
203.	Joseph........	Gloucester, N. J.
1101.	Joseph........	Green Co., Ala.
323.	Joseph........	Knoxville, Tenn.
113.	Joseph........	Manasquan, N. J.
385.	Joseph........	Manasquan, N. J.
97.	Joseph........	Mansfield, N. J.
339.	Joseph........	Mansfield, N. J.
651.	Joseph.............	New Jersey.
1064.	Joseph.............	New York.
169.	Joseph.........	North Carolina.
209.	Joseph.......	Nottingham, N. J.
412.	Joseph.............	Nova Scotia.
556.	Joseph.............	Portsmouth.
18.	Joseph.........	Portsmouth, R. I.
122.	Joseph.........	Portsmouth, R. I.
152.	Joseph.........	Portsmouth, R. I.
192.	Joseph.............	Providence.
91.	Joseph........	Providence, R. I.

INDEX.

28. Joseph .... Providence, R. I
143. Joseph .... Swansea, Mass.
1160. Joseph .... Texas.
148. Joseph .... Tiverton, R. I.
262. Joseph .... Tiverton, R. I.
269. Joseph .... Tiverton, R. I.
1379. Joseph, C............ Fall River.
1879. Joseph, D............ Fall River.
714. Joseph E.............. Chicago.
1861. Joseph, E............ Fall River.
1288. Joseph, E.......... Philadelphia.
2305. Joseph F............ Fall River.
1078. Joseph, H.......... Milford, Neb.
547. Joseph H............ New York.
1126. Joseph, L............... Alabama.
2011. Joseph, Lane.......... Alabama.
1919. Joseph, M............ Dunee, Ill.
1954. Joseph, R............ California.
1480. Joseph V............ Fall River.
1505. Joseph W............ Fall River.
2754. Joseph Wilson.... Philadelphia.
2403. Josephine .... Canada.
1946. Josephine .... Chicago.
2453. Josephine ....Sharpsburg, Tex.
2620. Josephine C......... Dundee, Ill.
295. Joshua ...... Delaware Co., N. Y.
417. Joshua .... Nova Scotia.
540. Joshua .... Pompey, N. Y.
797. Joshua, W.......... Horton, N. S.
359. Josiah .... Burlington Co., N. J.
1254. Josiah .... Emleytown, N. J.
1316. Josiah .... New Jersey.
499. Josiah .... Nova Scotia.
979. Josiah .... Nova Scotia.
2263. Josiah Brick .... New Jersey.
2189. Josiah, P............ New Jersey.
24. Joyce, (Hance).Shrewsbury, N. J.
815. Judah .... Cornwallis, N. S.
1941. Judith A., (Gray).Massachusetts.
750. Judith, (Durfee) ....Fall River.
2012. Julia .... Cunningham, Ala.
2917. Julia ....Goldsboro, N. C.
502. Julia .... Nova Scotia.
1817. Julia, Clara ...... Massachusetts.
2296. Julia H., (Robinson).Fall River.
2591. Julia, (Jermain).Goldsboro, N. C.
1528. JuliaRebecca ......Nova Scotia.
1195. Julia, S., (Gillet).... New Jersey.
2223. Julia Strawbridge.Ardmour, Pa.
1312. Julia, (Walker)....New Jersey.
1986. Juliet, R............ Los Angeles.
1776. Justinia .... Fall River.

K

2762. Kain Edward S.... Philadelphia.
1337. Kate .... New York.
2568. Kate Cole ....Wilmington, N. C.
2332. Keen Corinne, (Freeman)........
.... Pennsylvania.
2334. Keen Dora .... Philadelphia.
2332. Keen Florence.... Philadelphia.
2335. Keen Margaret.... Philadelphia.
2805. Kester Edith .... New Jersey.
2804. Kester Joseph...... New Jersey.
1629. Keziah .... Chelsea, Mass.
849. Knight H............ Fall River.
2411. Knight H............ Fall River.

L

1002. Laban .... Fall River.
739. Ladowick .... Fall River.
2010. Lane Beal............Alabama.
1418. Laura A., (Harrington) ....
.... Massachusetts.
2251. Laura, (Allaire) ....New Jersey.
1022. Laura, (Chandler) ...Fall River.
2362. Laura Evelina .... New Bedford.
840. Laura, (Kingsley) ....New York.
2398. Laura L., (Flemming) ..Canada.
2606. Laurence Leander .. Fall River.
2231. Lavinia Cook ......Philadelphia.
1285. Lavinia, (Cook) ....Philadelphia.
1575. Lavinia J............ Nova Scotia.
684. Lavinia, (Loomer)..Nova Scotia.
1239. Lavinia, (Price) ....
995. Lazarus .... Fall River.
1201. Leah ....
2118. Leah, (Shaw) ....
1005. Leander .... Fall River.
1783. Leander .... Fall River.
1696. Lee D.......... Galveston, Tex.
2913. Lee Edwin...... North Carolina.
2914. Lee Georgia......North Carolina.
938. Lefavour ...... Patterson, N. J.
864. Lemira, (Jones) ....Warren, R. I.
486. Lemuel .... Tiverton.
1536. Lemuel P.. .... Nova Scotia.
775. Lemuel Perry .... Nova Scotia.
2341. Leonora .... Fall River.
2483. Leonard I.... ....
1820. Le Valley Adelaide R............
.... Massachusetts.
1819. Le Valley Arthur F............
.... Massachusetts.
1822. Le Valley Benjaman W..........
.... Massachusetts.

INDEX.   337

1825.	Le Valley Emeline B............... Massachusetts.	2550.	Lovel Mary E., (Minnich)....... Philadelphia.
1824.	Le Valley James G............... Massachusetts.	2220.	Loyd William H....Ardmore, Pa.
		2221.	Loyd John S........ Ardmore, Pa.
1818.	Le Valley M. C., (Lovel).......... Massachusetts.	1230.	Lucia, (Adams) ....New Jersey.
		2458.	Lucile .... Sharpsburg, Tex.
1823.	Le Valley Sarah F............... Massachusetts.	826.	Lucilla ....Cornwallis, N. S.
1821.	Le Valley William P............. Massachusetts.	1072.	Lucina ...... Madison Co., N. Y.
		1716.	Lucy .... Fall River.
816.	Levi .... Nova Scotia.	1816.	Lucy .... Fall River.
2841.	Lewis Colonna, Washington, D. C.	2487.	Lucy .... Fall River.
2418.	Lewis Lambert .... Tonti, Ill.	494.	Lucy, (Borden) ......Fall River.
2433.	Lewis Mercer....Wallkill, N. Y.	678.	Lucy (Coward) ....New Jersey.
1556.	Lila L., (Smith) .... Somerville, Mass.	1767.	Lucy J., (Bennett) ..Providence.
		1493.	Lucy Jane, (Fisher) ..Fall River.
2387.	Lila Reeve ....New Brunswick.	2491.	Lucy, (Parsons) .... Fall River.
2271.	Lillian May .... New York.	449.	Lucy, (Rogers) ....Tiverton.
2414.	Lillian V............. Fall River.	1016.	Lugenia .... Fall River.
2158.	Lillie, (Chaffee)..... Edna, Kan.	1612.	Luther .... Phillipsburg, Can.
2614.	Lindon Cook .. Somerset, Mass.	440.	Luther .... Warren, R. I.
2749.	Linnard, Eliz. C.. Philadelphia.	1649.	Luther C............. Fall River.
2750.	Linnard Helen E..Philadelphia.	959.	Luther E............. Tiverton.
2137.	Linnard Joseph H..Philadelphia.	1697.	Luther G.... Minneapolis, Minn.
2138.	Linnard Theodore B............. Philadelphia.	1395.	Luther Irene, (Bliss) (Keep)....
		1394.	Luther Lorenzo ...... New York.
2173.	Lizzie A ......Jacobstown, N. J.	1396.	Luther Sarah B., (Newhall).....
2347.	Lizzie Drew .... Fall River.	998.	Lydia .... Fall River.
1904.	Lizzie S., (Chace).Massachusetts.	357.	Lydia ....Imlaystown, N. J.
2605.	Loban Gilbert ..Crystal Lake, Ill.	167.	Lydia, (Beck)....Botetourte, Vt.
2227.	Logan Elizabeth ....New Jersey.	452.	Lydia, (Borden) .....Fall River.
2225.	Logan Oliver E.... New Jersey.	917.	Lydia, (Coggeshall)..Fall River.
2228.	Logan Rena .... New Jersey.	215.	Lydia, (Ford)..Nottingham, N. J.
2226.	Logan Scott Jr...... New Jersey.	511.	Lydia, (Hathaway) ....
2818.	Longstreet George ..New Jersey.	455.	Lydia, (Hathaway) ..Fall River.
2816.	Longstreet Lillian.. New Jersey.	1120.	Lydia J., (Sheppard)...Alabama.
2817.	Longstreet Russel ..New Jersey.	963.	Lydia, (King).......... Tiverton.
841.	Lorena .... Quebec.	909.	Lydia, (Macomber)...Fall River.
2457.	Loretta .... Sharpsburg, Tex.	2121.	Lydia, (Nelson) ....
2270.	Loretta Jane .... New York.	542.	Lydia, (Shearman) ...Fall River.
1073.	Louis Willington .... Kalamazoo, Mich.	1087.	Lydia, (Slocum)......Fall River.
		1390.	Lydia Swan ....
2489.	Louisa .... Fall River.	1566.	Lydia, (Thomas) ....Nova Scotia.
2573.	Louisa, (Davis) .North Carolina.	1501.	Lydia W., (Davis) .... Fall River.
1467.	Louisa G., (Aldrich) .... Massachusetts.	2905.	Lynch Annie M.Goldsboro, N. C.
		2901.	Lynch Carrie B.Goldsboro, N. C.
2539.	Louisa M. ....Fall River.	2903.	Lynch Charles B.Goldsboro, N. C.
1718.	Louise .... Fall River.	2899.	Lynch George G.................. Goldsboro, N. C.
2369.	Louise Gould ...... Fall River.		
1899.	Louise, (Kornegay) .... North Carolina.	2900.	Lynch Herbert A................ Goldsboro, N. C.
1668.	Louise M......... Scituate, R. I.	2898.	Lynch James B..Goldsboro, N. C.
2715.	Love Annie ....Pattonsburg, Va.	2902.	Lynch Mary B..Goldsboro, N. C.
2716.	Love Charles ..Pattonsburg, Va.	2904.	Lynch Ruth B..Goldsboro, N. C.
2717.	Love Mary .... Pattonsburg, Va.	1018.	Lysander .... Fall River.

INDEX.

1443. Lysander .... Fall River.
1052. Lysander .... Tiverton.

M

2202. M. Augusta ....Jacobstown, N. J.
2405. Mabel H. .... Canada.
2570. Mabel M........Goldsboro, N. C.
2443. Male Phila .... Galveston, Tex.
935. Major .... Fall River.
1726. Major .... Fall River.
2026. Malbert T....... Cedartown, Ga.
1874. Malvina, (Hazlehurst) ....
.... Massachusetts.
1348. Malvina, (Seeley) ..Connecticut.
1349. Marcus H. .... New York.
162. Marcy, (Feamley) ..... Virginia.
1709. Marcy, (Perry) ....New Bedford.
1167. Margaret ......Calhoun Co., Ala.
1307. Margaret, (Atkinson). ....
.... New Jersey.
2930. Margaret D.... Goldsboro, N. C.
1576. Margaret E......... Nova Scotia.
1572. Margaret E., (Rand).Nova Scotia.
2272. Margaret Eline ......New York.
328. Margaret, (Keith) ....
.... Knoxville, Tenn.
1952. Margaret, (Pearce) ....Alabama.
1998. Margaretta R., (Peabody) ......
.... Alabama.
356. Margarite, (Allen)..New Jersey.
2386. Margery Ellen .... Nova Scotia.
1436. Maria B., (Jenny).New Bedford.
1523. Maria F............ Nova Scotia.
1430. Maria, (Hinckley) ...Fall River.
2560. Maria L., (McLeod) ....
.... North CaCrolina.
1098. Maria W., (Telfair) ....Alabama.
2788. Marian .... New York City.
732. Marietta, (Crocker) ..Fall River.
2434. Marion .... Walkill, N. Y.
2787. Marjorie V...... Colorao Springs.
2475. Marshall Edith G...Pomona, Cal.
2476. Marshall Horatio V.Pomona, Cal.
1859. Marshall W..........Fall River
441. Martin .... Gardners' Neck.
2384. Martha Adelia .....Nova Scotia.
1146. Martha, (Bryant) ....Tennessee.
286. Martha, (Dresser) ...Fall River.
1287. Martha E. ...... New Jersey.
1580. Martha E., (Tupper).Nova Scotia.
1714. Martha H. Cobb, ...
321. Martha, Hawkins......Virginia
817. Martha Jane ....Nova Scotia
1234. Martha, (Loree) ....

1907. Martha M., (Laurence) ....
.... Massachusetts.
1273. Martha, (Painter) ..New Jersey.
15. Mary ....
1199. Mary. ....
201. Mary .... Burlington, N. J.
1780. Mary ... . Fall River.
2790. Mary .... Fall River.
208. Mary .... Gloucester, N. J.
109. Mary .... New Jersey.
1517. Mary .... Nova Scotia.
119. Mary .... Portsmouth, R. I.
26. Mary ....Providence.
2888. Mary .... Wilmington, N. C.
1508. Mary A. ....Massachusetts.
1768. Mary A., (Borden) ....Fall River.
774. Mary A., (Gould) ....Nova Scotia.
1808. Mary A., (Waring) ..Fall River.
1406. Mary Ann .... Fall River.
813. Mary Ann .... Nova Scotia.
1049. Mary Ann, (Booth) ....Tiverton.
1381. Mary Ann, (Read) .. Fall River.
729. Mary, (Anthony) .... Fall River.
397. Mary, (Bailey) ......Fall River.
438. Mary, (Bradford) ....Providence.
1378. Mary, (Brow) .... Fall River.
2574. Mary C..........Goldsboro, N. C.
1156. Mary C., (Bacon).Corsicana, Tex.
846. Mary C., (Bennett)..........R. I.
1431. Mary C., (Brayton) ..Fall River.
1500. Mary C., (Crapo)....Fall River.
2929. Mary Carrow ..Goldsboro, N. C.
451. Mary, (Cook) .... Fall River.
13. Mary, (Cook) ..Providence. R. I.
529. Mary, (Crandall) ...... Tiverton.
1589. Mary E.......... Cornwallis, N. S.
930. Mary E ....
1105. Mary E., (Cheney) (Campbell)...
.... Alabama.
2415. Mary E. D. .... Fall River.
786. Mary E., (Dickey) ..Nova Scotia.
1741. Mary E., (Edmons) ....
2357. Mary E., (Howland)..Fall River.
1272. Mary E., (Scott) ....New Jersey.
1520. Mary Elizabeth.....Nova Scotia.
1693. Mary Elizabeth .... Texas.
2350. Mary Emma, (Pierce).Fall River.
2001. Mary Esther .... Alabama.
1036. Mary F., (Brownell)..Portsmouth.
1229. Mary, (Fennimore) ....
.... Beverley, N. J.
136. Mary, (Gifford) ....
.... Dartsmouth, Mass.
314. Mary, (Gladding) ....Bristol.

INDEX.

1996.	Mary Gray .... California.	2625.	McGiffert Mary E...Eutaw, Ala.
562.	Mary Hamblin	2892.	McLeod Georgia H., (Stover) ....
536.	Mary, (Hathaway.)		.... North Carolina.
2336.	Mary I. .... Fall River.	2893.	McLeod Mauds A...... ... ...
1424.	Mary J., (Hartley) ...Fall River.		.... North Carolina.
806.	Mary J., (Patterson).Nova Scotia.	2363.	Mert A......... Fort Pierre, S. D.
2261.	Mary J., (Smyth) ..New Jersey.	1380.	Melinda, (Eddy) ...... Fall River.
1109.	Mary James, (Grace) ..Alabama.	1003.	Melvin .... Fall River.
1681.	Mary Jane .... Hartford, Ct.	1024.	Merril H. ....
1511.	Mary Jane.......... Nova Scotia.	450.	Meribah, (Borden) ...Fall River.
779.	Mary, (Johnson) ..Nova Scotia.	265.	Meribah, (Barker)....Fall River.
93.	Mary, (King) ..Providence, R. I.	94.	Meribah, (Thornton) ....
151.	Mary, (Lawton)		.... Providence, R. I.
	.... Fall River, Mass.	19.	Mercy .... Portsmouth, R. I.
1922.	Mary M. E. L., (Wells).Elgin, Ill.	21.	Meribah, (Crawford) ....
2145.	Mary M., (Harris)..Philadelphia.		.... Portsmouth, R. I.
2302.	Mary M. Martin ..... Fall River.	2936.	Michaux Mary Louise ....
1928.	Mary M., (Reynolds) ...Oakland.		.... Goldsboro, N. C.
2500.	Mary Maria .... Fall River.	29353.	Michaux Sarah Borden ....
2537.	Mary Matilda .... Fall River.		.... Goldsboro, N. C.
342.	Mary, (McKean)	1685.	Milam .... Texas.
	.... .Philadelphia, Pa.	2446.	Milam Underwood...............
2851.	Mary McKee.......... New York.		.... Galveston, Tex.
1803.	Mary P............... Fall River.	2837.	Milbank Laurence.Yonkers, N. Y.
218.	Mary, (Pintard).	2838.	Milbank Lee B..Yonkers, N. Y.
	.... Shrewsbury, N. J.	2836.	Milbank Phila N..Yonkers, N. Y.
1255.	Mary R., (Emley) ..New Jersey.	2918.	Mildred B.......Goldsboro, N. C.
664.	Mary R., (Tilton) ...New Jersey.	2590.	Miller Bessie West ....
41.	Mary, (Potts) ..Portsmouth, R. I.		.... North Carolina.
131.	Mary, (Rodman) ..New Bedford.	2684.	Miller Cesna....Kansas City, Mo.
2897.	Mary S............. New Orleans.	2584.	Miller Charles B.Goldsboro, N. C.
834.	Mary S., (Jones) ..Rhode Island.	2934.	Miller Charles P ....
908.	Mary, S., (Petty) ...Fall River.		.... Goldsboro, N. C.
241.	Mary, (Sherman) ..Tiverton, R. I.	2933.	Miller Ethel B..Goldsboro, N. C.
879.	Mary, (Simmons).	2588.	Miller Frank Marvin ....
	.... Cape Cod,, Mass.		.... North Carolina.
1309.	Mary, (Taylor) .... New Jersey.	2586.	Miller Hugh Lee.North Carolina.
302.	Mary, Thomas)......Fall River.	2845.	Miller John Borden.............
101.	Mary, (Tindall)		.... Pasadena, Cal.
	.... Shrewsbury, N. J.	2583.	Miller John C...Goldsboro, N. C.
2525.	Mary Valentine ...... Fall River.	2682.	Miller Kirk ....Kansas City, Mo.
607.	Mary W., (Sheldon).Orange, N. J.	2585.	Miller Louise, (Michaux) ....
460.	Mary, (Weeks) ..... Fall River.	2587.	.... North Carolina.
2440.	Mary Willie, (Johnson)		Miller Mary B., (Southerland) ..
	.... Los Angeles.	2844.	.... North Carolina.
1757.	Mary, (Winslow) .Massachusetts.	2589.	Miller Philadelphia B............
1143.	Massey .... Tennessee.		Miller Robert Bascom...........
1545.	Matilda A........... Nova Scotia.	2683.	.... North Carolina.
2330.	Matthew .... New York.	2399.	Miller Walter..Kansas City, Mo.
6.	Matthew ......Portsmouth, R. I.	2872.	Milton Luther .... Canada.
118.	Matthew .... Portsmouth, R. I.		Minnich Charles H...............
128.	Matthew ..... Portsmouth, R. I.	2870.	.... Philadelphia.
27.	Matthew ...... Providence, R. I.	2869.	Minnich Clara Frank ....
1402.	Matthew C. D......... New York.	2562.	.... Philadelphia.
2639.	McGhee Matthew B... Alabama.		Minnich Marie L..Philadelphia.
			Minnie Bell, (Lynch) ....
			.... North Carolina.

2580. Minnie D....... Goldsboro, N. C.
2249. Minnie F., (Hill) ...New Jersey.
2472. Minnie Lee ......Galveston, Tex.
1103. Miranda, (Clark)....Eutaw, Ala.
1562. Miriam R., (Kincaid)..............
.... Massachusetts.
1163. Mitchel .... Calhoun Co., Ala.
2262. Morris Peak ....Colorado.
1563. Mortimer .... Chelsea, Mass.
2426. Munsill Ettie C., (Ulrich) ....
.... New Haven.
2427. Munsill Gail B........ Hartford.
2428. Munsill Marcus .... Hartford.
2849. Munsill Marcus Mills..........
.... Hartford, Ct.
2578. Murray .... Goldsboro, N. C.

N

1165. Nancy .... Alabama.
1046. Nancy .... Fall River.
2495. Nancy .... Fall River.
433. Nancy .... Scituate.
2127. Nancy, (Clark) ....
329. Nancy, (McWilliams). Tennessee.
856. Nancy, (Slocum) ....Fall River.
2028. Nannie, (Frey)) ...Newnan, Ga.
825. Naomi ....Quebec.
535. Nathan .... Pompey, N. Y.
1055. Nathan .... Pompey, N. Y.
446. Nathan .... Tiverton.
1781. Nathan D. .... Fall River.
1115. Nathan Lane .... California.
752. Nathaniel .... Fall River.
1943. Nathaniel .... Massachusetts.
842. Nathaniel .... Quebec.
255. Nathaniel........Scituate, R. I.
1468. Nathaniel B......... Fall River.
2366. Nathaniel B......... Fall River.
1960. Nathaniel Barnett ......Florida.
2447. Nellie Hamilton ....
.... San Antonio, Tex.
1209. Nelson ....
1633. Nelson .... Boston, Mass.
1422. Nelson C........... Fall River.
2604. Nettie, (Baldwin) ...... Chicago.
2320. Newhall Catherine W.California.
2322. Newhall Henry B.....California.
2321. Newhall Stedman.... California.
2619. Nina Susan Prentice.Dundee, Ill.
1414. Norman E........... Fall River.
1924. Norman T........... New York.

O

2696. Obenchain Alexandra ..Florida.
2072. Obenchain Alice M....Kentucky.
2070. Obenchain Caroline ..Kentucky.
2692. Obenchain Cecelia C..Kentucky.
2694. Obenchain Elizabeth S..Florida.
2695. Obenchain Fannie M....Florida.
2074. Obenchain Florence M.Kentucky.
2065. Obenchain Francis G....Florida.
2068. Obenchain James T...Tennessee.
2693. Obenchain Jenette B.....Florida.
06.2. Obenchain Letitia A..Kentucky.
2073. Obenchain Lura B....Kentucky.
2071. Obenchain Margaret..Kentucky.
2689. Obenchain Margery.. Kentucky.
2067. Obenchain Sallie S....Kentucky.
2691. Obenchain Thomas H.Kentucky.
2064. Obenchain William A.Kentucky.
2690. Obenchain William A.Kentucky.
1729. Oliver .... Fall River.
189. Oliver ....Johnston, R. I.
1564. Olivia Ann, (Rand), Nova Scotia.
1917. Olney .... Colorado.
1061. Olney......Jefferson Park, Colo.
1900. Omer Elton .... Fall River.
2781. Orville C............ New Jersey.
2358. Othneil .... Fall River.
1850. Otto Emerson ...... Nova Scotia.
1931. Owen D.......... Norwich, N. Y.

P

2777. Pagenstetcher, Albertina ....
.... New Jersey.
2779. Pagenstetcher, Chas. F. R.......
.... New Jersey.
2778. Pagenstetcher, Mabel F.........
.... New Jersey.
2336. Paine Ellen C., (Huling) ....
.... Massachusetts.
2204. Painter Bertha .... New Jersey.
2205. Painter Grant H.... New Jersey.
1039. Pardon S............ Fall River.
1975. Parham B...... Greensboro, Ala.
275. Parker .... Fall River.
468. Parker .... Fall River.
928. Parker .... Fall River.
943. Parker .... Fall River.
1305. Parker .... New Jersey.
1058. Parker .... Pompey, N. Y.
2245. Parker Edith M., (Loyd).Chicago.
2247. Parker Ethel Frances..Chicago.
2246. Parker Irene Borden....Chicago.
864. Paschal P........ Brazoria, Tex.

INDEX. 341

2449.	Paschal P..... San Antonio, Tex.	912.	Philadelphia, (Church)...........
282.	Patience .... Fall River.		.... Fall River.
517.	Patience, (Briggs) ...Fall River.	1679.	Philadelphia W., (Johnson) .....
394.	Patience, (Butter) ...Fall River.	1004.	Philander .... Fall River.
1725.	Patience, (Butler)...Fall River.	1858.	Philander .... Fall River.
538.	Patience, (Cook) .....Tiverton.	1885.	Philander W......... Fall River.
1735.	Patience, (Davies)...............	755.	Philip .... Fall River.
	.... Massachusetts.	1388.	Philip D............. Fall River.
243.	Patience, (Durfee)	2317.	Philip D............. Fall River.
	.... Fall River, Mass.	1427.	Philip H............. Fall River.
447.	Patience, (Durfee) ..Fall River.	1739.	Philip H............. Fall River.
462.	Patience, (Haggard) ..Newport.	1782.	Philip H............. Fall River.
955.	Patience, (Munroe) .Providence.	2961.	Philip Pearce .... Fall River.
1048.	Patience, (Ross) ....Connecticut.	1071.	Philura Ann ..Madison Co., N. Y.
403.	Patience, (Slade) ...... Tiverton.	1253.	Phoebe A., (Emley) ....
1158.	Patrick D............ Hope, Ark.	312.	Phoebe, (Anthony)..Portsmouth.
2581.	Paul .... Goldsboro, N. C.	744.	Phoebe Durfee ....Tiverton.
1588.	Pauline L......... Nova Scotia.	721.	Phoebe, (Durfee) ....Fall River.
287.	Peace .... Fall River.	2516.	Phoebe E., (Busby)....Fall River.
1027.	Peace, (Bassett) .... Fall River.	2360.	Phoebe Sarah .... Fall River.
246.	Peace, (Borden)	534.	Phoebe, (Wordell) ....New York.
	.... Fall River, Mass.	2827.	Pierce Madeline ..Massachusetts.
923.	Peace, (Gardner)....Fall River.	2948.	Preston Mary A..Bowling G., Ky.
474.	Peace, (Warren) .... Fall River.	1019.	Prince Sears .... Fall River.
539.	Peace, (Wordell) .... Fall River.	284.	Priscilla .... Tiverton.
2630.	Pearce Fannie G.....Selma, Ala.	1054.	Priscilla, (Grinnel) ....Tiverton.
1982.	Pearce Georgie....	525.	Priscilla, (Hicki) ....Fall River.
2628.	Pearce Joseph B...... Selma, Ala.	1498.	Priscilla, (Wilcox) ...Fall River.
1978.	Pearce Lovie .... Selma, Ala.	1736.	Prudence .... Fall River.
2632.	Pearce Lovie B......Selma, Ala.	950.	Prudence, (Borden) ..Fall River.
1980.	Pearce Penelope ...... Alabama.	1581.	Prudence L........Nova Scotia.
1932.	Pearl A., (Ball) .... Los Angeles.	975.	Prudence, (Philips).Valley Mills.
773.	Pearl, (Gould) .... Nova Scotia.	1166.	Pollie .... Calhoun Co., Ala.
461.	Peleg .... Fall River.	890.	Polly, (Davis).........Tiverton.
1038.	Peleg .... Fall River.	2257.	Pontin Alfaretta, (Ramage) ....
1759.	Peleg .... Fall River.		.... New Jersey.
2027.	Pelham, (Harper) ...... Florida.	2255.	Pontin Carrie, (Pagenstetcher)..
2430.	Penelope A., (Hamilton)		.... New Jersey.
	.... New York.	2254.	Pontin Georgiana, (Field) ....
459.	Penelope, (Brownell). Fall River.		.... New Jersey.
1352.	Percival E........ Connecticut.	2256.	Pontin Sarah E., (Bates) ....
2295.	Percival E........ Shelton, Ct.		.... New Jersey.
2808.	Percy .... New Jersey.	2253.	Pontin William Dykes ....
811.	Perry .... Cornwallis, N. S.		.... New Jersey.
401.	Perry .... Fall River.	343.	Potts Ann, (Cox) ....
248.	Perry .... Nova Scotia.		.... Bordentown, N. J.
416.	Perry .... Nova Scotia.	2111.	Powell Anne F...Chrisman, Tenn.
1531.	Perry A........... Nova Scotia.	2113.	Powell Carrie Evelyn..Kentucky.
650.	Peter .... Burlington, N. J.	2110.	Powell Charles E.... Kentucky.
352.	Peter ..... Burlington Co., N. J.	2091.	Powell Henry .... Kentucky.
2235.	Peter R...... Tunkhannock, Pa.	2094.	Powell Lilly ...., Kentucky.
629.	Petetiah .... Mansfield, N. J.	2093.	Powell Lucy .... Kentucky.
1208.	Petetiah ....New York.	2095.	Powell Luke .... Kentucky.
		2092.	Powell Mary, (Foster).Kentucky.
		2112.	Powell Minnie B..... Kentucky.

R

1258. Rachael, (Bean)....New Jersey.
2910. Rachael II.......Goldsboro, N. C.
679. Rachael, (Lawrence).New Jersey.
235. Rachel, (Cock)..................
.... Point Pleasant, N. J.
2597. Rachel Moye....Goldsboro, N. C.
1314. Randall .... New Jersey.
2240. Randall .... New Jersey.
2607. Raymond Davis .....Fall River
2497. Reason .... Fall River.
66. Rebecca ....
1510. Rebecca .... Fall River.
1881. Rebecca .... Fall River.
2119. Rebecca .... Fall River.
1202. Rebecca .... New York.
1150. Rebecca, (Alexander).Tennessee.
241. Rebecca, (Borden).Warren, R. I.
165. Rebecca, (Bronson) ....Virginia.
183. Rebecca, (Brown) ....
.... Bordentown, N. J.
1565. Rebecca, (Burbidge) ....
.... Nova Scotia.
1313. Rebecca, (Dolen) ....New Jersey.
1130. Rebecca K., (Grover)..Cairo, Ill.
772. Rebecca, (Lockhart) ....
.... Nova Scotia.
327. Rebecca, (Overstreet) ....
.... Overton, Tenn.
139. Rebecca, (Russell)...Dartmouth.
882. Rebecca, (Simmons) ....
..........Cape Cod. Mass.
225. Rebecca, (Wooley)..Shrewsbury.
1482. Rescome .... Fall River.
424. Rescome .... Westport, R. I.
832. Rescome.........Westport, R. I.
434. Rhoda .... Scituate.
855. Rhoda, (Cameron) ...Fall River.
746. Rhoda, (Cook) ...... Fall River.
1435. Rhoda, (Davis) ...... Fall River.
1293. Rhoda, (Gaskill) ...New Jersey.
213. Rhoda, (Robbins) ....
.... Nottingham, N. J.
475. Rhoda, (Warren) ....Fall River.
409. Rhoda, (Westgate) ....Tiverton.
762. Rhoda, (Wilcox) .Massachusetts.
1894. Rhodes .... Borden, Cal.
1114. Rhodes .... San Francisco, Cal.
2280. Rice Emma E., (Schofield)......
.... Michigan.
2283. Rice George Carleton ....
.... Michigan.
2281. Rice Mary M., (Parry) ....
.... Michigan.

180. Richard ...... Bordentown, N. J.
292. Richard .... Fall River.
393. Richard .... Fall River.
757. Richard .... Fall River.
728. Richard .... Fall River.
736. Richard .... Fall River.
1486. Richard .... Fall River.
36. Richard ...... Fall River, Mass.
237. Richard ...... Fall River, Mass.
238. Richard ...... Fall River, Mass.
171. Richard .... Freehold, N. J.
187. Richard .... Johnston, R. I.
344. Richard ....Johnston, R. I.
11'. Richard........Manasquan, N. J.
229. Richard ...... Manasquan, N. J.
63. Richard .... New Jersey.
381. Richard .... New Jersey.
692. Richard .... New Jersey.
1327. Richard .... New Jersey.
2242. Richard .... New Jersey.
548. Richard .... New York.
3. Richard ...... Portsmouth, R. I.
14. Richard .... Providence, R. I.
32. Richard .... Providence, R. I.
373. Richard ..... Shrewsbury, N. J.
699. Richard A..........New Jersey.
1399. Richard B............ Fall River.
2534. Richard D............ Fall River.
850. Richard E........ Scituate, R. I.
22. Richard Eversham ....
.... Burlington Co., N. J.
1573. Richard H.........Nova Scotia.
2335. Richard P............ Fall River.
675. Richard P..........New Jersey.
1297. Richard P.......... New Jersey.
2230. Richard T. Cook.. Philadelphia.
75. Richardson John ....
74. Richardson Thomas ....
73. Richardson William ....
1834. Richmond .... Los Angeles.
2708. Riffe Martha H., (Duncan)......
.... Virginia.
2953. Riffe Peter B......... Kentucky.
2707. Riffe Walter Scott .... Virginia.
1953. Rinnie L., (Borden) ..California.
436. Robert .... Scituate.
1554. Robert A.......... Moncton, N. B.
1737. Robert E............ Fall River.
2296. Robert H............ Shelton, Ct.
2862. Robert Henry .... Fall River.
1526. Robert Leard ..... Nova Scotia.
800. Robert, Remington...Fall River.
2442. Robert Stafford .Galveston, Tex.
2637. Robinson Henry B.....Alabama.

INDEX.

1809. Roger W............. Fall River.
2839. Rolden Lee ....Washington, D. C.
2438. Rolden, (Low)........New York.
1609. Romeo V............. Reno, Nev.
483. Rosannah, (Ames)..Sterling, Ct.
2928. Rowena M...... Goldsboro, N. C.
823. Roxanna ...... Cornwallis, N. S.
737. Roxanna .... Fall River.
776. Ruby, (Cox) ...... Nova Scotia.
1811. Ruel Channing ......Fall River.
1961. Ruffin G...... Moss Point, Miss.
1567. Rufus.............N. V. Scotia.
480. Ruhama, (Sprague) ..Fall River.
1835. Rupert .... Los Angeles.
576. Russell .... Portsmouth.
824. Ruth .... Cornwallis, N. S.
936. Ruth .... Fall River.
2826. Ruth .... Fort Pierre, S. D.
1050. Ruth, (Alford) ....New Bedford.
2531. Ruth Ann, (Waller) ..Fall River.
871. Ruth, (Borden) ......Fall River.
496. Ruth, (Davis) ...... Fall River.
272. Ruth, (Durfee) ..... Fall River.
481. Ruth, (Durfee) ...... Fall River.
526. Ruth, (Earl) .... Fall River.
346. Ruth, (Fraley)...Johnston, R. I.
554. Ruth, (Greene).........Warwick.
390. Ruth, (Harris) ...... Fall River.
251. Ruth, (Hart) .... Fall River.
306. Ruth, (Luke) .... Fall River.
519. Ruth, (Sabin) .... Fall River.
512. Ruth, (Stillwell).....Fall River.
965. Ruth, (Stranger) ...Providence.

S

67. Safety .... Bordentown, N. J.
2595. Sallie ....Goldsboro, N. C.
1144. Sallie .... Tennessee.
2595. Sallie ....D.... Goldsboro, N. C.
711. Sallie D., (Lawson) ....
.... Newburgh, N. Y.
1742. Sallie C., (Borden) ...Fall River.
1674. Samos D. W......... Fall River.
72. Samuel ......Bordentown, N. J.
428. Samuel .... Canada.
1723. Samuel .... Fall River.
1857. Samuel .... Fall River.
2504. Samuel .... Fall River.
2513. Samuel .... Fall River.
138. Samuel .... Fall River, Mass.
116. Samuel ....Manasquan, N. J.
228. Samuel........Manasquan, N. J.
10. Samuel .... Monmouth, N. J.

411. Samuel ....New Bedford, Mass.
204. Samuel .... New Jersey.
770. Samuel .... Nova Scotia.
1512. Samuel .... Nova Scotia.
837. Samuel .... Quebec.
852. Samuel .... Scituate.
377. Samuel .... Shrewsbury.
425. Samuel .... Tiverton.
423. Samuel .... Tiverton.
245. Samuel .... Tiverton, R. I.
884. Samuel .... Tiverton, R. I.
2166. Samuel A........... New Jersey.
253. Samuel Asa....Providence, R. I.
988. Samuel B.......... Nova Scotia.
1484. Samuel E............. Fall River.
2250. Samuel Morris .....New Jersey.
2863. Samuel Nelson ....Fall River.
829. Samuel Smith ......Nova Scotia.
802. Samuel. T..........Nova Scotia.
669. Samuel W .... New Jersey.
2640. Samuel Wheatly ....Washington.
2610. Sanford .... Fall River.
970. Sanford, (Arnold B)..Fall River.
45. Sarah..........Barbados. W. I.
1043. Sarah .... Fall River.
1464. Sarah .... Fall River.
2365. Sarah .... Fall River.
2498. Sarah .... Fall River.
153. Sarah .... Fall River, Mass.
501. Sarah .... Nova Scotia.
191. Sarah .... Providence, R. I.
144. Sarah .... Swansea, Mass.
2149. Sarah A............... Chicago.
2541. Sarah A............. Fall River.
2486. Sarah A............. Fall River.
1223. Sarah A........... Philadelphia.
1673. Sarah A. L. E........Fall River.
1769. Sarah A., (Norvel)..............
.... Massachusetts.
653. Sarah, (Allen) .....New Jersey.
2388. Sarah Allison ..New Brunswick.
937. Sarah Ann ....Fall River.
1241. Sarah, (Barton) .... New Jersey.
362. Sarah, (Black)............N. J.
537. Sarah, (Boomer) ... New York.
1122. Sarah C., (Burr) ..Talladega, Ala.
1275. Sarah, (Bussom) ...New Jersey.
492. Sarah, (Cook) .... Tiverton.
2005. Sarah Cordelia.........Alabama.
1389. Sarah D............. Fall River.
233. Sarah, (Debow) ....
.... Upper Freehold, N. J.
445. Sarah (Durfee) .... Tiverton.

1938. Sarah E., (Bennett) .... Massachusetts.
1801. Sarah F., (Brightman)....Mass.
2920. Sarah, Elizabeth.Goldsboro, N. C.
1332. Sarah F............. New Jersey.
268. Sarah, (Francis) ....Fall River.
1615. Sarah, (Gustin)..Massachusetts.
1721. Sarah, (Harris) ..... Fall River.
133. Sarah, (Hazzard).Newport, R. I.
29. Sarah, (Hodgson) .... Providence, R. I.
9. Sarah, (Holmes)....Portsmouth.
123. Sarah, (Howland).Tiverton, R. I.
2524. Sarah Jane .... Fall River.
2865. Sarah Jeanette.......Fall River.
1711. Sarah, (Jenny) ......Fall River.
2505. Sarah Josephine.....Fall River.
550. Sarah, (Kelley) ....
1830. Sarah L., (Miller) .... North Carolina.
2601. Sarah Louisa .... Providence.
725. Sarah, (Luther) .... Fall River.
1066. Sarah, McCormick)..New York.
330. Sarah, (McCoy) ......Knoxville.
956. Sarah, (M)hti)........Smithfield.
487. Sarah, (Negus) .... Tiverton.
1494. Sarah P............. Fall River.
778. Sarah, (Phinney)....Nova Scotia.
969. Sarah, (Sanford) (Young)........
753. Sarah, (Seabury) ..New Bedford.
563. Sarah, (Shearman) ..Fall River.
523. Sarah, (Simmons) .... Little Compton.
888. Sarah, (Sweet) .Somerset, Mass.
388. Sarah T., (DeGrauw)..New York.
120. Sarah, (Thurston).Newport, R. I.
1403. Sarah W., (Covel) ....Fall River.
1077. Sarepta, (Hoxie) ....Tonica, Ill.
2203. Scott Theodore .....New Jersey.
1470. Seabury, Caroline, (Lincoln).... .... Boston.
1472. Seabury, Charlotte A............... New Bedford.
1471. Seabury, Sarah L..New Bedford.
2290. Seeley Geneveve A..Connecticut.
2289. Seeley Henrietta M.Connecticut.
490. Seth .... Fall River.
1374. Seth .... Fall River.
2310. Seth A............. Fall River.
1924. Seymour Skiff, Cazenovia, N. Y.
1116. Sheldon .... Los Angeles, Cal.
1137. Sheldon Catherine J., (Franklin) .... New Jersey.
1136. Sheldon Mary F. (Woolsey) (Hopkins) ....

2153. Sherman .... Burlington, N. J.
2009. Shirley, Moss............Alabama.
989. Sidney D........... Nova Scotia.
1225. Sidney D........... Philadelphia.
2796. Sidney, Freeman .... Fall River.
1689. Sidney G.......Sharpsburg, Tex.
2455. Sidney Gail....Sharpsburg, Tex.
680. Sidney P....... Allentown, N. J.
2237. Sidney Parker ..... New Jersey.
1592. Sidney Pinco ..Cornwallis, N. S.
4617. Silah .... Bedford, Quebec.
836. Silas .... Quebec.
799. Silas H............. Nova Scotia.
1779. Silas Henry .... Fall River.
400. Simeon .... Fall River.
751. Simeon .... Fall River.
1465. Simeon .... Fall River.
1873. Simeon .... Fall River.
2364. Simeon .... Fall River.
2883. Slocumb, Charles D................ North Carolina.
2884. Slocumb, Louisa K................ North Carolina.
309. Smith .... Easton, N. J.
2393. Smith Donald B..Massachusetts.
2392. Smith Marion L..Massachusetts.
1525. Sophia A., (McLatch)............ Nova Scotia.
996. Sophronia .... Fall River.
1413. Spencer .... Fall River.
1774. Spencer .... Fall River.
2340. Spencer .... Fall River.
867. Squire ..... Minneapolis, Minn.
833. Squire .... Westport, R. I.
2618. Stella L., (Hemb) ..Dundee, Ill.
273. Stephen .... Fall River.
458. Stephen .... Fall River.
557. Stephen .... Fall River.
887. Stephen .... Fall River.
898. Stephen .... Fall River.
971. Stephen .... Fall River.
1722. Stephen .... Fall River.
1731. Stephen .... Fall River.
1765. Stephen .... Fall River.
145. Stephen .... Fall River, Mass.
289. Stephen .... New York.
521. Stephen .... Portsmouth.
301. Stephen ....Portsmouth, R. I.
263. Stephen .... Tiverton, R. I.
883. Stephen .... Tiverton, R. I.
2540. Stephen A............. Fall River.
2303. Stephen B............. Fall River.
1233. Stephen C. D........ Edna, Kan.
2965. Stover, Mary Louise .... North Carolina.

INDEX. 345

2705. Strickland, John Walter. ....
2701. Strickland, Laura L...
2704. Strickland, Lillie M. ....
2703. Strickland, Nellie E. ....
2699. Strickland, Margaret R., (Eskridge) ....
2077. Strickland Mary G., (Evans)....
.... Virginia.
2700. Strickland, Pandora O., (Bennett) ....
2076. Strickland Sallie E., (Riffe) .....
.... Virginia.
2698. Strickland Sarah Hester, (Bruce).
.... Forward to U. S.
2697. Strickland Virginia E., (Bruce).
2706. Strickland, William A. N. ......
2702. Strickland, Willie M ....
2075. Strickland Wm. E......Virginia.
2594. Sue, (Douglass).Charleston, S. C.
2746. Sullivan, Ethel V................
.... Washington, D. C.
2748. Sullivan, Theodore G.............
.... Washington S.
957. Susan .... Tiverton.
1006. Susan, (Brightman)..Fall River.
873. Susan, (Brown) ..... Fall River.
1361. Susan, (Brown)....New Jersey.
912. Susan D., (Brownell).Fall River.
1034. Susan Jane, (Tallman) ....
.... Portsmouth.
505. Susan, (Kennard), Malden, Mass.
2385. Susan Maria ......Nova Scotia.
575. Susan, (McCorrie)..Portsmouth.
551. Susan, (Meade) ..... New York.
830. Susan, (Sanford)....Fall River.
968. Susan, (Whitman)..Fall River.
267. Susanna, (Browneil) ..Fall River.
978. Susanna, (Thomas).Nova Scotia.
482. Susannah, (Boomer) .Fall River.
533. Susannah, (Boomer)..New York.
926. Susannah, (Coolidge), Fall River.
465. Susannah, (Hart) ...Fall River.
2759. Sutton, Benj. F.... Philadelphia.
2082. Sweetland Annie H., (Wiley) ....
.... Virginia.
1193. Sweetland, Caroline, (Wolcott)..
.... Greenup, Ky.
1185. Sweetland Charles Gould ......
.... California.
1183. Sweetland Elizabeth A., (Obenchain) ....
2084. Sweetland Elizabeth O., (Hill)..
.... Virginia.
2714. Sweetland, Essie H....Memphis.
2098. Sweetland George Lee.California.
2101. Sweetland Henry P..California.

1189. Sweetland, Henry P.... . ...
.... Sweetland, Cal.
1188. Sweetland, Isaac Van M........
.... Hamblin, W. Va.
1192. Sweetland, James Otis ....
.... Nevada Co., Cal.
2096. Sweetland Jefferson D..........
.... California.
2080. Sweetland John S......Virginia.
2102. Sweetland Laurence G.California.
2087. Sweetland Lewis Rolfe.........
.... Virginia.
2097. Sweetland Lura V., (Heath)....
.... California.
2085. Sweetland Maggie P., (Haile)...
.... Virginia.
1194. Sweetland, Margaret, (Powell)..
.... Greenup, Ky.
1187. Sweetland, Martha H., (Walker.)
2088. Sweetland Martha W., (Oxley)...
.... Virginia.
2081. Sweetland Mary H., (Love)......
.... Virginia.
1184. Sweetland Mary H. Strickland..
2713. Sweetland, Medora C., Memphis.
2712. Sweetland, Ross R....Memphis.
1191. Sweetland, Sallie E., (Powell)...
.... Kentucky.
2090. Sweetland Sallie R....Virginia.
1186. Sweetland, Samuel McF........
.... Memphis, Tenn.
2709. Sweetland, Signal W..Memphis.
2089. Sweetland Virginia W., (Sanford)
.... Virginia.
2100. Sweetland William A.California.
1190. Sweetland, William A..Virginia.
311. Sybil, (Monroe) .... Bristol.
2294. Sylvanus S.......... Connecticut.
2538. Sylvia .... Fall River.
1761. Sylvia, (Almy) ....Tiverton.
901. Sylvia, (Durfee) ...... Fall River.
456. Sylvia, (Weeks.) ....

T

660. Tacy, (Valentine) ....
.... Burlington, N. J.
1951. Telfair, Eliza V. ....
1949. Telfair, Helen, (Clark)..Florida.
1948. Telfair, John S........Alabama.
1950. Telfair, Kate, (Hallowes) ....
1947. Telfair, William H.....Alabama.
1291. Tenbroeck .... New Jersey.
673. Tenbroeck W........New Jersey.
2159. Theodore..........Edna, Kansas.
2354. Theodora ....Fall River.
2236. Theodore .... New Jersey

346 INDEX.

1221. Theodore .... Philadelphia.
2144. Theodore John .... Philadelphia.
1425. Theodore W......... Fall River.
353. Thomas .... Burlington Co., N. J.
1959. Thomas .... California.
821. Tabitha ...... Cornwallis, N. S.
291. Thomas .... Fall River.
392. Thomas .... Fall River.
454. Thomas .... Fall River.
509. Thomas .... Fall River.
570. Thomas .... Fall River.
624. Thomas .... Fall River.
745. Thomas .... Fall River.
875. Thomas .... Fall River.
880. Thomas .... Fall River.
892. Thomas .... Fall River.
1727. Thomas .... Fall River.
1756. Thomas .... Fall River.
2507. Thomas .... Fall River.
110. Thomas ..... Imlaystown, N. J.
89. Thomas .... Johnston, R. I.
117. Thomas .... Johnston, R. I.
572. Thomas .... Lowell, Mass.
202. Thomas ... Monmouth Co., N. J.
354. Thomas ... Monmouth Co., N. J.
690. Thomas .... New Jersey.
261. Thomas .... Newport, R. I.
1338. Thomas .... New York.
210. Thomas ...... Nottingham, N. J.
276. Thomas .... Nova Scotia.
795. Thomas .... Nova Scotia.
953. Thomas..Paris, Bourbon Co., Ky.
308. Thomas .... Portsmouth.
1096. Thomas .... Portsmouth.
40. Thomas ...... Portsmouth, R. I.
4. Thomas .... Providence, R. I.
31. Thomas .... Providence, R. I.
234. Thomas .... Rumson, N. J.
25. Thomas ...... Shrewsbury, N. J.
135. Thomas .... Tiverton, R. I.
1524. Thomas A........ Nova Scotia.
1920. Thomas C...... Cazenovia, N. Y.
1963. Thomas C...... Fernandina, Fla.
1968. Thomas C...... New Bern, Ala.
497. Thomas Cox .... Nova Scotia.
1740. Thomas E............. Fall River.
2520. Thomas E............. Fall River.
1065. Thomas F...... Cazenovia, N. Y.
2940. Thomas Fuller.Goldsboro, N. C.
1079. Thomas H............ Fall River.
863. Thomas H........ Galveston, Tex.
696. Thomas H........... New York.
1094. Thomas H...... New York City.
981. Thomas H...... Santa Ana, Cal.

1398. Thomas J............. Fall River.
649. Thomas J............. New Jersey.
1102. Thomas James ..... Mobile, Ala.
1534. Thomas James ..... Nova Scotia.
2493. Thomas Leonard .... Fall River.
2435. Thomas Paschal ....
.... Washington, D. C.
605. Thomas R...... New Berne, Ala.
1997. Thomas Richardson... Alabama.
1125. Thomas S............. Alabama.
2316. Thomas S............. Fall River.
2007. Thomas Sheppard ...... U. S. N.
383. Thomas T.......... New Jersey.
2860. Thompson, Marshall B.... Texas.
2178. Tilton Abbie B..... New Jersey.
2179. Tilton Beth .... New Jersey.
1261. Tilton Clark .... New Jersey.
1264. Tilton Elizabeth, (Wainwright).
2177. Tilton Ella B., (Borden)........
.... New Jersey.
2806. Tilton, Elsie M..... New Jersey.
1262. Tilton Emely ...... New Jersey.
2771. Tilton, Hazel........ New Jersey.
1265. Tilton John .... New Jersey.
2176. Tilton, John B...... New Jersey.
1263. Tilton Josiah B.... New Jersey.
2174. Tilton Joseph H..... New Jersey.
2807. Tilton, Martie Rose..New Jersey.
2175. Tilton Mary B., (Kester) ....
.... New Jersey.
2772. Tilton, Richard B... New Jersey.
1015. Timon .... Fall River.
1915. Timothy .... New York.
2653. Trammel Chesley.... Italy, Tex.
860. Turpin Bradford................
861. Turpin John M.................
379. Tylee .... Shrewsbury.

U

2847. Ulrich, Anna M..... New Haven.
2848. Ulrich, John M..... New Haven.
2846. Ulrich, Leslie B.... New Haven.

V

1917. Van Buren .... New York.
1171. Van Meter, Caroline E., (Adams)
.... Kentucky.
2059. Van Meter, Charles C..Kentucky.
1173. Van Meter, Chas. J.............
.... Bowling Green, Ky.
1175. Van Meter, Clinton C...........
.... Danville, Ky.
2685. Van Meter, Clinton C.Kentucky.

624. Van Meter, Elizabeth, (Carper) .... Virginia.
1178. Van Meter, Ellen M.... Virginia.
622. Van Meter, Hannah, (McFarran) .... Virginia.
626. Van Meter, Jacob .... Charleston, Va.
627. Van Meter, Joseph .... Fincastle, Va.
1170. Van Meter, Julia O., (Hobson).. .... Kentucky.
1179. Van Meter, Margaret J. Virginia.
2686. Van Meter, Mary Bell.Kentucky.
623. Van Meter, Mary, (Hedrich) .... Virginia.
1169. Van Meter, Mary J., (Cook) .... Virginia.
2060. Van Meter, Mary U., (Miller).... .... Missouri.
625. Van Meter, Placentia, (McFarran) .... Virginia.
1176. Van Meter, Robert L..Arkansas.
628. Van Meter, Sallie, (Sweetland) .... Kentucky.
1172. Van Meter, Samuel K. .... Bowling Green, Ky.
1181. Van Meter, Sarah E., (Helms)... .... Tennessee.
1174. Van Meter, Sarah F., (Clarkson) .... Bowling Green, Ky.
1180. Van Meter, William A........... .... Knoxville, Tenn.
2061. Van Meter, William S.Kentucky.
1168. Van Meter, William S..Virginia.
2814. Vernon .... New Jersey.
2931. Virginia W..... Goldsboro, N. C.

W

2200. W. Robert ....Jacobstown, N. J.
2183. Wainwright, Aaaron.New Jersey.
2184. Wainwright, Abbie B .... New Jersey.
2185. Wainwright, Anna R.... ... . New Jersey.
2182. Wainwright, Emma, (Buck)..... .... New Jersey.
552. Wait (Lawton) ....
283. Wait, (Simmons) ....Fall River.
1669. Waity A.......... Scituate, R. I.
1667. Waity M......... Scituate, R. I.
2004. Wallace J............. Alabama
2769. Walter A..... ..., Philadelphia.
2232. Walter Augustus .Philadelphia.
2274. Walter Charles ......New York.
1260. Walter E...... Jacobstown, N. J.
2939. Walter Eccles ...North Carolina.
2592. Walter Eugene..Goldsboro, N. C.

702. Walter H....... Red Bank, N. J.
582. Ward, Abigail .. Beaufort, N. C.
588. Ward, Alice.....Beaufort, N. C.
585. Ward, Comfort ..Beaufort, N. C.
589. Ward, Elizabeth..Beaufort, N. C.
587. Ward, Hannah..Beaufort, N. C.
586. Ward, (Hope, (Borden)........... .... Beaufort. N. C.
590. Ward, Rufus.....Beaufort, N. C.
584. Ward, Sheppard..Beaufort, N. C.
583. Ward, William ..Beaufort, N. C.
1365. Wardell Elizabeth...New Jersey.
1363. Wardell Harry .....New Jersey.
1364. Wardell Susan ......New Jersey.
2856. Weld, Eleanor M...Los Angeles.
2855. Weld, James Romaine .... Los Angeles.
2622. Wells George L........ Elgin, Ill.
2945. White, Katherine .... Bowling Green, Ky.
2722. Wiley, Edith .... Kentucky.
2726. Wiley, Elizabeth ..... Kentucky.
2723. Wiley, Frank .... Kentucky.
2724. Wiley, Laura .... Kentucky.
2720. Wiley, Mattie ....Kentucky.
2719. Wiley, Nettie......... Kentucky.
2718. Wiley, Robert .... Kentucky.
2721. Wiley, Sallie .... Kentucky.
2725. Wiley, Wyatt ....Kentucky.
1599. Wilford .... Cornwallis, N. S.
2613. Willard Peachy..Somerset, Mass.
1203. William. ....
46. William..Barbadoes, West Indies.
43. William .... Beaufort, N. C.
158. William .... Beaufort, N. C.
316. William .... Beaufort, N. C.
1613. William ...... Bedford ,Quebec.
358. William ...... Burlington, N. J.
1162. William ...... Calhoun Co., Ala.
431. William .... Canada.
1944. William .... Chicago.
801. William .... Cornwallis, N. S.
199. William .... Eversham, N. J.
270. William .... Fall River.
304. William .... Fall River.
473. William .... Fall River.
722. William .... Fall River.
945. William .... Fall River.
1084. William .... Fall River.
1384. William .... Fall River.
2514. William ....,..... Fall River.
146. William ...... Fall River, Mass.
154. William ...... Fall River, Mass.
1138. William .... Green Co., Ala.

INDEX.

351.	William	Hazleton, N. J.
188.	William	Johnston, R. I.
382.	William	Loyd, N. J.
652.	William	New Jersey.
1274.	William	New Jersey.
1362.	William	New Jersey.
601.	William	New Bern, Ala.
418.	William	Nova Scotia.
791.	William	Nova Scotia.
1514.	William	Nova Scotia.
1151.	William	Palestine, Tex.
1056.	William	Pompey, N. Y.
561.	William	Portsmouth.
92.	William	Providence, R. I.
1624.	William	Quebec, Canada.
854.	William	Scituate, R. I.
488.	William	Tiverton
324.	William	Washington Co., Ark.
2361.	William A	New Bedford.
1532.	William A	Nova Scotia.
1104.	William Alfred	Alabama.
2000.	William Alfred	Alabama.
2371.	William Arthur	Fall River.
2623.	William Austin.	New Berlin, N. Y.
2959.	William Ayres	
2799.	William Bradford	New Bedford.
2477.	William C	Fall River.
1606.	William C	Philipsburg, Canada.
2740.	William C	Surgeon U. S. A.
672.	William D	New Jersey.
1227.	William D	Philadelphia.
1075.	William Dean	N. Y.
2285.	William Duncan	Auburn, Neb.
2864.	William E	Fall River.
1473.	William G	Fall River.
1940.	William I	Boston, Mass.
1837.	William H	California.
1090.	William H	Fall River.
1670.	William H	Fall River.
1906.	William H	Fall River.
2379.	William H	Fall River.
2533.	William H	Fall River.
1832.	William H	Goldsboro, N. C.
1228.	William H	New Jersey.
1326.	William H	New Jersey.
1401.	William H. H	Fall River.
2596.	William Henry	Goldsboro, N. C.
2260.	William Henry	Nova Scotia.
1544.	William J	Nova Scotia.
1132.	William James	Jersey City.
1155.	William Joseph	Oxford, Fla.
1687.	William Joseph	Texas.
1927.	William L	New York.
1845.	William L	Nova Scotia.
1025.	William N	Fall River.
1869.	William N	Massachusetts.
2143.	William Noston	Philadelphia.
1032.	William O	Fall River.
1557.	William P	Chelsea, Mass.
1796.	William S	Fall River.
1080.	William W	Borden, Ind.
2025.	Willie C., (Tredaway)	Newnan, Ga.
2327.	Willie Owen	New York.
1657.	Wilson W	Scituate, R. I.
2409.	Windsor M	Boston.
1286.	Winfield E	New Jersey.
2108.	Wolcott, Alanson H.	Kentucky.
2104.	Wolcott, Albert S.	Kentucky.
2109.	Wolcott, Ella V	Kentucky.
2106.	Wolcott, Mattie W., (Foreman)	Kentucky.
2105.	Wolcott, Viola L., (Wilson)	Kentucky.
2103.	Wolcott William L.	Kentucky.
2018.	Woolsey Katherine, (Hamilton)	

1st Gen.	**Roger Willi**[ams]	
	MARY	
2nd Gen.	Mary. Freeborn	Providen[ce]
3rd Gen.	D 1766 Mary. Peleg. Roger. Dan'l. Pati[ence]	
	Elizabeth CARPENTER	
4th Gen.	7/12 1802 Dan'l. Robert. Silas. Peleg. Timo[thy]	
	HANAH WRIGHT	
5th Gen.	REUBEN. ESTHER. THANKF[UL]	
	HENRY WHEELER	
6th Gen.	ESTHER. THANKFUL. P[...]	
7th Gen.		

351. William ....Hazleton, N. J.	1670. William H.......... Fall River.
188. William .... Johnston, R. I.	1906. William H.......... Fall River.
382. William ....Loyd, N. J.	2379. William H.......... Fall River.
652. William .... New Jersey.	2533. William H.......... Fall River.
1274. William .... New Jersey.	1832. William H..... Goldsboro, N. C.
1362. William .... New Jersey.	1228. William H.......... New Jersey.
601. William ....New Bern, Ala.	1326. William H.......... New Jersey.
418. William .... Nova Scotia.	1401. William H. H....... Fall River.
791. William .... Nova Scotia.	2596. William Henry..Goldsboro, N. C.
1514. William .... Nova Scotia.	2260. William Henry .... New Jersey.
1151. William .... Palestine, Tex.	1544. William J........... Nova Scotia.
1056. William .... Pompey, N. Y.	1132. William James .....Jersey City.
561. William .... Portsmouth.	1155. William Joseph .... Oxford, Fla.
92. William ...., Providence, R. I.	1687. William Joseph .... Texas.
1624. William .... Quebec, Canada.	1927. William L............ New York.
854. William .... Scituate, R. I.	1845. William L......... Nova Scotia.
488. William .... Tiverton	1025. William N......... Fall River.
324. William...Washington Co., Ark.	1869. William N......... Massachusetts.
2361. William A........ New Bedford.	2143. William Noston ..Philadelphia.
1532. William A........ Nova Scotia.	1032. William O........... Fall River.
1104. William Alfred .... Alabama.	1557. William P........ Chelsea, Mass.
2000. William Alfred .... Alabama.	1796. William S......... Fall River.
2371. William Arthur...... Fall River.	1080. William W......... Borden, Ind.
2623. William Austin.New Berlin, N. Y.	2025. Willie C., (Tredaway) ....
2959. William Ayres................	.... Newnan, Ga.
2799. William Bradford ..New Bedford.	2327. Willie Owen .... New York.
2477. William C........... Fall River.	1657. Wilson W........ Scituate, R. I.
1606. William C..Philipsburg, Canada.	2409. Windsor M.............. Boston.
2740. William C...... Surgeon U. S. A.	1286. Winfield E......... New Jersey.
672. William D........... New Jersey.	2108. Wolcott, Alanson H...Kentucky.
1227. William D.......... Philadelphia.	2104. Wolcott, Albert S.... Kentucky.
1075. William Dean.............N. Y.	2109. Wolcott, Ella V...... Kentucky.
2285. William Duncan..Auburn, Neb.	2106. Wolcott, Mattie W., (Foreman)..
2864. William E........... Fall River.	.... Kentucky.
1473. William G........... Fall River.	2105. Wolcott, Viola L., (Wilson)......
1940. William I........ Boston, Mass.	.... Kentucky.
1837. William H........... California.	2103. Wolcott William L....Kentucky.
1090. William H........... Fall River.	2018. Woolsey Katherine, (Hamilton)

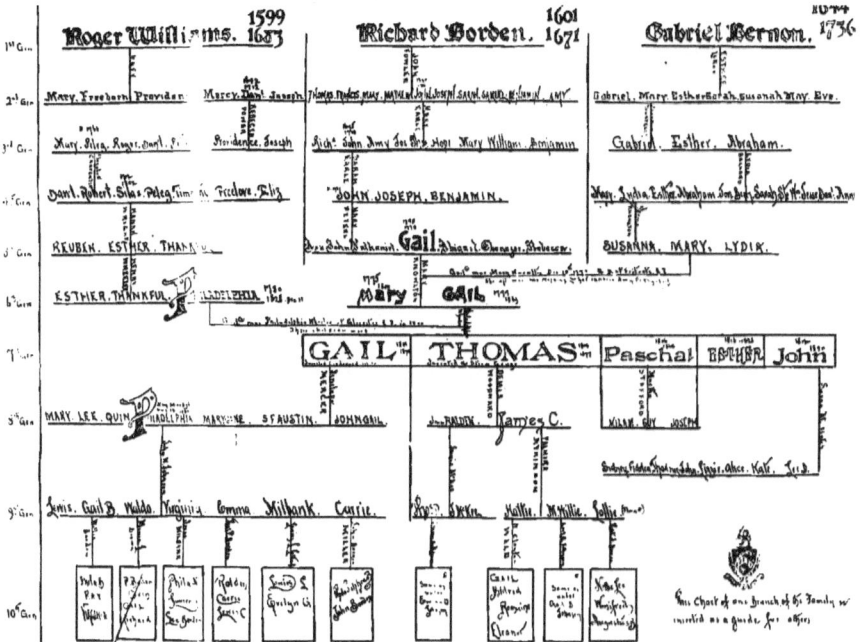

599   **1644**
683 **Bernon,** **1736**

n¹. Jaseph h. Susanah. May. Eve.

e. Joseph Abraham.
                    LYDIA BALLARD
. Eliz   1. Benj. Sarah. S^te W^m. Jesse. Dan¹. Anne.

.YDIA.

A. 1780
A. 1828 -Dec¹

www.ingramcontent.com/pod-product-compliance
Lightning Source LLC
Chambersburg PA
CBHW050844300426
44111CB00010B/1122